Statistical Engineering

Also Available from ASQ Quality Press

Applied Statistics for the Six Sigma Green Belt
Bhisham C. Gupta and H. Fred Walker

Applied Data Analysis for Process Improvement: A Practical Guide to Six Sigma Black Belt Statistics
James L. Lamprecht

Design of Experiments with MINITAB
Paul Mathews

Failure Mode and Effect Analysis: FMEA From Theory to Execution, Second Edition
D. H. Stamatis

Statistical Quality Control Using Excel, Second Edition
Steven M. Zimmerman, Ph.D. and Marjorie L Icenogle, Ph.D.

The Desk Reference of Statistical Quality Methods
Mark L. Crossley

Improving Performance Through Statistical Thinking
ASQ Statistics Division

Six Sigma for the Shop Floor: A Pocket Guide
Roderick A. Munro

Defining and Analyzing a Business Process: A Six Sigma Pocket Guide
Jeffrey N. Lowenthal

To request a complimentary catalog of ASQ Quality Press publications, call 800-248-1946, or visit our Web site at http://qualitypress.asq.org.

Statistical Engineering

An Algorithm for Reducing Variation
in Manufacturing Processes

Stefan H. Steiner and R. Jock MacKay

ASQ Quality Press
Milwaukee, Wisconsin

American Society for Quality, Quality Press, Milwaukee 53203
© 2005 by American Society for Quality
All rights reserved. Published 2005
Printed in the United States of America

12 11 10 09 08 07 06 05   5 4 3 2 1

Library of Congress Cataloging-in-Publication Data

Steiner, Stefan H., 1964–
   Statistical engineering : an algorithm for reducing variation in manufacturing
     processes / by Stefan H. Steiner and R. Jock MacKay.
     p. cm.
   Includes bibliographical references.
   ISBN 0-87389-646-7
   1. Production management.   2. Manufacturing processes.   I. MacKay, R. Jock.
II. Title.

TS155.S773   2005
658.5—dc22

2004030676

No part of this book may be reproduced in any form or by any means, electronic, mechanical, photocopying, recording, or otherwise, without the prior written permission of the publisher.

Publisher: William A. Tony
Acquisitions Editor: Annemieke Hytinen
Project Editor: Paul O'Mara
Production Administrator: Randall Benson

ASQ Mission: The American Society for Quality advances individual, organizational, and community excellence worldwide through learning, quality improvement, and knowledge exchange.

Attention Bookstores, Wholesalers, Schools, and Corporations: ASQ Quality Press books, videotapes, audiotapes, and software are available at quantity discounts with bulk purchases for business, educational, or instructional use. For information, please contact ASQ Quality Press at 800-248-1946, or write to ASQ Quality Press, P.O. Box 3005, Milwaukee, WI 53201-3005.

To place orders or to request a free copy of the ASQ Quality Press Publications Catalog, including ASQ membership information, call 800-248-1946. Visit our Web site at www.asq.org or http://qualitypress.asq.org.

Quality Press
600 N. Plankinton Avenue
Milwaukee, Wisconsin 53203
Call toll free 800-248-1946
Fax 414-272-1734
www.asq.org
http://qualitypress.asq.org
http://standardsgroup.asq.org
E-mail: authors@asq.org

∞ Printed on acid-free paper

To Anne Marie, Erik, and Emily—S.H.S

To Samm—R.J.M.

Contents

Acknowledgements xix

Preface xxi

Chapter 1   Introduction 1
   1.1 Truck Pull 2
   1.2 Engine Block Leaks 4
   1.3 Camshaft Lobe Runout 5
   1.4 Sand Core Strength 7
   1.5 Crankshaft Main Diameter 7
   1.6 Paint Film Build 8
   1.7 Refrigerator Frost Buildup 9

PART I   SETTING THE STAGE 11

Chapter 2   Describing Processes 13
   2.1 Language of Processes 13
   2.2 Causes of Variation 15
   2.3 Displaying and Quantifying Process Variation 18
   2.4 Models for Variation and the Effects of Causes 23

Chapter 3   Seven Approaches to Variation Reduction 29
   3.1 Fixing the Obvious Based on Knowledge of a Dominant Cause 30
   3.2 Desensitizing the Process to Variation in a Dominant Cause 32
   3.3 Feedforward Control Based on a Dominant Cause 33
   3.4 Feedback Control 34
   3.5 Making the Process Robust 36
   3.6 100% Inspection 38
   3.7 Moving the Process Center 39

Chapter 4 An Algorithm for Reducing Variation ... 41
- 4.1 The Statistical Engineering Variation Reduction Algorithm ... 41
- 4.2 How to Use the Algorithm Effectively ... 45

Chapter 5 Obtaining Process Knowledge Empirically ... 51
- 5.1 Question, Plan, Data, Analysis, and Conclusion (QPDAC) Framework ... 52
- 5.2 Examples ... 58
- 5.3 Summary ... 64

PART II GETTING STARTED ... 67

Chapter 6 Defining a Focused Problem ... 69
- 6.1 From a Project to Problems ... 69
- 6.2 The Problem Baseline ... 73
- 6.3 Planning and Conducting the Baseline Investigation ... 76
- 6.4 Examples ... 81
- 6.5 Completing the Define Focused Problem Stage ... 86

Chapter 7 Checking the Measurement System ... 89
- 7.1 The Measurement System and its Attributes ... 90
- 7.2 Estimating Measurement Variation ... 92
- 7.3 Estimating Measurement Bias ... 98
- 7.4 Improving a Measurement System ... 101
- 7.5 Completing the Check the Measurement System Stage ... 102

Chapter 8 Choosing a Working Variation Reduction Approach ... 105
- 8.1 Can We Find a Dominant Cause of Variation? ... 106
- 8.2 Can We Meet the Goal by Shifting the Process Center Without Reducing Variation? ... 108
- 8.3 Can We Reduce Variation by Changing One or More Fixed Inputs Without Knowledge of a Dominant Cause? ... 110
- 8.4 Does the Process Output Exhibit a Strong Pattern Over Time? ... 112
- 8.5 Summary ... 113

PART III FINDING A DOMINANT CAUSE OF VARIATION ... 115

Chapter 9 Finding a Dominant Cause Using the Method of Elimination ... 117
- 9.1 Families of Causes of Variation ... 118
- 9.2 Finding a Dominant Cause Using the Method of Elimination ... 121
- 9.3 Implementing the Method of Elimination ... 124

Chapter 10 Investigations to Compare Two Families of Variation **131**

    10.1 Stratification 131
    10.2 Comparing Two Time-Based Families 135
    10.3 Comparing Upstream and Downstream Families 138
    10.4 Comparing Assembly and Component Families 141
    10.5 Comments 144

Chapter 11 Investigations to Compare Three or More Families of Variation **149**

    11.1 Multivari Investigations: Comparing Time- and Location-Based Families 149
    11.2 Comparing Families Defined by Processing Steps 163
    11.3 Comparing Component Families 167

Chapter 12 Investigations Based on Single Causes **179**

    12.1 Group Comparison: Comparing Parts With Binary Output 179
    12.2 Investigating the Relationship Between Inputs and a Continuous Output 183

Chapter 13 Verifying a Dominant Cause **193**

    13.1 Verifying a Single Suspect Dominant Cause 194
    13.2 Issues With Single Suspect Verification Experiments 197
    13.3 Verifying a Dominant Cause From a Short List of Suspects 200
    13.4 Further Issues and Comments 205

PART IV ASSESSING FEASIBILITY AND IMPLEMENTING A VARIATION REDUCTION APPROACH **211**

Chapter 14 Revisiting the Choice of Variation Reduction Approach **213**

    14.1 Fixing the Obvious: Implementing an Available Solution 214
    14.2 Compensating for Variation in the Dominant Cause 217
    14.3 Reformulating the Problem in Terms of a Dominant Cause 219
    14.4 Continuing the Search for a More Specific Dominant Cause 223
    14.5 Dealing With Dominant Causes That Involve Two Inputs 225
    14.6 Summary 225

Chapter 15 Moving the Process Center **227**

    15.1 Examples of Moving the Process Center 228
    15.2 Assessing and Planning a Process Center Adjustment 238

Chapter 16 Desensitizing a Process to Variation in a Dominant Cause **241**

    16.1 Examples of Desensitization 242
    16.2 Assessing and Planning Process Desensitization 255

Chapter 17 Feedforward Control Based on a Dominant Cause 259
17.1 Examples of Feedforward Control 260
17.2 Assessing and Planning Feedforward Control 266

Chapter 18 Feedback Control 269
18.1 Examples of Feedback Control 270
18.2 Assessing and Planning Feedback Control 280

Chapter 19 Making a Process Robust 285
19.1 Examples of Process Robustness 286
19.2 Assessing and Planning Process Robustness 297

Chapter 20 100% Inspection 301
20.1 Examples of 100% Inspection 302
20.2 Assessing and Planning 100% Inspection 304

Chapter 21 Validating a Solution and Holding the Gains 307
21.1 Validating a Solution 307
21.2 Holding the Gains Over the Long Term 310
21.3 Completing the Implement and Validate Stage 312

References *313*

Index *319*

Material on Enclosed CD-ROM

CASE STUDIES	**CD–1**
Case Study I Brake Rotor Balance	**CD–3**
Case Study II Rod Thickness	**CD–13**
Case Study III Crankshaft Main Diameter	**CD–23**

EXERCISES	**CD–39**
Exercises	**CD–41**
Chapter 2	CD–41
Chapter 3	CD–43
Chapter 5	CD–43
Chapter 6	CD–44
Chapter 7	CD–45
Chapter 9	CD–48
Chapter 10	CD–49
Chapter 11	CD–53
Chapter 12	CD–56
Chapter 13	CD–57
Chapter 14	CD–59
Chapter 15	CD–61
Chapter 16	CD–63
Chapter 17	CD–66
Chapter 18	CD–67
Chapter 19	CD–68
Chapter 21	CD–75

Exercise Solutions **CD–77**
    Chapter 2 .... CD–77
    Chapter 3 .... CD–83
    Chapter 5 .... CD–84
    Chapter 6 .... CD–86
    Chapter 7 .... CD–90
    Chapter 9 .... CD–100
    Chapter 10 .... CD–101
    Chapter 11 .... CD–109
    Chapter 12 .... CD–118
    Chapter 13 .... CD–121
    Chapter 14 .... CD–126
    Chapter 15 .... CD–128
    Chapter 16 .... CD–134
    Chapter 17 .... CD–144
    Chapter 18 .... CD–152
    Chapter 19 .... CD–156
    Chapter 21 .... CD–170

SUPPLEMENTS **CD–177**

Chapter 2 Supplement   Describing Processes **CD–179**
    S2.1 Pareto Principle CD–179
    S2.2 Defining Dominant Cause CD–180
    S2.3 Dominant Cause Involving Two (or More) Inputs CD–180
    S2.4 Classifying Causes of Variation CD–182
    S2.5 Process Capability CD–183
    S2.6 Relating the Two Kinds of Variation CD–184
    S2.7 Variation within Groups and Group to Group CD–185
    S2.8 Gaussian Model CD–187

Chapter 4 Supplement   An Algorithm for Reducing Variation **CD–191**
    S4.1 Comparison of Variation Reduction Algorithms CD–191

Chapter 5 Supplement   Obtaining Process Knowledge Empirically **CD–193**
    S5.1 Attributes CD–193
    S5.2 Sampling Protocols and Their Effects CD–196
    S5.3 Study, Sample, and Measurement Errors CD–198
    S5.4 Outliers CD–198

Chapter 6 Supplement   Defining a Focused Problem **CD–201**
    S6.1 Confidence Intervals for the Process Baseline Attributes CD–201
    S6.2 Process Stability and Quantifying Baseline Performance CD–203

Chapter 7 Supplement   Checking the Measurement System CD–205

    S7.1 Assessing a Binary Measurement System CD–205
    S7.2 Assessing a Destructive Measurement System CD–209
    S7.3 Repeatability, Reproducibility, Linearity, and Stability CD–211
    S7.4 Gage R&R and Isoplot Investigations
         to Assess Measurement Variation CD–214
    S7.5 Interpreting Measurement Variation CD–220
    S7.6 Effect of Measurement Variation CD–220
    S7.7 Finding a Dominant Cause of Measurement Variation CD–221

Chapter 9 Supplement   Finding a Dominant Cause Using the Method of Elimination CD–225

    S9.1 Comparison of Strategies for Finding a Dominant Cause CD–225
    S9.2 No Single Dominant Cause CD–227

Chapter 10 Supplement   Investigations to Compare Two Families of Variation CD–229

    S10.1 Analysis of Variance CD–229
    S10.2 Possible Explanations for Observed Changes in Variation CD–232
    S10.3 Why Formal Hypothesis Tests Are Not Recommended CD–234
    S10.4 Other Plans for Comparing Upstream and Downstream
          Families CD–235
    S10.5 Applying Leverage in Comparing Assembly
          and Component Families CD–237

Chapter 11 Supplement   Investigations to Compare Three or More Families of Variation CD–239

    S11.1 Analysis of Variance for Multivari Investigations CD–239
    S11.2 Handling Families Expected to Have a Haphazard Effect CD–246
    S11.3 Regression Analysis for Variation Transmission Investigations ... CD–250
    S11.4 Other Component-Swapping Investigations CD–253
    S11.5 Component-Swapping with Three Groups or Subassemblies CD–254

Chapter 12 Supplement   Investigations Based on Single Causes CD–257

    S12.1 Matrix Scatter Plots and Draftsman Plots CD–257
    S12.2 Group Comparison Versus Paired Comparison CD–259
    S12.3 Regression Extensions CD–260

Chapter 13 Supplement   Verifying a Dominant Cause CD–263

    S13.1 Understanding Repeats and Replicates CD–263
    S13.2 Randomization, Replication, and Blocking CD–264
    S13.3 Variables Search CD–265

Chapter 15 Supplement   Moving the Process Center CD–267

    S15.1 Fractional Factorial Experiments CD–267

Chapter 16 Supplement Desensitizing a Process to Variation in a Dominant Cause **CD–277**
  S16.1 Mathematical Representation of Process Desensitization CD–277
  S16.2 Fractional Factorial Experiments for Desensitization CD–278
  S16.3 Further Analysis for the Eddy Current Measurement Example ... CD–281

Chapter 17 Supplement Feedforward Control Based on a Dominant Cause **CD–285**
  S17.1 Simulating the Benefit of Selective Fitting CD–285
  S17.2 More on Making Predictions CD–287

Chapter 18 Supplement Feedback Control **CD–289**
  S18.1 Alternative Simple Feedback Controllers CD–289
  S18.2 Simulating the Benefit of Feedback Control CD–290
  S18.3 Properties of the Exponential Weighted Moving Average CD–292
  S18.4 Connection to PID Controllers CD–295

Chapter 19 Supplement Making a Process Robust **CD–297**
  S19.1 Taguchi's Methodology CD–297

APPENDICES USING MINITAB CD–299

Appendix A Data Storage and Manipulation **CD–301**
  A.1 Row/Column Data Storage Format CD–301
  A.2 Making Patterned Data CD–303
  A.3 Calculating Derived Characteristics CD–304
  A.4 Selecting a Data Subset CD–304
  A.5 Stacking Columns CD–305
  A.6 MINITAB Macros CD–307

Appendix B Numerical Summaries **CD–311**
  B.1 Numerical Summaries for Continuous Characteristics CD–311
  B.2 Numerical Summaries for Discrete Characteristics CD–313

Appendix C Graphical Summaries **CD–317**
  C.1 Histogram CD–317
  C.2 Run Chart CD–318
  C.3 Box Plot CD–319
  C.4 Scatter Plot CD–320
  C.5 Multivari Chart CD–326
  C.6 Exponential Smoothing CD–330

Appendix D Analysis of Variance (ANOVA) **CD–333**
    D.1 One-Way ANOVA CD–333
    D.2 ANOVA for Two or More Inputs CD–336

Appendix E Regression Models and Analysis **CD–339**
    E.1 Regression with a Single Input CD–339
    E.2 Regression with Multiple Inputs CD–344

Appendix F Planning and Analysis of Designed Experiments **CD–347**
    F.1 Setting Up the Experimental Design CD–347
    F.2 Analysis of the Experimental Results CD–352

Figures and Tables

Figure 1.1	Truck pull (as found).	3
Figure 1.2	Truck pull (as desired).	3
Figure 1.3	Proportion of leaking blocks versus month of production.	4
Figure 1.4	Camshaft.	5
Figure 1.5	Lobe deviation from ideal.	6
Figure 1.6	Histogram of camshaft lobe BC runout.	6
Figure 1.7	Current and initial desired core strengths (pounds).	7
Figure 1.8	Histogram of crankshaft diameters for the front position of the first main.	8
Figure 1.9	Histogram of film build.	9
Figure 2.1	Exhaust manifold.	13
Figure 2.2	Exhaust manifold process map.	14
Figure 2.3	A dominant cause with continuous input and binary output.	17
Figure 2.4	A dominant cause with continuous output.	17
Table 2.1	A dominant cause with binary input and output.	17
Figure 2.5	Two examples of a dominant cause involving two inputs with a binary output.	18
Figure 2.6	Histogram of camshaft lobe angle errors.	19
Figure 2.7	Histogram of camshaft lobe BC runout.	20
Figure 2.8	Box plots of camshaft lobe angle errors by lobe position.	21
Figure 2.9	Run chart for angle error at lobe 12.	21
Figure 2.10	Stream one V6 piston diameter at operation 270 by hour.	22
Figure 2.11	A Gaussian output model.	23
Figure 2.12	Stack height.	24
Table 2.2	Model for each component.	24
Figure 2.13	Reduction in variation if we remove a cause contributing a given proportion of the total variation.	26
Figure 2.14	Pareto chart of potential gain.	26
Figure 3.1	Reducing variation in a dominant cause.	30
Figure 3.2	Aligning the process center stratified by a dominant cause.	31
Figure 3.3	Desensitizing a process to variation in a dominant cause.	32

Figure 3.4	Reducing output variation using feedforward control.	33
Figure 3.5	Feedback control.	35
Figure 3.6	Variation in flow rate over time.	35
Figure 3.7	Making the process robust to cause variation.	37
Figure 3.8	100% inspection.	38
Figure 3.9	Moving the process center.	39
Figure 4.1	Statistical engineering variation reduction algorithm.	42
Figure 5.1	Target population to data.	57
Figure 5.2	Histogram and run chart of V6 piston diameter.	58
Figure 5.3	Part of a V6 piston machining process.	59
Figure 5.4	Hypothetical target population scatter plot comparing two measurement systems.	59
Table 5.1	Piston gage comparison results.	60
Figure 5.5	Scatter plot comparing sample diameters at the final and Operation 270 gages.	61
Figure 5.6	Fascia molding and painting process.	61
Figure 5.7	Schematic of the study units.	62
Table 5.2	Summary of ghosting scores.	63
Table 6.1	Leak classification for 100 leaking blocks.	70
Figure 6.1	A connecting rod.	71
Figure 6.2	Pareto chart for rod scrap by operation.	72
Figure 6.3	Hierarchy of project, problems, and questions.	73
Table 6.2	Problem types and corresponding baseline measures.	74
Figure 6.4	Baseline histogram of rod thickness.	74
Figure 6.5	Rod thickness by day.	75
Table 6.3	Relative precision for estimating a standard deviation as a function of sample size.	77
Table 6.4	Relative precision (%) for estimating a proportion p with 95% confidence.	77
Figure 6.6	Row/column format storage of rod thickness baseline data.	78
Figure 6.7	Artificial example showing a thickness trend.	79
Figure 6.8	Artificial example showing a large thickness day effect.	79
Figure 6.9	Average warranty cost versus truck pull class.	82
Figure 6.10	Truck pull baseline histogram.	82
Figure 6.11	Box plots of truck pull by day.	83
Figure 6.12	Histogram and box plot of right caster from baseline.	84
Figure 6.13	Correlation between in-plant and subjective pump noise measurement systems.	85
Figure 6.14	Baseline histogram for noisy pumps.	86
Figure 7.1	Front and rear camshaft journal locations.	90
Figure 7.2	Representation of measurement variation (standard deviation) and bias.	91

Figure 7.3	Baseline performance for journal 1 front diameter.	93
Figure 7.4	Camshaft diameter measurement investigation data in row/column format.	94
Figure 7.5	Diameter and (diameter minus part average) by part number.	95
Figure 7.6	Diameter and (diameter minus part average) by week.	95
Table 7.1	Journal diameter CMM data.	99
Figure 7.7	Measurement error stratified by part number and week.	100
Figure 8.1	Two cases for shifting the process center.	108
Figure 8.2	Baseline histogram of sand core strength.	108
Figure 8.3	Wheel bearing seal failure times.	109
Figure 8.4	Baseline histogram of camshaft lobe BC runout.	110
Figure 8.5	Variation in flow rate over time.	112
Figure 8.6	Flowchart to help choose a working approach before a dominant cause is known.	114
Figure 9.1	The piston machining process map.	118
Figure 9.2	Diagnostic tree for piston diameter variation.	119
Figure 9.3	Diagnostic tree for average label height.	119
Figure 9.4	Partition of the causes of oil consumption.	120
Figure 9.5	Diagnostic tree for V6 piston diameter variation.	121
Figure 9.6	Diagnostic tree for V6 piston diameter problem after Operation 270 investigation.	122
Figure 9.7	Final diagnostic tree for excessive variation in V6 piston diameter.	123
Table 9.1	First-time reject rate by test stand.	126
Figure 9.8	Diagnostic tree for hypothetical search.	127
Figure 9.9	Box plot of diameter by Operation 270 stream from hypothetical investigation.	127
Figure 9.10	Box plot of diameter by Operation 270 stream from hypothetical investigation showing full extent of baseline variation.	128
Table 10.1	Scrap stratified by cavity.	132
Figure 10.1	Rod thickness stratified by position.	133
Figure 10.2	Camshaft journal runout by lobe.	134
Figure 10.3	Diagnostic tree for camshaft lobe runout.	134
Figure 10.4	Part-to-part and shift-to-shift families.	135
Figure 10.5	Box plots of BC runout by day.	135
Figure 10.6	Box plots of camshaft journal runout by day for Lobe 8 (left) and Lobe 12 (right).	136
Figure 10.7	Plastic scale model of the engine block.	137
Figure 10.8	Diagnostic tree for excess porosity.	137
Figure 10.9	Multivari chart of block porosity versus hour.	138
Figure 10.10	Camshaft process map.	139
Figure 10.11	Lobe 12 BC runout after heat treatment and final.	139
Figure 10.12	Lobe 12 BC runout before and after heat treatment.	140

Figure 10.13	Concentration diagram showing location of rail damage.	141
Figure 10.14	Closing effort by velocity.	143
Figure 10.15	Hypothetical results for disassembly-reassembly investigation.	143
Figure 10.16	Multivari chart with less than full extent of variation.	146
Figure 10.17	Box plot of diameter by Operation 270 stream from hypothetical investigation.	147
Figure 11.1	Cavity-to-cavity, mold-to-mold, part-to-part, and hour-to-hour families.	150
Figure 11.2	Plot of side shift by day.	151
Figure 11.3	Diagnostic tree for side shift.	152
Figure 11.4	Histogram of side shift values from the multivari investigation.	153
Figure 11.5	Multivari chart of side shift by time.	153
Figure 11.6	Multivari charts for side shift versus pattern on left, side shift versus group on right.	154
Figure 11.7	Multivari chart of side shift versus pattern and time.	154
Figure 11.8	Histogram of journal diameter for second baseline investigation.	156
Figure 11.9	Camshaft production process map.	156
Figure 11.10	Diameter variation in the multivari.	157
Figure 11.11	Multivari charts of diameter by position, grinder, and batch/hour.	158
Figure 11.12	Multivari charts of diameter by grinder versus position and batch/hour.	158
Figure 11.13	Multivari charts of diameter by group.	159
Figure 11.14	Multivari chart from first fascia cratering multivari investigation.	160
Figure 11.15	Multivari chart from second fascia cratering multivari investigation.	161
Figure 11.16	Piston machining process map.	163
Figure 11.17	Scatter plots of outgoing versus ingoing diameter (by operation).	164
Figure 11.18	Painting process map.	166
Figure 11.19	Diagnostic tree for headrest failure example.	169
Figure 11.20	Diagnostic tree for power window buzz noise.	172
Figure 11.21	Illustration of swapping the group of components labeled G1.	175
Figure 12.1	Box plots of wall thickness by block type at locations 3 and 4.	180
Figure 12.2	Plot of primary seal fit by group (left) and by plastisol amount (right).	181
Figure 12.3	Plot of crossbar dimension versus barrel temperature and hydraulic pressure.	184
Figure 12.4	Diagnostic tree for search for the cause of excessive caster variation.	185
Table 12.1	Component dimensions ranked by residual standard deviation.	186
Figure 12.5	Scatter plots of right caster versus the component characteristics.	187
Table 12.2	Pair wise component dimensions ranked by residual standard deviation.	188
Figure 12.6	Scatter plot of sand scrap proportion versus moisture and compactness.	189
Figure 12.7	Sand scrap proportion versus sand temperature with quadratic fit.	189
Figure 13.1	Runs, replicates, and repeats for an experiment with a single suspect at two levels.	195

Table 13.1	Valve lifter clearance experiment plan and results.	195
Figure 13.2	Oil consumption by lifter clearance level.	196
Figure 13.3	Crossbar dimension verification experiment results.	197
Table 13.2	A two-level factorial experiment with three suspects and eight treatments.	201
Table 13.3	Suspects and levels for brake rotor verification experiment.	201
Table 13.4	Brake rotor verification experiment results.	202
Figure 13.4	Weight by treatment for the brake rotor verification experiment.	202
Figure 13.5	Pareto chart of the effects for brake rotor verification experiment.	203
Table 13.5	Balance weight averages by thickness variation and tooling.	204
Figure 13.6	Interaction between tooling and core thickness variation.	204
Figure 13.7	Window leaks verification experiment results.	206
Table 13.6	Designed experiments terminology.	207
Figure 14.1	Right caster daily averages by alignment machine.	215
Figure 14.2	Plot of primary seal fit by plastisol amount.	216
Figure 14.3	Scatter plot of crossbar dimension versus barrel temperature.	218
Figure 14.4	Scatter plot of flushness versus left front pad height.	220
Figure 14.5	Dot plot of seal strength for leakers and nonleakers.	221
Figure 14.6	Plot of final base circle BC runout by heat treatment spindle.	224
Figure 14.7	Flowchart to help decide how to proceed after finding a dominant cause.	226
Figure 15.1	Changing the process center.	227
Figure 15.2	Battery seal group comparison results.	228
Table 15.1	Heat seal experiment candidates and levels.	229
Table 15.2	Treatments and seal strength for battery seal experiment.	229
Figure 15.3	Seal strength by treatment combination.	230
Figure 15.4	Pareto chart of effects for battery seal experiment.	230
Figure 15.5	Main effect plot for melt temperature.	231
Figure 15.6	Box plots of dip bump score by core wash solution.	232
Table 15.3	Boss shrink defect scores.	233
Table 15.4	Experimental design and results for differential carrier experiment.	234
Figure 15.7	Pareto chart of the effects for piston shrink defect experiment.	236
Figure 15.8	Main effects of significant inputs for shrink defect experiment.	237
Figure 15.9	The effect of aligning substream centers on output variation.	239
Figure 16.1	Original (left) and new (right) relationship between the dominant cause and output characteristic.	241
Table 16.1	Treatments for the engine block porosity desensitization experiment.	243
Figure 16.2	Porosity by treatment for high and low pouring temperatures.	243
Figure 16.3	Scrap rate by lubricant score.	244
Table 16.2	Candidate levels and scrap rates for low and high level of lubricant amount.	245

Figure 16.4	Scrap rate for low and high level of lubricant amount versus treatment.	245
Figure 16.5	Interaction plots for oil pan scrap desensitization experiment.	246
Figure 16.6	Interaction plots of die temperature and binder force by lubricant amount (supplier C only).	247
Table 16.3	Eight-run fractional factorial refrigerator desensitization experiment.	248
Table 16.4	Plan and data for refrigerator frost buildup desensitization experiment.	249
Figure 16.7	Temperature on cooling plate by treatment.	250
Figure 16.8	Interaction plots of candidates and the environment cause output in cooling plate temperature.	250
Figure 16.9	Box plots of eddy current hardness measurements by day.	251
Figure 16.10	Scatter plot of eddy current versus Brinell hardness measurements.	252
Table 16.5	Candidate levels for eddy current measurement experiment.	253
Figure 16.11	Plot of eddy current versus Brinell hardness for freq. = 200, temp. = 35, and gain = 30.	254
Figure 17.1	Feedforward control schematic.	259
Table 17.1	Potato chip spots data.	260
Figure 17.2	Scatter plot of dark spot score versus sugar concentration (%).	261
Figure 17.3	Baseline histogram of imbalance.	262
Figure 17.4	Vectoring to reduce imbalance.	262
Figure 17.5	Histogram of simulated imbalances for vectored assemblies.	263
Table 17.2	Results of simulating selective fitting.	264
Figure 18.1	Feedback control schematic.	269
Figure 18.2	Box plot of tightness by front and axle cable batch.	270
Table 18.1	Parking brake tightness adjuster experiment results.	271
Figure 18.3	Diameter at stream 1 at Operation 270 by minute.	272
Figure 18.4	Diameter at stream 1 of Operation 270 by minute.	273
Figure 18.5	Piston diameters versus time.	275
Figure 18.6	Run chart of flow rate.	276
Figure 18.7	EWMA smoothing of paint flow rate.	277
Figure 18.8	Right caster angle over time.	279
Figure 18.9	Right caster average by shift.	279
Figure 18.10	A machining process with persistent shifts due to tooling.	280
Table 19.1	Candidates and levels for burn robustness experiment.	286
Table 19.2	Experimental plan and data for burn robustness experiment.	287
Figure 19.1	Burn by treatment with added vertical jitter.	287
Figure 19.2	Pareto plot of effects on average burn score.	288
Figure 19.3	Main effect plot for back pressure.	288
Table 19.3	Candidates and levels for iron silicon concentration robustness experiment.	289
Table 19.4	Treatments and results for iron silicon concentration robustness experiment.	290
Figure 19.4	Silicon concentration by treatment combination.	291

Figure 19.5	Pareto chart of the effects on log(s) for iron silicon experiment.	292
Figure 19.6	Main effects for iron silicon concentration robustness experiment.	292
Table 19.5	Candidates selected for the pinskip robustness experiment.	293
Table 19.6	Electroplating pinskip experimental plan and data.	294
Figure 19.7	Pareto analysis of effects in pinskips experiment.	295
Figure 19.8	Cube plot for tank 1 concentration, tank 2 concentration, and tank 2 temperature.	295
Figure 19.9	Number of defective grills versus run order.	296
Figure 20.1	Reducing variation by adding inspection limits.	301
Figure 21.1	Histogram of crossbar dimension in the validation investigation.	309
Figure 21.2	Porosity scrap rate by month.	309

Acknowledgments

This book is the product of many years' experience helping manufacturing organizations reduce variation. We want to thank all of our clients and colleagues at the Institute for Improvement in Quality and Productivity (IIQP). The IIQP is a not-for-profit organization, started in 1985, committed to the development, communication, and application of methods for quality and productivity improvement.

Former and current IIQP colleagues include Bovas Abraham, Dennis Beecroft, Jerry Lawless, Jack Robinson, Jim Whitney, and Clif Young, among others. The numerous corporate members of the institute over the years included AT Plastics, Bell Canada, BF Goodrich Canada, BUDD Canada, Campbell Soup Company, Canada Post Corporation, C-I-L, Continental Can Canada, Dofasco, Epton Industries, GSW Water Products Company, General Motors Canada, Lear Seating Canada, Lear Seigler Industries, Metal Koting Continuous Colour Coat, Nortel Networks, Quebec and Ontario Paper Co., Research in Motion (RIM), Stelco Steel, The Woodbridge Group of Companies, Uniroyal Goodrich Canada, Wescast Industries, and Xerox Canada. Thanks also to other companies we have worked with, including Fisher and Paykel, Focus Automation Systems, Ford Motor Company, Imperial Oil of Canada, Seagram Americas, Stackpole, Toyota Motor Manufacturing Canada, and many others. We learned from you all.

We have also benefited from helping to organize and attending the Continuous Improvement Symposiums held by General Motors of Canada each year since 1987. The symposiums provide an opportunity for employees and suppliers to showcase process improvement success stories. Some of the examples in the book come from these symposiums.

We want to single out Mike Brajac, Pete Peters, and Mark Smith at General Motors of Canada, with whom we have had a long-standing relationship. You are the best problem solvers we know. We also thank Wescast Industries and especially Harry Schiestel, who taught us a lot about foundries and how to solve problems.

We have had outstanding relationships with our colleagues at the University of Waterloo who have contributed, perhaps unknowingly, in many ways to the book. We especially thank Winston Cherry and Wayne Oldford for their penetrating questions and stimulating discussions, and Dennis Beecroft, who taught us there is more to variation reduction than statistics.

Also, our appreciation goes to the undergraduate and graduate engineering and statistics students taking the inaugural offering of the University of Waterloo course STAT 435/835, Statistical Methods for Process Improvement, in the Winter 2004 term. The students used a draft of the book and provided many helpful comments and suggestions to improve the presentation.

In the early stages of writing this book, we were fortunate to be on sabbatical. During fall 2003, Jock received support from CSIRO in Melbourne, Australia, with thanks to Richard Jarrett and Geoff Robinson. Stefan spent a year starting in May 2002 at the University of Auckland, New Zealand, with thanks to Chris Wild.

We were guided through production by the staff of ASQ Quality Press and Kinetic Publishing Services. Special thanks to Paul O'Mara, Laura Varela, and John Ferguson.

MINITAB is a trademark of Minitab in the United States and other countries and is used herein with the owner's permission. Microsoft Word and Microsoft Excel are trademarks of Microsoft.

Last, but certainly not least, we would like to acknowledge the wonderful support from our families over the long road required for writing this book.

Preface

Reducing the variation in process outputs is a key part of process improvement. If you have picked up this book, you probably do not need to be convinced of the truth of this statement. For mass-produced components and assemblies, reducing variation can simultaneously reduce overall cost, improve function, and increase customer satisfaction with the product. Excess variation can have dire consequences, leading to scrap and rework, the need for added inspection, customer returns, impairment of function, and a reduction in reliability and durability.

We have structured the book around an algorithm for reducing process variation that we call *Statistical Engineering*. The algorithm is designed to solve chronic problems on existing high- to medium-volume manufacturing and assembly processes. The algorithm will not apply to urgent, short-term sporadic problems such as what to do when a defective item is found. Instead, we look at the problem of reducing the frequency of such defective items.

The fundamental basis for the algorithm is the belief that we will discover cost-effective changes to the process that will reduce variation if we increase our knowledge of how and why a process behaves as it does. A key way to increase process knowledge is to learn empirically—that is, to learn by observation and experimentation. We discuss in detail a framework for planning and analyzing empirical investigations, known by its acronym QPDAC (Question, Plan, Data, Analysis, Conclusion). We use the QPDAC framework at many stages of the Statistical Engineering algorithm to help plan, analyze, and interpret the results of appropriate investigations.

Using the algorithm, you are guided through a series of empirical investigations to a cost-effective solution to the problem. The purpose and plan for each investigation depends on:

- The stage of the algorithm
- The accrued knowledge from earlier investigations
- Other engineering and process knowledge

We classify all effective ways to reduce variation into seven approaches. A unique aspect of the algorithm forces early consideration of the feasibility of each of the approaches. Selecting a working approach helps generate effective and efficient solutions. The choice of approach affects the process knowledge required and hence how we proceed.

Some of the variation reduction approaches (but not all) require knowledge of a dominant cause of variation. We present a low-cost strategy for finding a dominant cause based on families of causes and the method of elimination. The method of elimination uses a series of simple investigations, each of which is designed to eliminate a large number of possibilities from those remaining.

We illustrate all aspects of the algorithm with many examples adapted from our experience. In some cases, we have disguised the data to protect confidentiality; in others, we have taken some liberties with what actually happened to make a point.

Throughout the book, we use the statistical software package MINITAB for all calculations and most displays. To apply the Statistical Engineering algorithm, the user requires a software package capable of making basic plots and finding simple numerical summaries. By referring to the procedures in the software, we avoid algebraic expressions for the most part. We describe the calculations with words and let the software deal with the numerical implementation.

We avoid formal or complicated statistical analysis procedures. Whenever possible, we use graphical displays to guide us to the correct interpretation of the data. We assume that the reader has been exposed to basic statistical concepts and tools, such as standard deviations, averages, histograms, run charts, box plots, scatter plots, process maps, and flowcharts. We provide detailed explanations of more sophisticated analysis tools as needed, including multivari charts, analysis of variance (ANOVA), regression, and designed experiments. We include appendices to explain how to use MINITAB to produce the analysis for all of the methods discussed.

Important issues surround the management of process improvement projects. These are the same issues that arise in any project management exercise. Priorities must be set, plans and schedules made, resources provided, and so on. We do not deal with these issues in detail; rather, we focus on the algorithm, the variation reduction approaches, and the tools required to achieve variation reduction.

The Statistical Engineering algorithm is not meant to replace global improvement systems such as Six Sigma. It is focused on and designed for process improvement in high- to medium-volume manufacturing processes. We suggest that the algorithm, strategies, and methods be incorporated into a general improvement system and used where appropriate.

TARGET AUDIENCE

The primary audience of this book is people who are involved in the improvement of manufacturing processes. They include:

- Process engineers with responsibility for reducing variation, decreasing costs, improving quality, and so on
- Six Sigma Green Belts, Black Belts, and Master Black Belts
- Trainers in process improvement methods
- Academics and students interested in quality and productivity improvement
- Teachers and students of courses in engineering statistics

WHY THIS BOOK IS NEEDED AND HOW IT WILL BENEFIT THE READER

This book is unusual because it focuses directly on the goal of variation reduction through the use of the Statistical Engineering algorithm and its associated approaches. The book is not a collection of statistical analysis tools and methods useful in achieving the goal. Having a lot of tools at your disposal does not help if you do not have a good idea what to do next or when to use a particular tool.

This book will benefit the reader in many ways. In particular, it will help you learn:

- A structured way to address variation reduction problems through a series of process investigations, each depending on what has been learned to that point
- The seven variation reduction approaches
- How to conduct empirical investigations in a sound manner to get reliable conclusions
- The method of elimination for finding a dominant cause of variation
- The appropriate use of statistical tools to support the structured algorithm
- How to assess feasibility and implement each of the seven variation reduction approaches

We present numerous examples and three case studies to convince you that the algorithm works.

HOW TO USE THIS BOOK EFFECTIVELY

While reading the text, we suggest you start an improvement project, or at least think about your own process problems. The more analogies you can draw between our examples and your processes, the better you will understand the material.

We advise the reader to try the exercises and explore the data sets to help gain confidence in the use of the approaches and methods.

STRUCTURE OF THE BOOK

This book is structured around the Statistical Engineering variation reduction algorithm. To limit length and cost, we have divided the book into two parts. In the printed text, we present the algorithm and the concepts and tools needed to use it effectively. On the enclosed CD-ROM, we provide chapter supplements, case studies, exercises, data sets, and appendices. In the chapter supplements, we give more technical details, discuss some important complications and competing methods, and give references for further reading. The printed text can stand alone, but we believe you will find the supplements helpful.

There are four major parts in the printed text:

Part I: Setting the Stage—We start by introducing the language of processes, such as outputs, fixed and varying inputs, dominant cause, and so on. Then we present the variation reduction algorithm, the seven variation reduction approaches, and a framework (QPDAC) for learning empirically by investigating the process.

Part II: Getting Started—We look at how to focus, define, and quantify a problem. We also look at methods for assessing the measurement system and for choosing a working variation reduction approach to guide further investigations.

Part III: Finding a Dominant Cause of Variation—We describe the method of elimination for finding a dominant cause that uses families of causes of variation. We provide a number of investigation plans and analysis methods to help eliminate possible causes. We introduce experimental plans to verify that we have found a dominant cause.

Part IV: Assessing Feasibility and Implementing a Variation Reduction Approach—We return to the choice of variation reduction approach in light of the results of a search for the dominant cause or a decision to skip such a search. In separate chapters, we discuss assessing the feasibility of and implementing each approach. We finally consider validating the solution and look at methods for preserving the gains.

The enclosed CD-ROM contains:

- The chapter supplements
- Three case studies
- Exercises with solutions
- All data from the examples and exercises (in both Microsoft Excel and MINITAB worksheet format)
- Appendices that will help you use MINITAB

In summary, the outstanding features of the book are:

- A structured algorithm for reducing variation in processes
- A classification of potential solutions into seven variation reduction approaches
- An emphasis on planning for data collection and simple analysis methods
- Use of the method of elimination to economically find a dominant cause of variation
- Many examples (with more than 100 datasets available) to illustrate all stages of the algorithm
- Separation of the "how to" (main text) from the supplementary material (CD-ROM)

- Demonstration of the use of MINITAB to help with the implementation of statistical tools, allowing greater focus on the interpretation of the data
- More than 65 exercises designed to reinforce the ideas and tools

We encourage readers to send us feedback regarding their use of Statistical Engineering and the related tools and methods. Our e-mail addresses are shsteiner@uwaterloo.ca and rjmackay@uwaterloo.ca.

Stefan Steiner and Jock MacKay
Waterloo, Ontario, Canada
November 2004

1
Introduction

Problems are only opportunities in work clothes.
—Henry J. Kaiser, 1882–1967

This book presents a systematic algorithm for reducing variation. The algorithm is tailored to high- to medium-volume manufacturing processes where it is feasible to measure the values of selected process inputs and outputs.

We use the word *variation* to mean both the deviation of the output from a target value and the changing value of the output from part to part. For example, in a machining process that produces V6 pistons, the target value for the diameter is 101.591 millimeters. The measured diameter in millimeters of three successive pistons is

101.593, 101.589, 101.597

We can see variation in both senses since none of the pistons has the target diameter and all have different diameters. We will formulate problems by defining appropriate performance measures that capture the nature of the variation that we want to reduce. Excessive variation leads to poor performance, low customer satisfaction, scrap and rework, complex downstream control plans, and so on. If we can resolve such problems, we can reduce costs and improve quality and performance.

The fundamental basis for the algorithm is our belief that *by increasing knowledge of how and why a process behaves as it does, we will discover cost-effective changes to the process that will reduce variation.* One way to increase process knowledge is to learn empirically, that is, to learn by observation and experimentation. Statistics is the discipline that teaches us how to learn empirically. Statistics provides the answers to questions such as "How should we plan our process investigation?" and "How do we interpret the data that we have?" The algorithm we propose relies heavily on statistical methods and tools combined with existing engineering knowledge and theory. Using the algorithm, we will plan and carry out one or more investigations to learn about process behavior.

In most cases, to be cost-effective, the proposed changes involve better process control or alterations to process settings rather than fundamental design changes or replacing process equipment. We have classified these low-cost changes into a set of generic variation reduction approaches and have structured the algorithm to force early consideration of an approach.

The specific objectives of the book are to help you to:

- Think strategically about how to achieve cost-effective variation reduction.
- Reduce variation by following a step-by-step algorithm.
- Understand sources of variation and their role in process improvement.
- Learn how to better use empirical methods; that is, learn effective and efficient ways to plan, execute, and analyze the results of a process investigation.

The purpose of this chapter is to provide examples of the types of problems we can address using the proposed algorithm. Here we discuss the problems only, not the path to their solution. We revisit these examples later in the book. We hope these problems will motivate you to read further. If you can draw analogies between your own processes and problems and those described, then we are confident that you can achieve great benefits by applying the algorithm, approaches, and methods found in the rest of the book.

1.1 TRUCK PULL

Front wheel alignment on light trucks is a set of characteristics that affect the handling of a vehicle and the life of its tires. One component of the alignment is called *pull*. Pull is an important characteristic because it indicates how well the truck will track on a standard highway. A driver can feel pull—a value close to target will produce a more drivable vehicle.

Pull is a torque, measured in Newton-meters, and is a function of right and left front wheel camber and caster angles. For the vehicles discussed here, the relation between pull and the alignment angles is

Pull = 0.23*(right caster − left caster) + 0.13*(right camber − left camber)

The alignment characteristics are measured on every truck assembled. In order to improve customer satisfaction, the manufacturer decided to reduce variation in pull around the target value. The performance of the process over a two-month period at the start of the project is shown in Figure 1.1. The histogram is based on pull values from 28,258 trucks.

The histogram was created using MINITAB and the data given in the file *truck pull baseline*. You can find this file (in both a MINITAB worksheet and an Excel spreadsheet) on the enclosed CD-ROM. See the appendices for more information on using MINITAB.

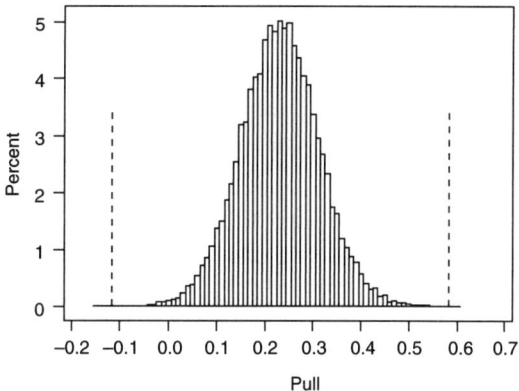

Figure 1.1 Truck pull (as found).

The dashed vertical lines on Figure 1.1 show the specification limits for pull, set at –0.12 and 0.58 Newton-meters. The target value is 0.23 Newton-meters. We see that every truck has pull within the specification limits (in fact, there are a few pull values outside these limits that are not visible in the histogram because of the large number of data points). Any truck with pull that does not meet the specifications is repaired and remeasured before shipment.

The goal of the project was to reduce the variation in pull around the target so that the histogram would look like that in Figure 1.2. If this goal can be achieved, the process will produce a greater proportion of trucks with pull close to the target value; hence, there will be greater overall customer satisfaction. As well, the proportion of trucks needing rework will be smaller, thus reducing cost.

This is a problem in reducing variation in pull from truck to truck. The process is currently centered on the target. As shown by Figure 1.1, adjusting the process center to increase (or decrease) pull on every truck will make the process worse, because then more

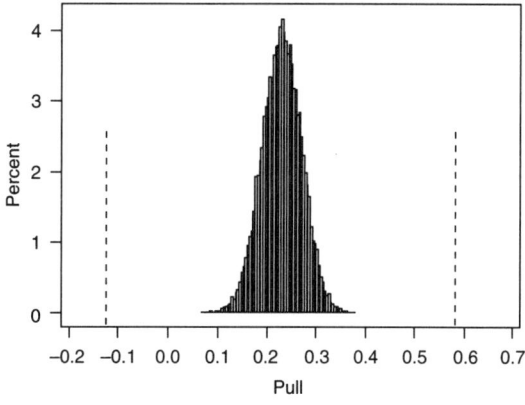

Figure 1.2 Truck pull (as desired).

4 Chapter One: Introduction

trucks will have pull outside the specification limits and fewer trucks will have pull near the target value.

We have found that excessive variation from part to part is a widespread problem. For example, many measurement systems exhibit substantive variation if the same part is repeatedly measured, especially over a long period. This variation leads to poor process control, disputes between customers and suppliers, the scrapping of good parts, and so on. Often we understand enough about a process to have an easy method to keep the center of the process near the target value. We rarely have such an adjustment to reduce the variation from one part to the next.

1.2 ENGINE BLOCK LEAKS

A cast-iron foundry manufactured engine blocks, which were then machined at an engine assembly facility. The engine plant pressure tested each block for leakage in the oil and water passages and scrapped leaking blocks. The cost of the scrap, including the wasted machining, was several hundred thousand dollars per year. This cost was assigned to the foundry because it was accepted that the leaks were generated in the casting process. The foundry management established a team to reduce the frequency of block leakers. Figure 1.3 is a run chart of the proportion of leaking blocks over several months' production prior to the start of the project. The team's goal was to reduce the proportion of leakers to less than 1%.

This example is typical of many processes that generate scrap, rework, and returns due to defects. A painting process may produce visual defects such as dirt, craters, and so forth. A molding process can generate defects such as porosity, shrinks, or inclusions; an assembly process can deliver parts that fail to function; and so on. In all these examples, a part either has the defect or it does not. The target value is no defect. The goal of the project is to reduce the frequency of the defect—that is, to reduce the variation in the output by making more parts without defects.

Problems defined in terms of a binary output rather than a continuous output can be difficult because it is harder to learn about the process. In most applications, the defective rate

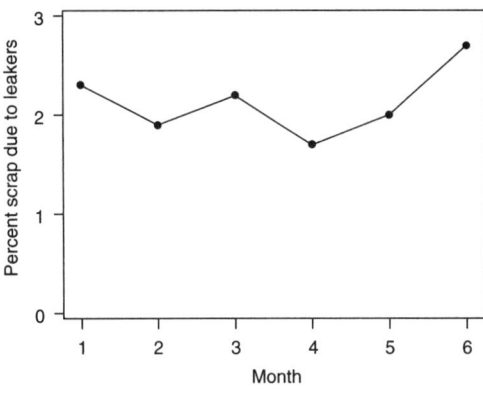

Figure 1.3 Proportion of leaking blocks versus month of production.

at the beginning of the project is likely to be low, so that investigations require large sample sizes to find a few defectives. However, we can sometimes translate a problem defined in terms of a binary output into one with a continuous output. For example, the cause of some of the leaks was the underlying wall thickness, a continuous characteristic. If an internal wall gets too thin, then a leak will occur. For these leaks, we can translate reducing the proportion of leakers into reducing variation in wall thickness about the target value. We expect to be able to learn about wall thickness variation more easily. When possible, we will translate a problem with a binary output into one with a continuous output. This is not always possible. For example, leaks due to sand inclusions cannot (easily) be directly associated with an underlying continuous characteristic, and we are forced to deal with the binary output and large sample sizes.

1.3 CAMSHAFT LOBE RUNOUT

The geometry of the lobes of a camshaft, as shown in Figure 1.4, is critical in the functioning of an engine. The rotation of the camshaft lobes drives the opening and closing of the engine valves. The displayed camshaft has 12 lobes, three of which are indicated by the white arrows.

Viewed from the side, the base of the lobe (±60° from the centerline) is ideally an arc of a circle. Figure 1.5 is a trace of the deviation from ideal, circle (in millimeters) over the base of one lobe.

Of the six measured critical characteristics related to lobe geometry, *base circle runout* was historically the most problematic and thus was chosen as the focus of a variation reduction exercise. Base circle (BC) runout is a positive measure of the deviation (maximum – minimum over the ±60° arc) of the actual lobe geometry from the ideal circle. A value of 0.000 millimeters means that the base is exactly circular. The maximum allowable BC runout is 0.040 millimeters, or 40 microns.

A sample of 108 parts, 9 per day, was collected over 12 days, and the BC runout for each of the 12 lobes was measured on each camshaft. The 1296 runout measurements (recorded in microns) and some other geometric characteristics of the lobes are available in the file *camshaft lobe runout baseline*. A histogram of the runout values over all lobes is shown in Figure 1.6.

The BC runout for all lobes was well below the specification limit, but because of the effects of this critical characteristic on engine performance, management initiated a

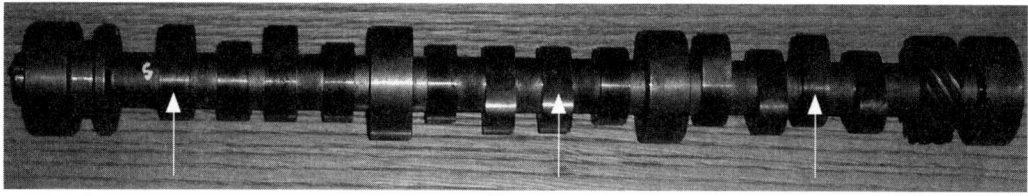

Figure 1.4   Camshaft.

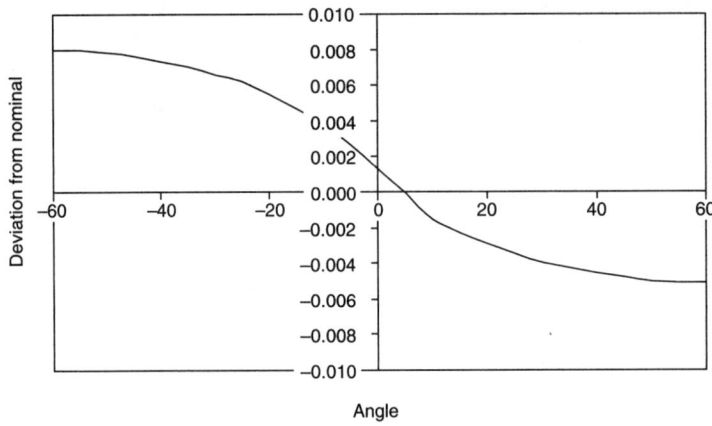

Figure 1.5 Lobe deviation from ideal.

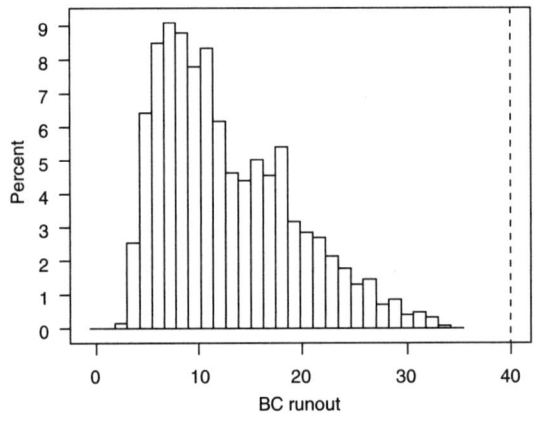

Figure 1.6 Histogram of camshaft lobe BC runout (dashed line gives the specification limit).

project to improve the process. Here the goal was to change the process so that the right tail of the histogram would be shifted to the left and a greater proportion of lobes would have runout close to zero. Because runout must be positive, adjusting the process to lower runout on all lobes was not feasible. In this context, we take variation reduction to mean that we will make a higher proportion of lobes with runout close to the target value of 0 microns.

There are many other process characteristics with a one-sided specification and a physical lower bound. Examples include flatness, porosity, taper, and so on. We can improve the process performance defined in terms of such characteristics by increasing the concentration of values near the target value. That is, we reduce the variation in the process output about the target value.

1.4 SAND CORE STRENGTH

In a cast-iron foundry, there was breakage due to handling of the molded sand cores that create the cavities within the casting. The loss of cores added cost and threatened the financial viability of the overall production of castings. The strength of cores was measured using a destructive measurement system. A sample of 100 cores (five samples spread out over a single day of five consecutive shots of the four-cavity mold) was measured to demonstrate the current process performance. The data are given in the file *sand core strength baseline*. The histograms in Figure 1.7 show the initial performance and the project goal in which the strength would be increased by three pounds on each core.

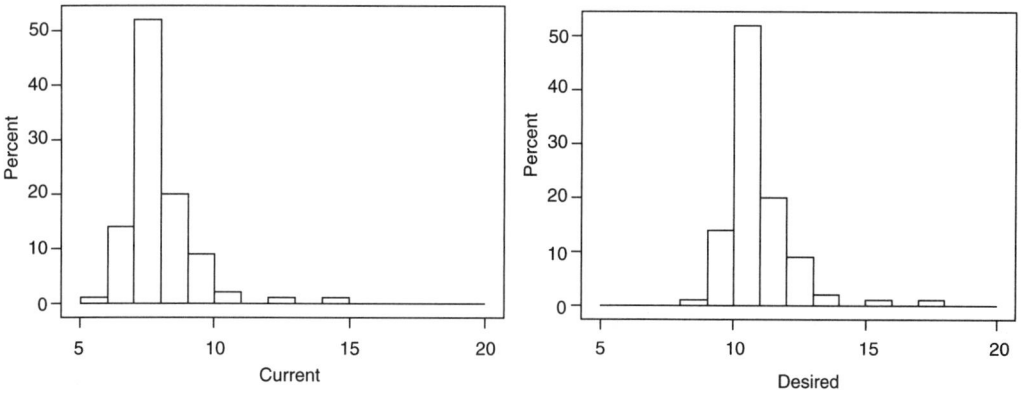

Figure 1.7   Current and initial desired core strengths (pounds).

There was no obvious target for core strength. The specification for the core was that it not be broken. The team planned to increase core strength by increasing the concentration of resin in the core-making process. On first glance, this is an example of a problem where we can improve a process by making a one-time adjustment that increases or decreases the output characteristic on all parts, that is, by changing the process center. Later, the team discovered that overly strong cores caused casting defects. They changed the goal to eliminate cores with low strengths. In other words, the new goal was to reduce variation in strength from core to core.

It is often easy to find a low-cost solution to shift the process center. The challenge is to avoid side effects—in which we replace one problem by another, as in the sand core strength example.

1.5 CRANKSHAFT MAIN DIAMETER

In a process to machine crankshaft main journals, there was excess diameter variation. The histogram of the diameter at the front position of the first main at the start of the project is shown in Figure 1.8. Note the diameter was recorded as the deviation from the target value, measured in microns. The specification limits were ±4 microns.

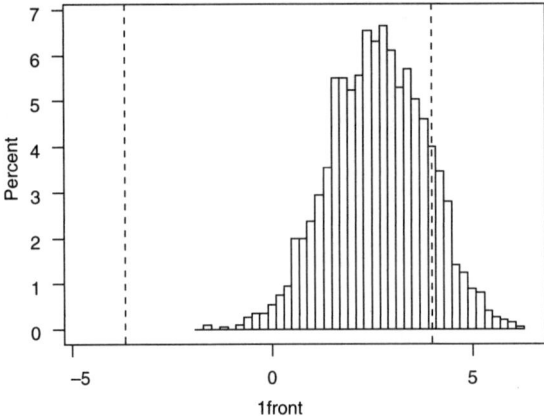

Figure 1.8 Histogram of crankshaft diameters for the front position of the first main.

The process was off target, and it seemed a simple matter to adjust the upstream grinders to reduce the average diameter closer to zero. Again, there were negative consequences. The process operators knew that crankshafts with oversized journals could be reworked but those with undersized journals were scrapped. They had deliberately centered the process above the target to avoid costly scrap. Here the goal is first to reduce piece-to-piece variation and shrink the width of the histogram. Then they could adjust the center of the process to the target without the risk of scrapped parts.

1.6 PAINT FILM BUILD

In a painting operation, there was excessive variation in film build (paint thickness) from vehicle to vehicle at particular locations. As a consequence, to meet the minimum film build specification of 15 thousandths of an inch, the target was kept well above the specification at 17. The film build of five consecutive cars was measured at five locations on the front door every hour for 16 hours. The data are given in the file *paint film build multivari*. A histogram of the film build values is given in Figure 1.9. We see that all film build measurements are above the minimum specification limit. However, running the process above target results in high paint usage and creates visual defects such as runs on occasion. The paint shop management initiated a project with the goal of reducing variation in film build. With lower variation, the film build target could be reduced, resulting in cost savings due to reduced paint usage and rework.

Variation reduction projects are often linked with productivity goals. They may involve cost reduction, as in the paint film build example. In other situations, variation reduction may lead to increased throughput. For example, by virtually eliminating the need for rework due to imbalance in a brake rotor production process, the team was able to increase the daily volume.

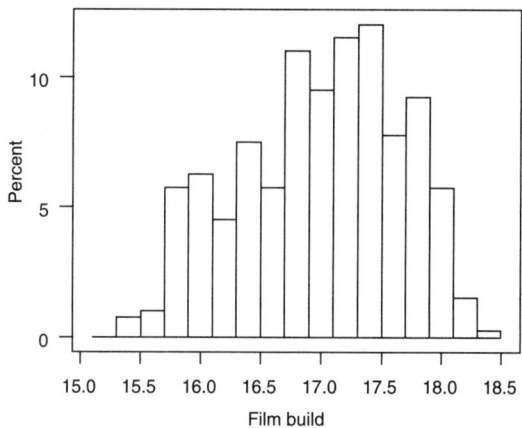

Figure 1.9   Histogram of film build.

1.7 REFRIGERATOR FROST BUILDUP

The manufacturer of a so-called frost-free refrigerator designed for temperate climates expanded its market to several tropical countries. Almost immediately, there were many customer complaints about the frost buildup inside the refrigerator. These complaints had been nonexistent in the traditional market. A team was formed to reduce the number of complaints. The manufacturer feared that the opportunity for market expansion could easily be lost.

There was no clear baseline to specify the problem because there were relatively few refrigerators in service in the new market, and there were difficulties tracking which refrigerators were in use and which were still to be sold. The goal of the team was to deliver the promised frost-free refrigerators to the customers in tropical climates, as measured by a reduction in the number of complaints.

This problem is different in several ways from those described earlier. First, at the start of the project, there was no objective way to measure the output characteristic of interest. Whether frost built up or not could be determined only during customer usage. Second, the circumstances under which the problem occurred were better understood, because there had been no similar complaints of frost buildup in the traditional temperate climate market. In fact, the team was quickly able to identify the likely causes of the problem. The ambient temperature and humidity were higher in the new markets, and also customer usage was different. Some of the new customers opened and closed the refrigerator more often than expected and also added large volumes of warm items to the refrigerator all at once. These suspected causes of frost buildup were not under the control of the manufacturer. They dominated the effects of inputs that changed during manufacturing and assembly.

Many problems have known or identified causes. In an automotive painting operation, process engineers discovered that a visual defect on the roof of the car could be eliminated by reducing the thickness (film build) of the clear coat, the last layer applied in the process.

However, there was a heavy price to pay for this supposed solution due to poorer overall appearance of the painted surface. Using the algorithm given in this book, the engineers found a solution that allowed film build to be increased without the appearance defect occurring.

 **Key Points**

- We can describe different features of variation in the specification of a problem:

    –Part-to-part variation

    –Deviation from target

    –Defect rate

- We can generate low-cost ideas for variation reduction if we better understand how and why a process varies.

- We can gather this knowledge through a series of process investigations supplemented by existing knowledge and theory.

PART I
Setting the Stage

If I had to reduce my message to management to just a few words, I'd say it all had to do with reducing variation.

—W. Edwards Deming, 1900–1993

In this first part of the book, we explore the meaning and consequence of process variation, provide tools for quantifying variation in an understandable way, and discuss various sources of variation. We also present an algorithm for reducing variation, centered around seven variation reduction approaches. We explain the seven variation reduction approaches and give examples. Finally, as the key to improvement is acquiring new process knowledge, we discuss the QPDAC (Question, Plan, Data, Analysis, Conclusion) framework, used to plan and analyze process investigations.

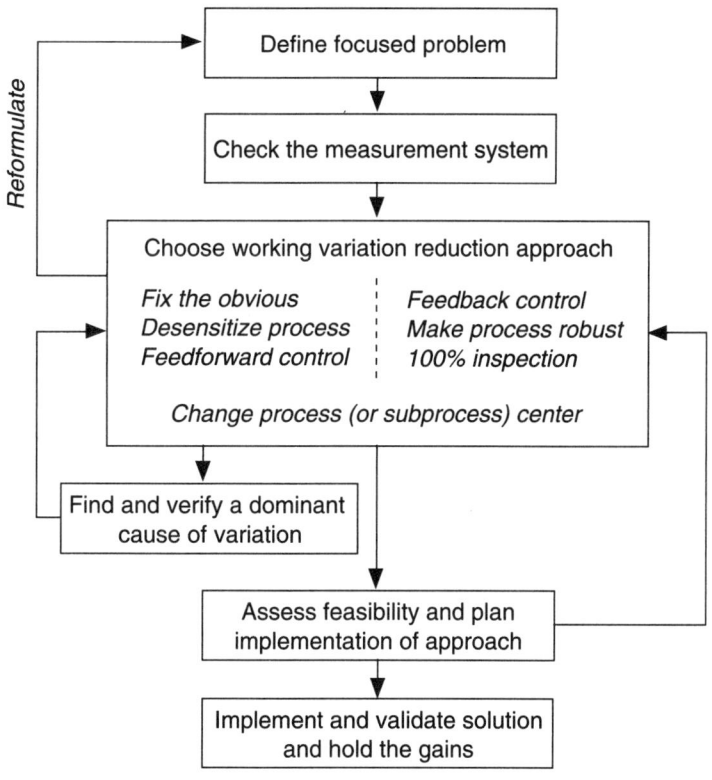

Statistical Engineering variation reduction algorithm.

2
Describing Processes

Failure to understand variation is a central problem of management.

—Lloyd S. Nelson

This book is about how to reduce variation in process output. Increased costs and decreased customer satisfaction are two of the negative consequences of excess variation. In this chapter, we provide some common language and tools to describe processes and variation.

2.1 LANGUAGE OF PROCESSES

First, we introduce some terms to describe a process and its behavior. You will likely be familiar with these terms; we introduce them here so that we can use them without confusion throughout the book. For a running example, we look at a process used to manufacture cast iron exhaust manifolds, as shown in Figure 2.1. In the casting process, scrap iron is melted and doctored to adjust its chemistry and physical properties such as temperature.

Figure 2.1 Exhaust manifold.

Sand molds are formed to determine the external shape of the manifold. Cores are molded to create internal space in the casting. A core is placed in the mold and the molten iron is poured. After cooling, the sand is shaken out, and the result is a rough casting. The casting is finished by machining various surfaces and drilling holes. Throughout, operators make measurements and process adjustments. They also inspect the castings for defects at several points.

A process can be divided into *subprocesses* and is almost always a part of a larger process. In the casting example, the melting of the iron, the creation of the mold, and the core-making are all examples of subprocesses. The manufacturing process for the manifold sits inside a system that includes the design process for both the part and the manufacturing process; the sales, order, and billing processes; the delivery process; and so on.

A *process map* or *flowchart* is a good tool to describe a process, especially if we choose an appropriate level of detail. For example, we can represent the major subprocesses or operations within the casting process by the simple flowchart in Figure 2.2.

This flowchart shows the major subprocesses and the order in which they occur. We can describe the subprocesses in much finer detail if we choose. The flowchart clarifies which operations in the process are parallel and which are sequential. The chart also shows the boundaries that we have selected for the process. We could have chosen to include more upstream or downstream subprocesses as part of the manufacturing process. The selection of process boundaries, like the choice of the level of detail, is driven by a tradeoff between presenting facts and providing information. Too much detail can obscure the information in the chart. We take these issues into consideration when designing a useful flowchart. See Harrington (1987) for a detailed discussion of chart selection, construction, and use.

One important feature of a process is that it is repeatable; each time it operates, the process produces a *unit*. In the example, each finished casting is a unit. For a manufacturing process, we can think of a unit as a part. For a measurement process, a unit is the act of making a measurement.

We use the word *characteristic* to describe a feature or quality of a unit. For the casting process, characteristics of a manifold include its hardness, its dimensions, the pouring temperature of the iron, the properties of the sand in the cores used for that casting, and so on. The *customers* are the people or organizations that use the process units. Here, the customers include the assemblers of the engine and the ultimate users of the vehicles that

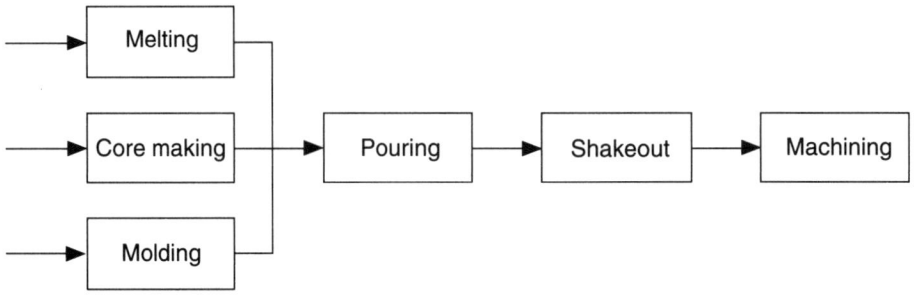

Figure 2.2 Exhaust manifold process map.

incorporate the engines. *Output characteristics,* or more briefly *outputs,* are characteristics of interest to customers. The properties of the casting that describe its performance in the engine and ease of assembly in the engine plant are outputs. We call characteristics of the casting such as the pouring temperature of the iron and the properties of the sand cores *input characteristics,* or *inputs* for short. We also classify characteristics in terms of the values they can take.

Classification	Description	Examples
binary	only two possible values	defective or not, high or low
discrete	integer values	number of defects
ordinal	ordered categories	score: low, medium, high
categorical	unordered categories	supplier: A, B, C
continuous	a range of real numbers	length, hardness, time

The *suppliers* are the people or organizations that control the subprocesses that determine many of the inputs. In the casting process, some suppliers are the providers of the sand for the cores and other raw materials, the equipment manufacturers, and so on. To complete the picture, we call the people involved in the process its *owners.* In the example, the owners are the process operators and managers. The owners are interested in the cost associated with each unit as well as many other input and output characteristics.

We classify input characteristics as *fixed* or *varying.* An input is fixed if it changes only when we deliberately make it change. For example, the target value for the pouring temperature of the iron is a fixed input. The instruction to the operator to sample five parts every hour, measure the hole locations, and make an appropriate adjustment is a fixed input. The fixed inputs may be changed by the process owners. An input is varying if its value changes from unit to unit or time to time without deliberate intervention. For example, the dimensions of the cores change from casting to casting. Other varying inputs, such as pouring temperature and raw material characteristics, change more slowly over time.

Note that the target value or set point for pouring temperature is a fixed input, but the actual pouring temperature is a varying input. We need to keep this idea clear because sometimes inputs that we consider fixed actually vary. For example, two operators may interpret the same set of instructions in different ways. The instructions are a fixed input, but the implementation of the instructions is a varying input.

2.2 CAUSES OF VARIATION

To reduce variation, we must change one or more fixed inputs of the process. In Chapter 3, we consider some approaches to variation reduction in which we first identify a dominant cause of the variation and then change fixed inputs to reduce the effects of these causes. In this section, we make clear this notion of a dominant cause. We continue with the casting and machining example to illustrate the points.

In Chapter 1, we saw that most problems involved reducing the unit-to-unit variation. We define a cause with respect to this kind of variation.

A *cause of variation in a process output* is a varying input with the property that if all other (varying) inputs were held constant, then the output changes when the input changes. Part of this definition is conceptual since it is not possible to hold all other inputs constant for all units.

In the example, suppose the output is the hardness of the casting measured at a particular location. Hardness varies from casting to casting. The concentration of carbon in the iron when it is poured into the mold is a cause of hardness variation. When this concentration changes, all other inputs being constant, the hardness of the casting will change. There are many causes of hardness variation.

Now for our first controversial statement. A fixed input cannot be a cause of variation. For example, the design of the product or process is not the cause of variation since the design is a fixed input. Since we are interested in problems defined in terms of variation, we will never say that the design is the cause of the problem. As you will see, we will change one or more fixed inputs to solve the problem, but these are not the causes.

We use a simple mathematical model to describe a cause by specifying the values of all varying inputs (again, this is only conceptually possible for a real process) for any one unit produced by the process. We write the functional model

$$output = f(input1, input2, ...)$$

With this model, *input1*, for example, is a cause if the *output* changes when the value of *input1* changes while all other inputs are held constant. Note that the function $f(\ )$ depends implicitly on the values of the fixed inputs.

The model is useful because it helps us understand what is meant by the effect of a cause. The *effect* of the cause (or varying input) is the change in the output produced by a change in the input. The effect depends on the size of the change in the input, the initial value of the input, and perhaps the values of the other inputs. A cause has a *large effect* if a relatively small change in the input produces a relatively large change in the output. We define a small change in the input and a large change in the output relative to the variation we see in these characteristics under regular operation of the process. We simplify the language if we call all varying inputs *causes*, even those with no effect.

For any process output, there are likely to be a large number of causes, each with an effect. We assume that the Pareto principle[1] (Juran et al., 1979) will apply and that only a few causes will have large effects. We call these causes *dominant*. We base our strategies and approaches to reducing variation on the assumption that there will be only one or two dominant causes. We justify this focus on dominant causes more fully in Section 2.4.

We can model the effect of a dominant cause as

$$output = f(dominant\ cause) + noise,$$

where $f(dominant\ cause)$ is a function that captures the effect of the dominant cause and the term *noise* captures the effect of all the other inputs. For a dominant cause, the range in $f(dominant\ cause)$ is greater than the output variation due to the noise.[2]

A dominant cause can be a single input or involve two or more inputs in a variety of ways. Figure 2.3 shows a dominant cause for a binary output that is either good (G) or bad (B). The horizontal line represents the normal range of values of the input. Small values of the input to the left of the dotted line correspond to good output, large values to bad output.

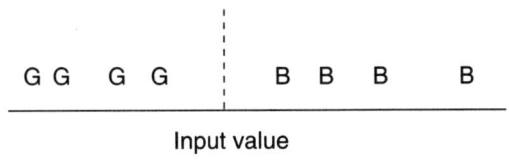

Figure 2.3 A dominant cause with continuous input and binary output.

Figure 2.4 shows another example of a dominant cause. The left-hand plot in the figure shows a continuous input and the right-hand plot a discrete input. In either instance, we see the full range of variation for both the input and output on their respective axes.

If both the input and output are binary (or discrete with a few possible values), we can depict a dominant cause using a table of percentages such as Table 2.1.

We hope to find a single dominant cause of variation such as shown in these examples. We may fail for several reasons. First there may be no single dominant cause; instead, we may find several causes, each with a relatively large effect. Second, we may find that the dominant cause involves two (or more) inputs.[3] Figure 2.5 shows two examples of a dominant

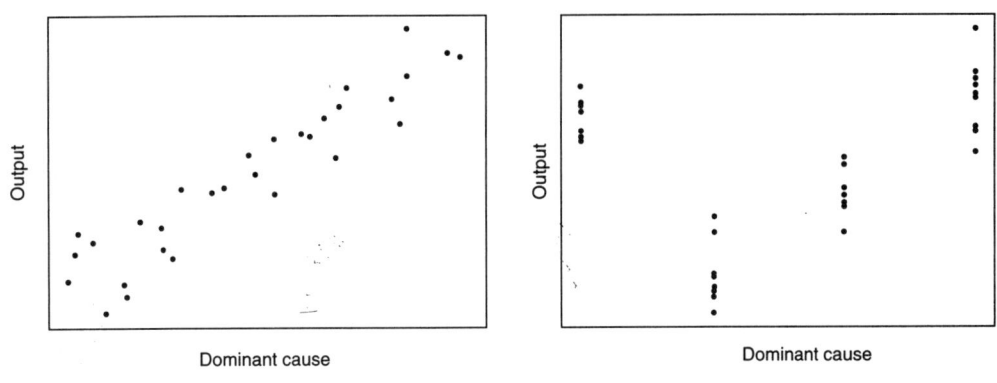

Figure 2.4 A dominant cause with continuous output.

Table 2.1 A dominant cause with binary input and output.

	Output pass	Output fail
Input low	80%	20%
Input high	97%	3%

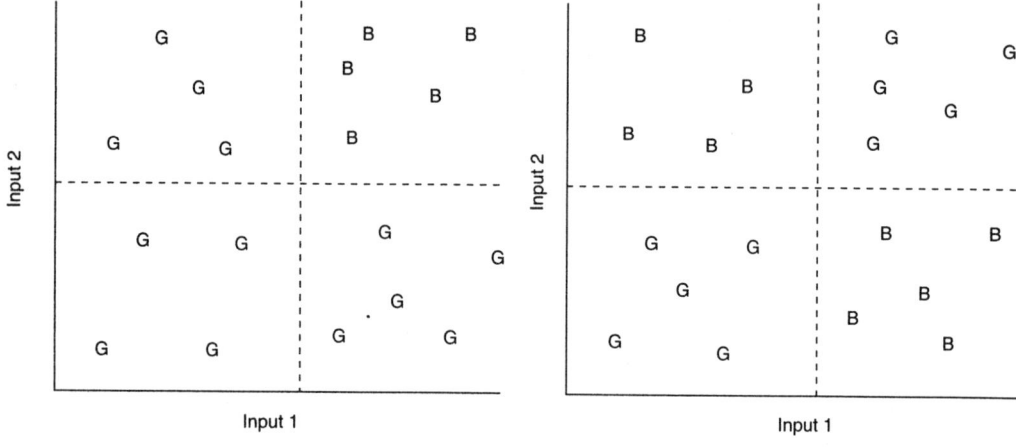

Figure 2.5 Two examples of a dominant cause involving two inputs with a binary output.

cause involving two continuous inputs, and a binary output where we denote good output as *G* and bad output as *B*. In the example presented in the left panel, both inputs must be simultaneously large in order to get bad output. In the right panel, good output results when the two input values are both large or both small. The latter case may arise if the output is a clearance between two assembled components whose critical dimensions are described by inputs 1 and 2.

Causes can be classified in many ways.[4] For us, the key issue is the size of the effects. We want to find dominant causes that contribute substantially to the variation in the output.

2.3 DISPLAYING AND QUANTIFYING PROCESS VARIATION

We will need to quantify process variation at several stages in the variation reduction algorithm. For example, we may set the goal to reduce the variation by 50%. We require a numerical measure of variation to define this goal explicitly.

We use the camshaft manufacturing process described in Chapter 1 to illustrate. The angle of the lobe axis is one important characteristic of the camshaft lobes. The specifications are ±400 thousandths of a degree measured as the deviation of the lobe axis from the nominal direction. A project team collected a sample of 108 camshafts, 9 per day over 12 days, and measured the angle error for each of the 12 lobes. The 1296 measurements of angle error and several other outputs are available in the file *camshaft lobe runout baseline*.

For a continuous output with two-sided specifications such as the angle error, we use the average and standard deviation (*stdev*) to summarize process behavior. From MINITAB, we get the following output.

```
Variable        N        Mean      Median     TrMean      StDev    SE Mean
angle        1296      -21.30     -18.00     -20.04      71.50       1.99

Variable           Minimum     Maximum          Q1         Q3
angle              -241.00      155.00      -67.00      30.00
```

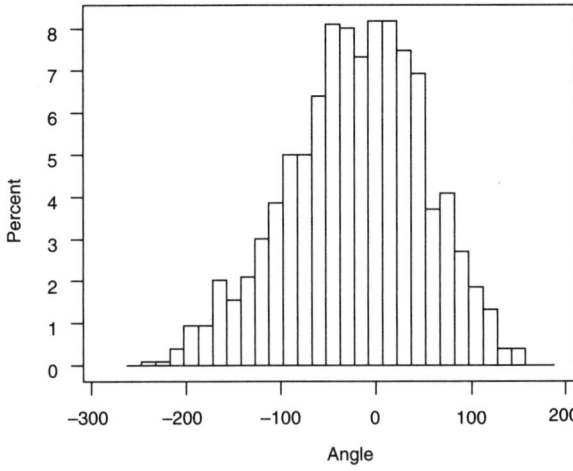

Figure 2.6 Histogram of camshaft lobe angle errors.

The average is –21.3 and the standard deviation is 71.5, as given in the columns Mean and StDev, respectively. See Appendix B for an interpretation of the other summary measures.

For the camshaft example, we can interpret the average and standard deviation using the histogram of angle error given in Figure 2.6.

The average is the point on the horizontal axis where the histogram would balance if we could cut it out of the page. In Figure 2.6, since the histogram is roughly symmetric about zero, the average (the balance point) is close to 0°. The width of the histogram is approximately six standard deviations when the histogram has a bell shape. Here, the width of the histogram is 400 thousandths of a degree and 6 *stdev* = 429. For a bell-shaped histogram, almost all of the characteristic values will fall within the range

$$average \pm 3\ stdev$$

From this argument, we see that the standard deviation is a measure of the unit-to-unit variation. The average is a measure of the process center. The distance from the average to the target is a measure of how well the process is targeted and hence a measure of the off-target variation. The average and standard deviation are sometimes combined in a capability ratio,[5] which can describe both kinds of variation simultaneously.

In the example, the target value for angle error is 0°. If the collected data represent the long-term behavior of the process, we cannot reduce variation significantly by better centering the process, that is, by adjusting the average to the target. Here, we must reduce the standard deviation to get a substantive reduction in the variation.[6]

Why is the standard deviation so large? Since the standard deviation measures the variation in angle error from lobe to lobe, there must be changes in varying inputs from lobe to lobe in the sample that explain the angle error variation. To reduce the standard deviation, we may first try to identify dominant causes; that is, varying inputs making major contributions to the standard deviation. Then we can try to eliminate the effects of these causes. We make this idea clear in the next section.

20 Part One: Setting the Stage

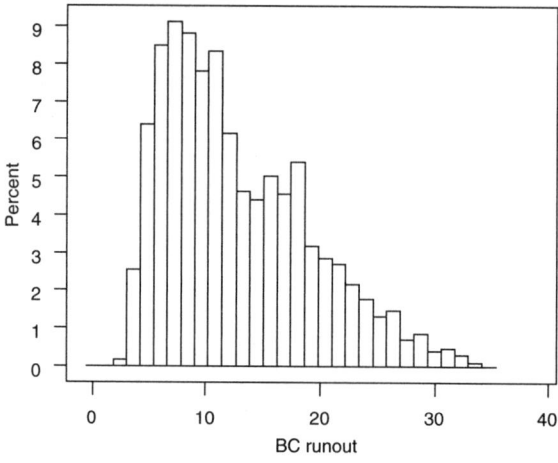

Figure 2.7 Histogram of camshaft lobe BC runout.

In Chapter 1, we described a problem in which the goal was to reduce variation in base circle (BC) runout, another characteristic of a camshaft. In that case, the target value was zero and the histogram of runout values is shown in Figure 2.7.

The average and standard deviations (given as follows by MINITAB) are roughly 12.6 and 6.4 microns, respectively.

```
Variable       N       Mean    Median    TrMean     StDev    SE Mean
BC runout    1296     12.643   11.100    12.271     6.389    0.177

Variable    Minimum   Maximum              Q1         Q3
BC runout    2.600    33.900             7.425     17.000
```

The histogram is not bell-shaped. The average ± 3 standard deviations is (–6.6, 31.8). We see that some values are above 31.8 and none are close to –6.6. We no longer have the interpretation that almost all of the values fall within three standard deviations of the average. However, we can still interpret the standard deviation as about 1/6 of the range since $(33.9 - 2.6) \div 6 = 5.2$. Here, to reduce variation around the target, we need to shift the average to the left and reduce the standard deviation. If we identify the dominant cause, we may find a low-cost way to reduce the variation.

Somewhat surprisingly, adjustment of the process center to the target can play a role in reducing unit-to-unit variation. To see how this happens, consider again the angle error data. Recall that there are 12 lobes on each camshaft with the positions numbered 1 to 12. We can use a box plot (see Appendix C), as given in Figure 2.8, to compare the performance of angle error from lobe position to position.

The lobe averages range from roughly 45 (lobe 2) to –90 (lobe 12). If we adjusted the process on each lobe separately so that the angle error average was on target for each lobe, then the overall standard deviation would be reduced to 61.6 from 71.5. You can see this result qualitatively by imagining all of the boxes in Figure 2.8 being shifted vertically to

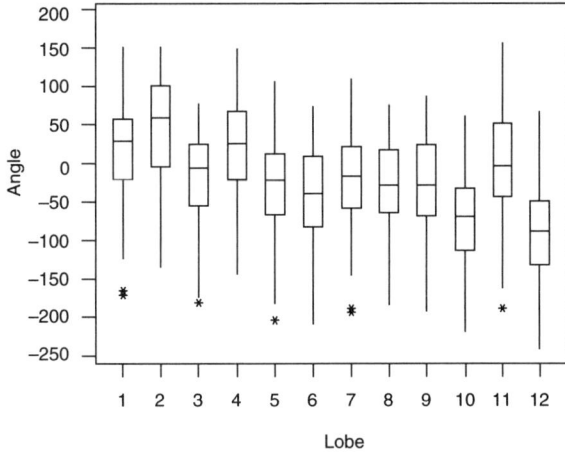

Figure 2.8 Box plots of camshaft lobe angle errors by lobe position.

have a centerline on zero. Then the range of values shown by the whiskers on the box plots and the overall standard deviation would be smaller.[7]

Data summaries such as the average, standard deviation, histogram, and box plots do not show how the process output varies over time. To show this behavior, we use a run chart, a simple plot of the output values against the order or time of collection. The run chart for the angle-error data from lobe 12 is shown in Figure 2.9.

The run chart shows how the process output varies over time. We may see cycles and smooth patterns on such a plot. The time structure of the output variation is important when we define the problem baseline (Chapter 6) and when we consider feedback controllers (Chapter 18). In Figure 2.9, we see most of the full range of variation from one camshaft to the next. There is no obvious longer-term pattern. In the production of V6 pistons, the diameter was recorded for one process stream at Operation 270 every minute for 200 minutes.

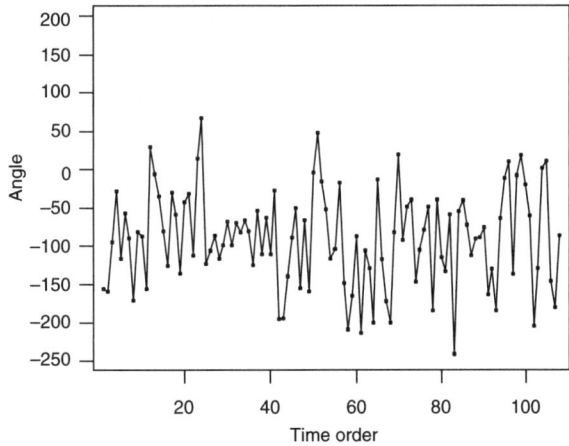

Figure 2.9 Run chart for angle error at lobe 12.

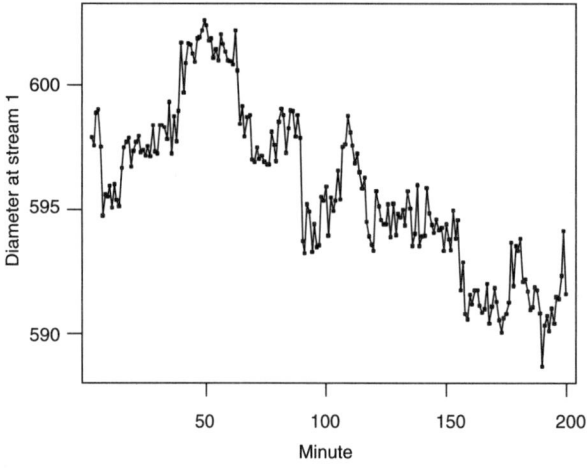

Figure 2.10 Stream one V6 piston diameter at operation 270 by hour.

The run chart is given in Figure 2.10. Here, the diameter shifts occasionally but relative to the long term, the output variation is small over the short term.

We can apply these numerical summaries and plots to data for a continuous output. If the output is binary (typically defective or not) or categorical with only a few values, we use proportions to quantify process variation. For example, in the process that produces manifolds described in Section 2.1, there are two manifolds produced in each mold. During an investigation, a team collected a sample of 40 castings, 20 from each cavity labeled A and B. The team classified each casting as defective (1) or not (0). We summarize the variation using the proportion defective. From MINITAB we get

```
Tabulated Statistics: output, cavity
  Rows: output     Columns: cavity
              A        B      All
  0          20       18       38
  1           0        2        2
  All        20       20       40
```

In this case, $2 \div 40 = 5\%$ of the castings in the sample are defective. The corresponding percentages are 0% and 10% for cavities A and B, respectively. We do not find histograms or bar charts helpful for these data. We can sometimes use a run chart to display clusters of defectives over time.

We have presented a number of charts and statistics for quantifying and describing process variation. We construct these summaries from a sample of units collected over time from the process. We must be careful that we collect data that gives us an accurate picture of the long-term process performance. For example, we would not have seen the structure of the variation over time in Figure 2.10 as clearly if we had sampled 200 pistons over a few hours rather than over 200 hours. We deal with this critical idea technically in Chapter 5 and practically throughout the book whenever we describe a process investigation. The bottom line is that we will not make progress without careful thought on how we collect process data.

2.4 MODELS FOR VARIATION AND THE EFFECTS OF CAUSES

Throughout the book, we use a simple model to connect the causes and the output. The model will help us identify dominant (and unimportant) causes of variation and to quantify their contributions to the variation in the output. As well, we can use the model to understand how to combine contributions from various causes and to understand how variation is transmitted through a process. The functional model, *output* = f (*input1, input2, ...*), which we introduced in Section 2.2, is useful to explain the idea of a cause and its effect but is too complex to apply in practice.

The basis of the model is a mathematical representation of a histogram that specifies the relative frequency with which different output values occur. To avoid confusion with the measured values, we denote the output in the model by an uppercase letter, typically Y. We associate a smooth curve, an idealized histogram, with Y as shown in Figure 2.11. In this case, the model is a Gaussian curve[8] (also called a bell or normal curve) that can be specified by two *parameters,* the mean (or center of the symmetric curve) and the standard deviation, a measure of the spread of the values. The mean and standard deviation (*sd*) associated with the model are directly analogous to the average and *stdev* of the histogram. The deviation between the mean and the target represents the off-target variation and the standard deviation describes the variation from unit to unit.

When we apply the model, we estimate the mean and standard deviation using the values of the output characteristic in a sample of units from the process.

The model behaves as expected if we rescale the output. Suppose the output value Y is changed by first multiplying by one constant (b) and adding another (a) so that

$$Y_{new} = a + bY_{old}$$

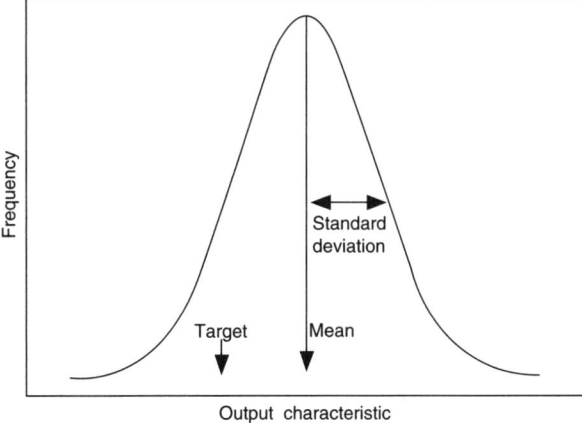

Figure 2.11 A Gaussian output model.

Then, in the model, the mean and standard deviation are changed to

$$\text{mean}(Y_{new}) = a + b\,\text{mean}(Y_{old}), \quad sd(Y_{new}) = |b|\,sd(Y_{old})$$

For example, suppose the model for a temperature output has mean 1234°F and standard deviation 56.7°F. If we change from Fahrenheit to Celsius (°C = 0.556°F − 17.778), then the new model has mean 668°C and standard deviation 31.5°C. Note that if we multiply Y by negative one, the standard deviation in the model does not change.

We want to decompose the model for the output into pieces corresponding to the effects of causes. Consider the following example. Suppose that two components A and B are stacked and the height of the assembly y is the output of interest. This is illustrated in Figure 2.12.

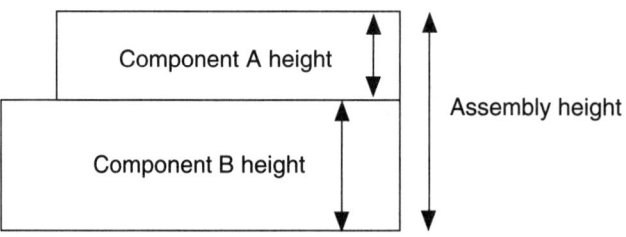

Figure 2.12 Stack height.

We have a model for the height of each component as described in Table 2.2. The means and standard deviations are based on data collected from the process that produces each component.

Here X_A and X_B represent the effect of component A and component B on the stack height. As such, we can construct a model for stack height from the two models of component heights, namely

$$Y = X_A + X_B$$

Adding the models does not mean that we add the idealized histograms. Fortunately, there are simple rules to describe how to combine the means and standard deviations. For the means, we get

$$\text{mean}(Y) = \text{mean}(X_A) + \text{mean}(X_B)$$

Table 2.2 Model for each component.

Component	Model	Mean	Standard deviation
A	X_A	12.212	0.045
B	X_B	17.567	0.033

In words, the mean of the sum is the sum of the means. This formula has important consequences when we adjust a process center to target by changing a fixed input. For example, we can use the model to predict the effect on the output if we shift the mean of component A.

For the standard deviation, we have

$$sd(Y) = \sqrt{sd(X_A)^2 + sd(X_B)^2} \tag{2.1}$$

In words, the standard deviation of the sum is the square root of the sum of the squares of the standard deviations of the terms in the sum. There is a demonstration in the exercises to convince you that this key formula is true. Equation (2.1) applies when the two component heights vary independently.

Using the model, the mean and standard deviation of the assembly height are

$$mean(Y) = 12.212 + 17.567 = 29.779, \quad sd(Y) = \sqrt{(0.045)^2 + (0.033)^2} = 0.056$$

Equation (2.1) has many important consequences. The standard deviation of the sum is much less than the sum of the standard deviations. This is good news when you are building up assemblies, because the overall variation will be less than the sum of the component variation. However, it is bad news when it comes to reducing variation. To see why, suppose that (for a price) we contemplate reducing the standard deviation of the height of component B by 50% from 0.033 to 0.016. We can use the model and Equation (2.1) to predict the impact on the variation of the assembly height. The effect on the standard deviation of the assembly height is surprisingly small; the standard deviation becomes 0.048, a 15% reduction.

More generally, suppose we consider only two sources of variation, one attributed to a particular cause and the second to all other causes. To be specific, suppose that the specific cause (actually a group of causes) is the measurement system, so that all other causes are responsible for the variation in the true value of the outputs. If the effects are additive and independent, at least approximately, the overall standard deviation is

$$sd(\text{total}) = \sqrt{sd(\text{due to measurement})^2 + sd(\text{due to rest})^2}$$

Now suppose that the standard deviation due to the measurement system is 30% of the overall standard deviation, that is, the ratio of sd(due to measurement) to sd(total) is 0.3. What is the percentage reduction in overall variation if the variation due to the measurement system is eliminated?

We constructed Figure 2.13 to give the percent reduction in overall standard deviation if we could eliminate completely the contribution of an identified cause. From the plot, we see that when the ratio of the sd(due to cause) to the sd(total) is 0.3, the potential gain is about 5%. In other words, if the ratio of measurement system variation to total variation is 0.3, you can reduce the overall standard deviation by at most 5% if you replace the current measurement system with one that is perfect.

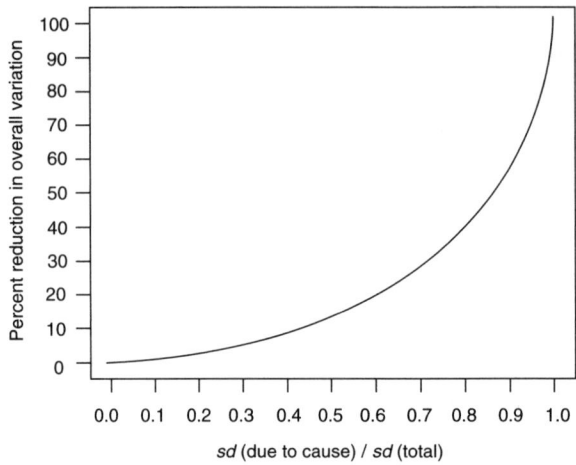

Figure 2.13 Reduction in variation if we remove a cause contributing a given proportion of the total variation.

In the stack height example, the ratio of $sd(X_B)$ to $sd(Y)$ is 0.59, and thus, by eliminating the variation in component B completely we can reduce the overall standard deviation by about 20%. Similarly, the ratio of $sd(X_A)$ to $sd(Y)$ is 0.80, and by eliminating the variation in component A, we reduce the overall standard deviation by about 41%.

The message from Figure 2.13 is that we need to address dominant causes of variation in order to make a significant reduction in variation. If we rank the potential gains (in terms of percent reduction in overall standard deviation) by eliminating the contributions of various causes, we expect to see a Pareto chart like Figure 2.14. There is little opportunity for gain in identifying causes with small contributions to the overall standard deviation. For this reason, the proposed variation reduction algorithm (see Chapter 4) focuses on finding and dealing with dominant causes.

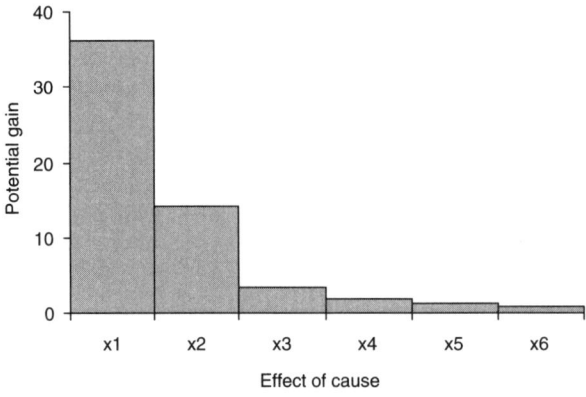

Figure 2.14 Pareto chart of potential gain.

We can also use a model to describe how variation is transmitted through the process. Suppose the scatter plot of the output versus some specific input is given by the left-hand side plot in Figure 2.4. We adopt the model

$$Y = a + bX + noise,$$

where Y represents the output, $a + bX$ the effect of the identified cause, X, and *noise* the rest of the causes, presumably unidentified. The parameters a and b describe the linear relationship between Y and X. From the model we can see that

$$sd(Y) = \sqrt{b^2 sd(X)^2 + sd(noise)^2}$$

In this example, we denote *sd* (due to *X*) by $b*sd(X)$. Using the model, we can predict how much of the standard deviation in Y is due to changes in X. For example if we could hold X fixed (that is, make its standard deviation 0), then the standard deviation of Y will be reduced to standard deviation of the noise. We can estimate this residual standard deviation by fitting a line to the scatter plot in Figure 2.4 (see chapters 10 to 12). We can also contemplate changing a fixed input to change the slope b; again, we can use the model to predict the effect of this change.

We can construct similar models for binary outputs. These are more complex and prove less useful so we omit them here.

Key Points

- Process descriptions should be kept as simple as possible while still capturing the needed information.
- Variation is an attribute of the process, not of single units produced by the process.
- We can quantify and describe process variation using summaries such as average, standard deviation, histograms, and run charts.
- Reducing variation in a cause has a substantial effect on the overall variation only if the cause is dominant.
- To determine the effect of various sources of variation on the overall variation, we have

$$sd(\text{overall}) = \sqrt{sd(\text{due to a specific cause})^2 + sd(\text{due to all other causes})^2}$$

End Notes (see the Chapter 2 Supplement on the CD-ROM)

1. The Pareto principle as applied to variation reduction problems is explored in more detail in the supplement.
2. We give a more formal derivation of the definition of a dominant cause. Some possible input-output relationships are also explored.
3. In the chapter supplement, we define what it means for a dominant cause to involve two (or more) inputs and explore the consequences.
4. We can classify causes of variation in many ways. For example, in Statistical Process Control (SPC), we speak of special and common causes. Taguchi (1986) calls causes controlled or noise and subdivides these classes even further. We look at these classification systems and their relationship to variation reduction.
5. There is an alphabet soup of capability rations—such as P_{pk}, C_{pk}, and so on—used to quantify process performance and its relationship to specifications. We give a brief discussion of these indices that are sometimes used to set the goal for a variation reduction project.
6. We have described two kinds of variation, deviation of the average from the target and variation from unit to unit. We provide a formula that relates the two kinds of variation to the "variation from the target," the single measure of variation that is likely related to cost.
7. In the camshaft lobe BC runout example, we showed how aligning the average angle error for each lobe could reduce the overall standard deviation. We give a key formula widely used in the analysis of variation to connect the overall standard deviation to the variation within groups and group to group.
8. The Gaussian distribution is widely applicable. We describe some of its key properties.

 Exercises are included on the accompanying CD-ROM

3
Seven Approaches to Variation Reduction

A fool can learn from his own experiences; the wise learn from the experience of others.

—Democritus, 460–370 B.C.

There are many ways to change fixed inputs in the process to reduce variation in an output characteristic. In this chapter, we classify methods for variation reduction into seven generic approaches (MacKay and Steiner, 1997–98):

1. Fixing the obvious based on knowledge of a dominant cause of variation
2. Desensitizing the process to variation in a dominant cause
3. Feedforward control based on a dominant cause
4. Feedback control
5. Making the process robust to cause variation
6. 100% inspection
7. Moving the process center closer to the target

We must identify a dominant cause of the variation for the first three approaches but not the final four.

To implement any of the approaches, we need to change one or more fixed inputs. The possible changes include:

- Changing a set point (for example, machine settings, specifications for an input, supplier, and so on)
- Adding or changing a process step (for example, adding inspection, replacing a gage, retraining an operator, rewriting instructions, and so on)

- Changing the control plan (for example, adding a feedback or feedforward controller, changing the current controller, and so on)
- Changing the product design

These changes can be implemented anywhere in the process including at suppliers. We need to be careful since changing any fixed input may add significant operating costs or produce serious *side effects* defined in terms of other process outputs.

In this chapter, we introduce and discuss the seven variation reduction approaches. For each approach, we discuss how and when it works and the potential difficulties of implementation. We provide details and further examples in chapters 14 through 20.

The variation reduction approaches are an integral part of the proposed algorithm. The algorithm is the topic of Chapter 4. The ultimate aim of the algorithm is to lead to process improvement by implementing one (or more) of the seven variation reduction approaches. We provide further comparison and discussion of how to choose an approach in chapters 8 and 14.

3.1 FIXING THE OBVIOUS BASED ON KNOWLEDGE OF A DOMINANT CAUSE

After finding a dominant cause of variation, we can sometimes reduce variation in the output by implementing an obvious fix. For instance, we might identify temperature as a dominant cause and reduce temperature variation by more frequent adjustment. The effect of reducing variation in a (continuous) dominant cause is illustrated by the scatter plots in Figure 3.1. The vertical dashed lines specify the range in the input (the dominant cause here) and the horizontal dashed lines the corresponding range in the output. If we reduce the variation in the input, as shown in the right panel of Figure 3.1, the variation in the output will be substantially reduced.

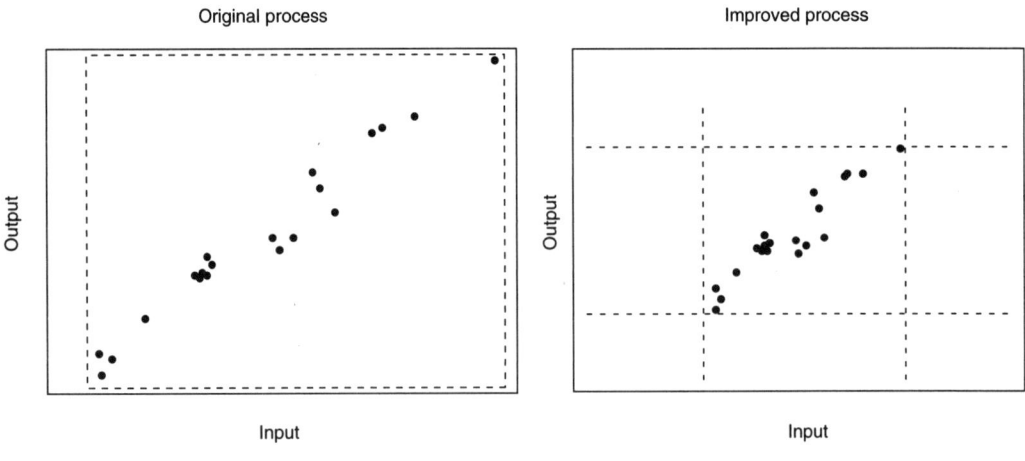

Figure 3.1 Reducing variation in a dominant cause.

We can sometimes eliminate variation in the dominant cause. In a drilling process, there was a frequent problem with holes drilled in the wrong location. Plates were drilled on a multispindle drill press. The defects were not found until assembly. Setting a plate upside-down or rotating it from the correct orientation before drilling was the dominant cause. The process engineer redesigned the fixture holding the plate so that it was impossible to mount the plate in the press incorrectly. The variation in the dominant cause was eliminated.

Figure 3.1 shows a continuously varying dominant cause. We can also reduce variation in a discrete cause, as shown in Figure 3.2. The dashed horizontal lines show the range of output. In the left panel of Figure 3.2, we see that machine number is a dominant cause of variation in the output. There are large differences in the average output level (process center) for different machines. In the right panel of Figure 3.2, we see that we can reduce variation by aligning the average output for each machine. There may be an obvious method for making such an adjustment.

For example, if a dominant cause is the difference in how two operators control the process, then we can reduce variation by retraining the operators and clarifying the control plan. In another example, a team identified the difference between two suppliers of valve lifters as the dominant cause of high oil consumption in truck engines. The team could eliminate most of the problem by switching to a single supplier or by establishing procedures to reduce differences between the suppliers.

In an engine assembly plant, the original problem was the high frequency of rejects for excessive noise at the valve-train test stands. The team discovered that there were large differences in average measured noise level for each of the three parallel test stands, even though the engines were haphazardly assigned to the stands for testing. After calibrating each stand, they found the average noise levels were roughly equal. The real challenge for the team was to keep the problem from recurring.

There are two conditions necessary for the fix-the-obvious approach to work. First, we must be able to identify a dominant cause. We waste valuable resources and make little or

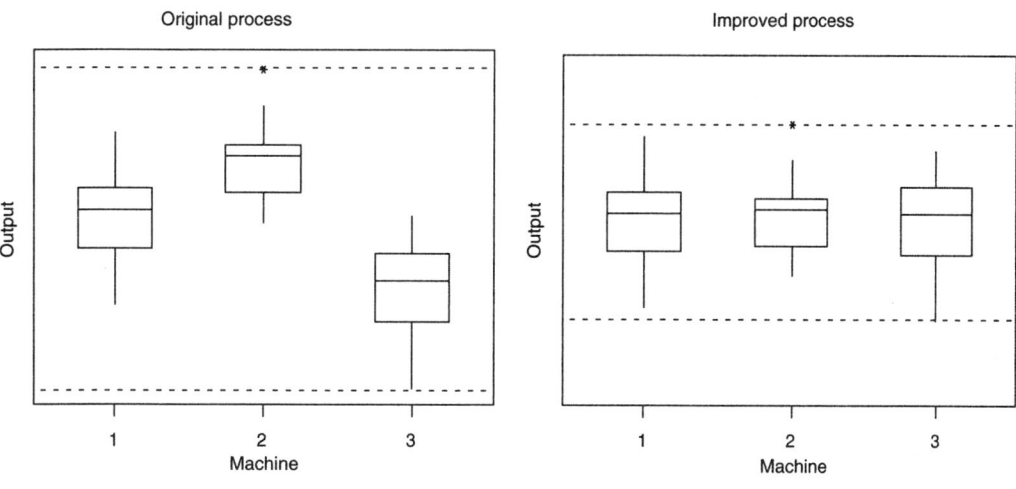

Figure 3.2 Aligning the process center stratified by a dominant cause.

no progress fixing problems based on a nondominant cause. Second, based on process knowledge, we must be able to address the cause with an obvious solution whose feasibility is without question.

3.2 DESENSITIZING THE PROCESS TO VARIATION IN A DOMINANT CAUSE

If we find a dominant cause as shown in the left panel of Figure 3.3, we can reduce variation in the output by flattening the relationship between the input (dominant cause) and the output as shown on the right panel of the figure. The dashed horizontal lines show the range of output values for the original and improved processes. We reduce the sensitivity of the output to variation in the input by changing some fixed inputs. We do not reduce variation in the dominant cause.

We find process desensitization useful when the variation in a dominant cause is hard or impossible to control. For example, we discussed a frost buildup problem in refrigerators in Chapter 1. The dominant causes of the variation in frost buildup were environmental and usage inputs, such as ambient temperature and the average number of times per hour that the door was opened. You can imagine that the instruction "Open this fridge at most once per hour" to reduce variation in a dominant cause would not be effective here. Instead, the manufacturer decided to investigate a number of design changes (that is, changes of fixed inputs) to reduce the sensitivity of the refrigerators to varying usage inputs. The changes investigated were directed by the knowledge of the causes of the problem.

For a second example, a team was charged with reducing scrap due to porosity found on a machined surface of an engine block. The porosity was due to gas bubbles trapped near the surface of the block when the molten iron solidified. The team decided to search for a dominant cause and discovered that low iron-pouring temperature resulted in a high frequency of porosity scrap. Looking for a more specific cause, they found that low-pouring

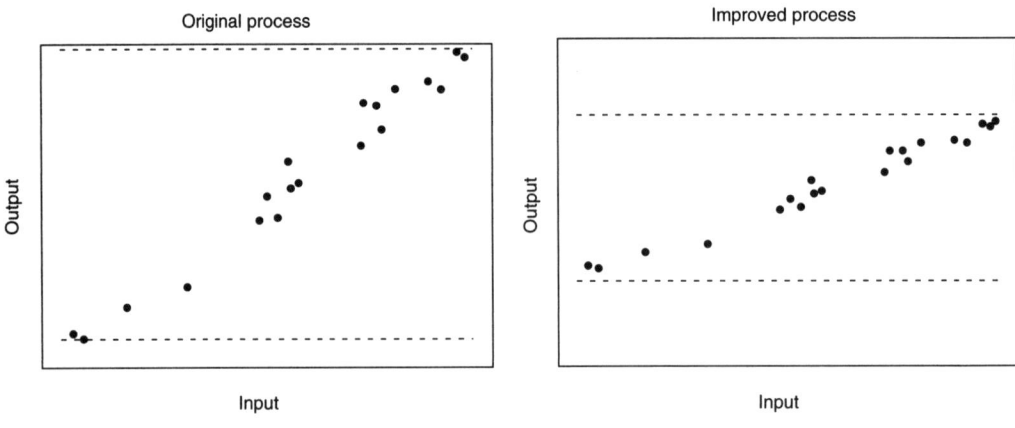

Figure 3.3 Desensitizing a process to variation in a dominant cause (dashed horizontal lines show the range of output values).

temperature was due to unscheduled and scheduled (for example, lunch) downtime at the pouring operation, where there was no method for maintaining iron temperature. The cost of adding controls was prohibitive. The team looked for ways to desensitize the porosity level to pouring-temperature variation. They discovered that changing the wash on the sand cores substantially reduced the sensitivity of the porosity to low pouring temperature. The foundry adopted this low-cost solution.

Process desensitization is a desirable approach since, once it is complete, no further action is required. First, we must find a dominant cause of variation. We can use knowledge of this cause to help select fixed inputs that we might change to desensitize the process. In the engine block porosity example, once the team understood that low pouring temperature was a dominant cause of porosity, they were led to consider changing the core wash and core sand composition. Without knowledge of a dominant cause, it is unlikely that they would have thought about changing these fixed inputs.

It is difficult to predict when desensitization will work. This is its great weakness. We require a great deal of process knowledge to make an output less sensitive to variation in a dominant cause. Which fixed inputs should we change? We require expensive designed experiments to determine these inputs (if any exist) and their appropriate settings. The experiments may fail to determine process settings that lead to improvement. Also, the new process settings may lead to increased operating costs or negative side effects.

3.3 FEEDFORWARD CONTROL BASED ON A DOMINANT CAUSE

With feedforward control, we adjust the process based on measurements made on a dominant cause of variation, anticipating the effect on the output. Suppose we have identified a dominant cause as in the left panel of Figure 3.4. If we have a high value for the input, we predict that we will get a high value for the output, so we adjust the process center downward to compensate.

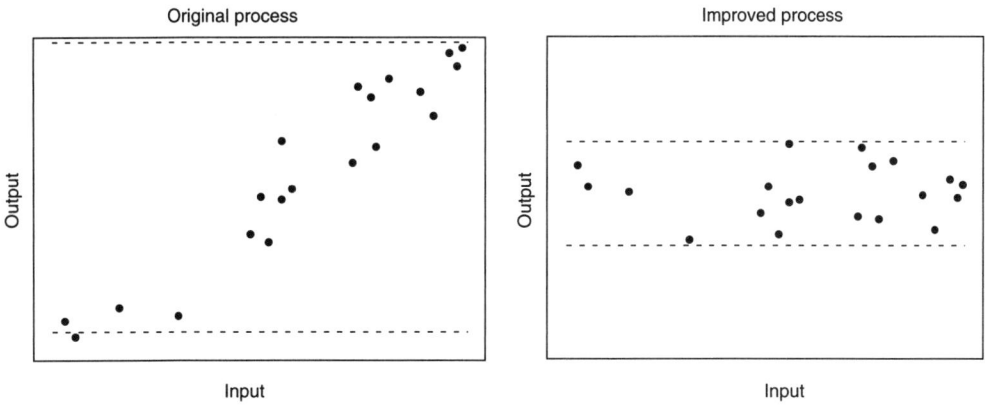

Figure 3.4 Reducing output variation using feedforward control.

Feedforward control is a form of process desensitization, as discussed earlier. However, with feedforward control, we achieve the desensitization by active adjustment as the process operates, rather than with a one-time change in the settings of some fixed inputs. We can see how feedforward control reduces the variation in the output in the right panel of Figure 3.4.

In the development of the truck alignment process discussed in Chapter 1, the geometry of the frame was a dominant cause of variation in the alignment characteristics. To reduce variation, the frame manufacturer measured each frame and recorded the geometric information with a bar code. The assembly plant read the information and manufactured a special component to adjust the alignment characteristics to remove the effect of frame geometry. This approach was relatively expensive but led to a dramatic reduction in the variation of the alignment characteristics.

Selective fitting is another example of feedforward control. Components are sorted and matched to produce good assemblies. This adds cost and complexity to the assembly process. In a problem to reduce steering wheel vibration, a team discovered that a dominant cause was an imbalance in the transmission. The problem was reformulated to reduce imbalance in the transmission. Then the team identified two mating components that were dominant causes of the transmission imbalance. They improved the balance of the transmission by measuring and vectoring the two transmission components, that is, assembling the components so that their individual contributions to imbalance tend to cancel. As a second step, they sorted transmissions on the basis of imbalance and used the better-balanced transmissions in the most sensitive vehicle models. Both changes are examples of feedforward control.

Feedforward control works under the following conditions. First, we must identify a dominant cause. Second, the relationship between the cause and the output must be well known and stable over time. Third, we must be able to measure the cause in a timely way. Finally, there must be a way to adjust the process to compensate for the anticipated effect of the cause. As a result, to implement feedforward control, we may need to find a way to adjust the process center.

If the dominant cause changes slowly and the conditions are met, we can use feedforward control with only occasional adjustments. For example, in a batch process, if a dominant cause is a raw material characteristic that is constant over a batch, we can use the value of the cause to set up the process to remove or reduce its effect. Here, we make an adjustment only once per batch.

There are substantial costs and risks associated with feedforward control. Costs arise because we need to identify a dominant cause, measure the cause, and repeatedly adjust the process. There is a danger of overadjustment if there is measurement error in determining the input (the dominant cause) value, if the relationship between the input and the output is not well understood, or if the adjustment scheme is imperfect. In addition, repeated process adjustment may be impractical or costly and may introduce undesirable side effects.

3.4 FEEDBACK CONTROL

The idea behind feedback control is to monitor the output characteristic and to predict future behavior from current and past observations. If the predicted output value is far from the target, we make an adjustment to the process. In Figure 3.5, the panel on the left

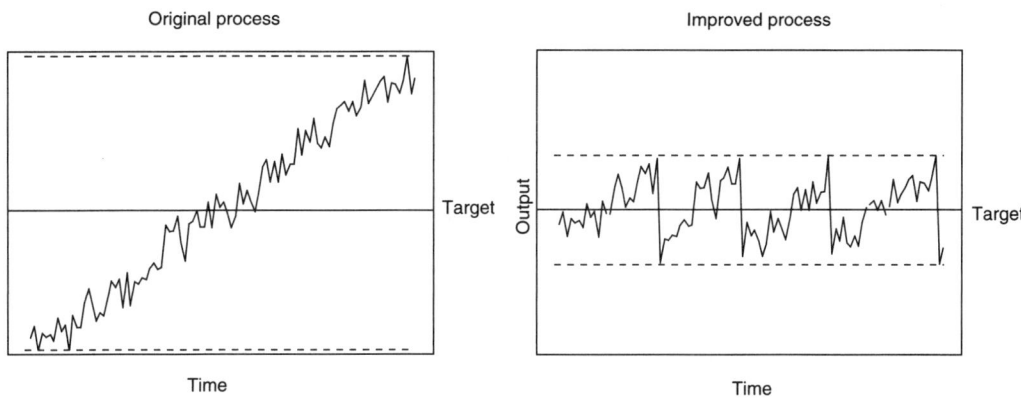

Figure 3.5 Feedback control.

shows the output of a process without feedback adjustment. The dashed horizontal lines indicate the range of output values. At any particular time, we can predict that the output will be larger in the future. If we predict that the output is far from the target, we adjust the process towards the target. The right panel of Figure 3.5 shows the dramatic reduction in variation of the output due to this approach. Here, using the feedback control procedure, we adjust the process whenever the output falls outside the adjustment limit (not shown, but near the top dashed horizontal line in the right panel of Figure 3.5). The size of adjustment is based on the deviation from the target of the last observed output value.

A team wanted to reduce variation in the film thickness in the painting of fascias. Too much paint resulted in defects such as sags and runs, and too little paint created appearance defects. The team discovered that a dominant cause of thickness variation was the flow rate of the paint. They reformulated the film thickness problem in terms of flow rate. The team then identified a dominant cause of variation in flow rate that was addressed by making equipment modifications. After these changes, the flow rate continued to show a pattern in the variation over time as shown in Figure 3.6.

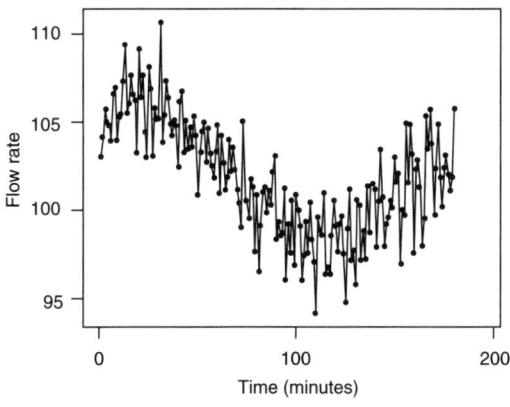

Figure 3.6 Variation in flow rate over time.

The team defined a feedback control scheme that involved considering an adjustment of flow rate every minute. If the flow rate fell outside prescribed limits, a valve was adjusted to move the process center to the target value of 103. The feedback scheme produced a further substantial reduction in film thickness variation.

As a second example, in a project to reduce scrap due to variation in the diameter of crankshaft journals, the team discovered that an automated 100% inspection gage drifted substantially over time. They could not determine the cause of the drift, so they decided to use a feedback controller to adjust the gage when appropriate. A reference part was measured several times just after the gage had been cleaned and calibrated. The team used these measurements to establish a centerline and adjustment limits for the gage. Every four hours the reference part was measured, and if the measured value fell outside the limits, the operators recalibrated the gage.

The use of setup procedures based on first-off measurements is another example of feedback control. For example, an operator used the measured output of the first few units produced after a tooling change to adjust the machine. Once a good setup was achieved, no further process measurements were taken or adjustments made until the next tool change.

We can apply feedback control successfully when three conditions are satisfied. First, the process must exhibit a strong time structure in the output variation as discussed in Chapter 2. Examples include drift due to tool wear and stratification due to batch-to-batch variation. Second, there must be an adjustment procedure to move the process center. Finally, the time to measure the output and adjust the process must be small relative to the rate of change of the process.

The major advantage of feedback control is that it requires no knowledge of the dominant cause of variation. We use only the measured output values. A drawback is the high cost of process measurements and adjustments. Finally, due to the feedback nature of the control, there is an inherent time delay. To identify when an adjustment is required, we must first observe some output values that are significantly different from the target value. Thus, feedback control is always reactive.

3.5 MAKING THE PROCESS ROBUST

To make the process robust, we first determine a performance measure, such as the standard deviation of the output values for 20 consecutive parts. Next we change one or more fixed inputs to see if the performance measure has improved. In Figure 3.7, we use a box plot to show performance, and the improvement due to changing the fixed inputs is clear. The horizontal dashed lines show the range of output values before the change in fixed inputs.

The challenge with this approach is to identify which fixed inputs to change. For example, to measure the concentration of silicon in cast iron, the operator poured coins from a sample of molten iron, prepared the coins by grinding the surface, and then used a spectrometer to determine silicon concentration. The repeatability, defined as the variation when the same batch of iron was measured by the same operator in a short time, was unacceptably high. To reduce this variation, the team decided to adopt the

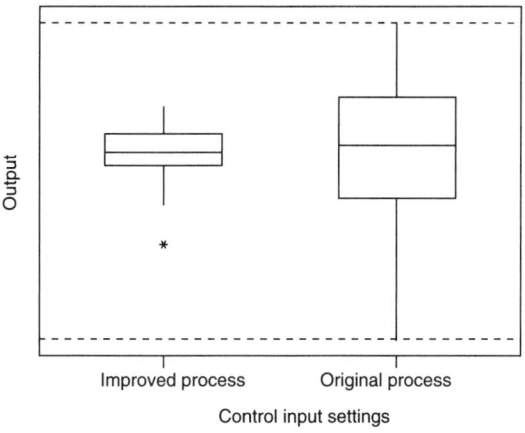

Figure 3.7 Making the process robust.

robustness approach. First they selected a run and a performance measure, the standard deviation of the measured silicon concentration in five coins poured from the same batch of iron. Second, they selected a number of fixed inputs, including mold shape, sample polishing method, and sample temperature. They organized a large experiment in which all of the selected inputs were changed in an organized fashion. For each combination of the inputs, they prepared and measured five coins and then calculated the performance measure. They found new settings for several inputs that reduced the repeatability variation.

As a second example, in a painting operation (discussed in Chapter 1), there was excessive variation in film build (paint thickness) from vehicle to vehicle at particular locations. As a consequence, to meet the minimum film build specification, the process operators kept the process center well above the lower specification. This resulted in high paint usage and occasional visual defects such as runs. To reduce the variation, the project team used the standard deviation of film build over five consecutive panels (it was too expensive to use cars for this investigation) to measure performance. Then they varied five fixed inputs, including some process parameters and paint properties, to explore settings to reduce the variation. The team discovered new settings of the fixed inputs that reduced the panel-to-panel variation by a factor of two. The average film build was then reduced with substantial savings in paint and rework costs.

Making the process robust will be successful if we can identify fixed inputs that can be changed to improve the performance measure. This approach is similar to desensitizing the process to variation in a dominant cause. In both, the goal is to make the process less sensitive to variation in the dominant cause. However, with the robustness approach, we select and change fixed inputs without first identifying the dominant cause. Without such knowledge, we find it more difficult to determine which fixed inputs to change and by how much. There is considerable risk with the robustness approach that significant resources will be used in a fruitless search for better process settings.

Note that we use the terms *robustness* and *desensitization* to label different approaches. Many authors use them interchangeably.

3.6 100% INSPECTION

100% inspection is the simplest variation reduction approach. We reduce output variation by identifying and then scrapping or reworking all items that have values of the output beyond selected inspection limits. We illustrate the effect of 100% inspection in Figure 3.8. We will scrap or rework (and reinspect) all units with output values outside the vertical dashed lines. Assuming no inspection or measurement error, the new full range of output is given by the dashed lines. The more the inspection limits are tightened, the greater the reduction in variation in the output characteristic.

In the production of cast-iron exhaust manifolds, a rare defect is a blocked port. The defect was not detected until the engine was assembled and tested. The team assigned to eliminate the defect had great trouble determining a dominant cause because the defect occurred so rarely. The team abandoned the search for the cause and, instead, designed and built a new device to automatically inspect all ports on every casting produced. Management agreed that the extra cost of the inspection was justified so that their customers could be assured that no manifolds with blocked ports would be shipped.

In the manufacture of a forced-air furnace, there was an emergency limit switch that shut down the furnace if the plenum got too hot. The heart of the switch was a plastic base in which a number of components were mounted to detect high temperatures and mechanically break the circuit. There was 100% inspection of the assembled switches. The scrap cost was high because the component and assembly costs could not be recovered from a scrapped switch. Using Pareto analysis, the team discovered that the most frequent reason for scrapping a switch was a broken tab in the plastic base. They also found that the tab was broken before any components were assembled into the base. The team decided to institute 100% inspection on the bases to scrap those with a broken tab before the components were installed. This reduced scrap costs by 50% and more than justified the inspection cost.

100% inspection is possible if the output characteristic can be determined for every unit in advance of shipping the product to a customer. Inspection has a number of negative features. The cost of reducing variation by tightening the inspection limits may be high due to increased rework and scrap costs and lost capacity. Also, the cost of inspection may be high if a new gage or additional labor is required. The reduction in variation will be less than

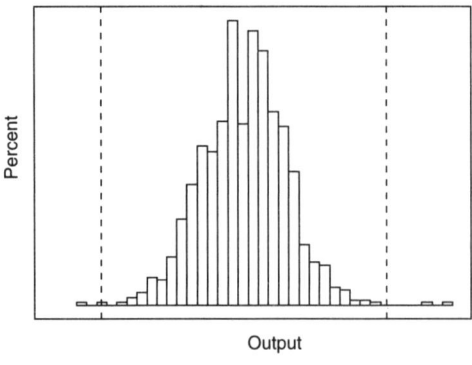

Figure 3.8 100% inspection.

anticipated if there are measurement or inspection errors. As a result, given the propensity of people to make inspection errors, most successful applications use automated inspection.

Inspection on the output has a negative reputation among quality professionals. However, we often apply 100% inspection on a dominant cause. This is called *source inspection,* one form of error proofing (Shingo, 1986).

3.7 MOVING THE PROCESS CENTER

This variation reduction approach is different from the others because we use it to change the process center to reduce off-target rather than unit-to-unit variation. We may need to move the process center closer to target as in Figure 3.9, or we may need to align sub-process centers as in Figure 3.2.

A team was charged with reducing the amount of shrinkage in a molded plastic casing. The casing was part of an assembly, and if there was excessive shrinkage, there was interference between the casing and the enclosed cable. Shrinkage was defined as a percentage change in casing length after a fixed curing time. The goal of the project was to reduce the average shrinkage. The team selected 15 fixed inputs, both raw material components and processing parameters. They planned an experiment where, for each combination of the input values, they measured the average shrinkage in a 500-meter length of casing. The results suggested two fixed inputs that could be changed to significantly reduce average shrinkage. The team did not determine the cause of shrinkage under the original process conditions.

In another example, a major source of scrap in the casting of aluminum differential carriers was a shrink defect resulting from the aluminum pulling away from the mold as it solidified. To reduce the rate of shrink defects, a team invented a five-point score that rated carriers from bad (score 1) to good (score 5). The goal was to increase the average score. Nine fixed inputs were varied in an experiment to determine levels that maximized the average score over 20 carriers. Again, the team made no attempt to determine a dominant cause of the shrink defect.

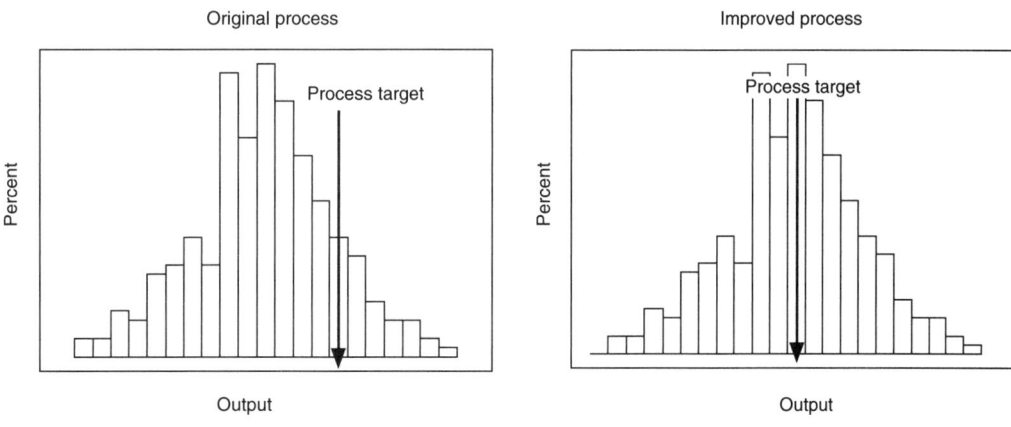

Figure 3.9   Moving the process center.

To apply the Move Process Center approach, we must discover a fixed input or inputs that we can change to shift the process center. Typically, we do this with an experiment in which we investigate several fixed inputs simultaneously.

There are risks associated with adjusting the process center. First, we may not find a fixed input that changes the process center and so waste the cost of the experiment. Second, changing the fixed input may add expense or produce negative side effects.

To apply feedback or feedforward control, we need a fixed input that we can change to adjust the output center a prescribed distance. We call such an input a *process center adjuster*, or just an *adjuster*. We use the same tools to find an adjuster or a fixed input to move the process center. With the Move the Process Center approach, we hope to make a single shift. With feedback or feedforward control, we plan to adjust the process often.

In many circumstances, we already know of a fixed input that will change the process center, but we may need to calibrate the size of the effect in order to target the process properly. We can use an experiment to determine the effect of the adjuster.

For many problems, moving the process center is identical to making the process robust. In the shrinkage example, since the output has a physical lower limit of zero, shifting the average and reducing the variation about zero are the same problem and the two approaches are the same. In the shrink defect example, the two approaches are again identical. In each case, we are led to look for fixed inputs that we can change to reduce the defect rate without finding a dominant cause.

Key Points

- To reduce variation, we must change fixed inputs. Without changing one or more fixed inputs, the process performance will not improve.
- We consider seven variation reduction approaches:
  1. Fixing the obvious using knowledge of a dominant cause of variation
  2. Desensitizing the process to variation in a dominant cause
  3. Feedforward control based on a dominant cause
  4. Feedback control
  5. Making the process robust
  6. 100% inspection
  7. Moving the process center
- Finding a dominant cause of variation is an important step in the first three approaches.
- The appropriate choice of variation reduction approach depends on the problem definition, the current state of knowledge about the process, and costs.

 Exercises are included on the accompanying CD-ROM

4

An Algorithm for Reducing Variation

Begin with the end in mind.

—Stephen Covey

In this chapter we provide an algorithm to address variation reduction problems such as those described in Chapter 1. The algorithm is structured around the seven variation reduction approaches introduced in Chapter 3.

We believe that variation reduction problems (or any problems) are best addressed using a step-by-step method—that is, an algorithm. There are many advantages to adopting an algorithm. Some people and teams are natural problem solvers that can follow their own instincts. Most of us can use some guidance. We find the algorithm useful because it:

- Is easy to teach (systematic and structured).
- Can be managed.
- Helps to avoid silly mistakes or oversights.
- Provides documentation of progress and success (or failure).
- Helps to ensure that all possible solution approaches have been considered.
- Makes most people, especially when working in teams, better problem solvers.

On the enclosed CD-ROM, we describe in detail three case studies that are successful applications of the algorithm. Here we outline the algorithm and discuss a range of implementation issues. In later chapters, we describe each stage in detail.

4.1 THE STATISTICAL ENGINEERING VARIATION REDUCTION ALGORITHM

We propose the algorithm given in Figure 4.1 to reduce variation in high- to medium-volume manufacturing processes.[1]

42 Part One: Setting the Stage

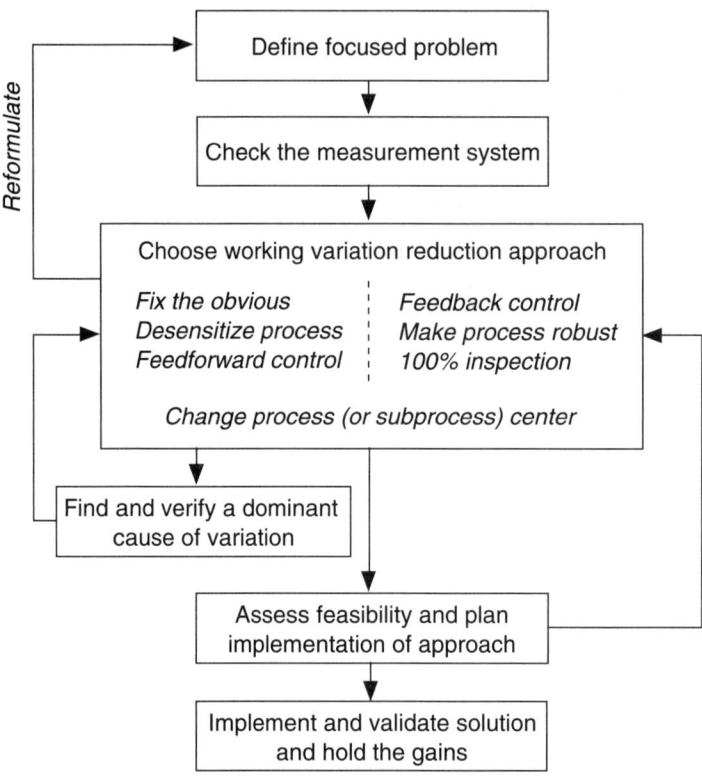

Figure 4.1 Statistical engineering variation reduction algorithm.

Process managers provide the information necessary to start the algorithm. This information should include a specified process, an improvement goal often expressed in monetary terms, and a schedule. The managers also enable people to carry out the work. For convenience, we refer to these people as the process improvement team or the team for short.

In the first stage of the algorithm, Define Focused Problem, the team narrows the process boundaries and selects the particular output characteristic(s) needed to specify the problem. For example, suppose the management statement of the problem is to reduce scrap costs for a particular process by 50% within a month. The focused statement may be to reduce a particular category of scrap, or scrap at a particular processing step. The team determines the nature of the excess variation (off-target, unit to unit) and establishes an appropriate baseline measure of the process performance. They then set a goal for the project in terms of the baseline performance that is consistent with the management goal.

One special feature of Statistical Engineering is the explicit and frequent use of the information provided by the baseline investigation. We establish a baseline to help:

- Set the goal in terms of a particular output.
- Search for the dominant cause.

- Assess the selected variation reduction approach.
- Validate the solution.

We cover the Define Focused Problem stage in detail in Chapter 6.

In the second stage, Check the Measurement System, the team ensures that the measurement system used for the selected output is not home to the dominant cause of variation. They also need to determine if the measurement system is adequate for use in later stages of the algorithm. We assess the measurement system by planning and executing an investigation to determine how much of the baseline variation can be attributed to the measurement system.

If we find that the measurement system is the home of the dominant cause of variation or is not adequate to support future investigations, we reformulate the problem into a new problem to address the variation in the measurement system. We can address this new problem with the algorithm. The process of interest is now the measurement system. We provide several examples of solving measurement system problems throughout the book.

We have seen many failures because a team tried to solve a problem without an adequate measurement system. Less often, we have seen teams become bogged down trying to improve a measurement system that is not the home of a dominant cause and is perfectly adequate for use in further investigations. We cover the stage Check the Measurement System in detail in Chapter 7.

At the next stage of the algorithm, Choose Working Variation Reduction Approach, the team looks ahead to decide how to reduce the variation. We believe that making this step explicit distinguishes this algorithm from the many others that have a similar intent. We have specified seven approaches to reducing variation, described in Chapter 3. At this stage of the algorithm, the team should assess each approach using:

- The nature of the problem and process
- The knowledge they have
- The knowledge required to implement the approach
- The likelihood and cost of obtaining the required process knowledge
- The likelihood of successful implementation
- The probable cost of implementation

The goal is to select a working approach and then direct efforts to determine if this approach is feasible. If it is, the team proceeds to the implementation stage; if not, they reconsider the approaches.

We have divided the approaches into two groups based on whether their implementation requires the identification of a dominant cause of variation. The approaches requiring the identification of a dominant cause are:

- Fixing the obvious (Chapter 14): Use knowledge of a dominant cause to implement an obvious solution.
- Desensitizing the process (Chapter 16): Change one or more fixed inputs to reduce the sensitivity of the output to changes in a dominant cause.

- Implementing feedforward control (Chapter 17): Use an adjustment scheme based on measured values of a dominant cause to anticipate and reduce its impact on the output.

The approaches not requiring the identification of a dominant cause are:

- Implementing feedback control (Chapter 18): Use an adjustment scheme based on measured values of the output to anticipate and reduce the impact of future changes.

- Making the process robust to cause variation (Chapter 19): Change one or more fixed inputs to reduce the baseline measure of variation.

- Using 100% inspection (Chapter 20): Use an inspection scheme to select units with less variation in the output.

- Moving the process center (Chapter 15): Change one or more fixed inputs to shift the process center.

The team must decide at this stage if they will search for a dominant cause. If the answer is yes, then they proceed with the search. If not, they should look at each of the other approaches that are not cause-based and select one.

In most applications of the algorithm, the team will decide to search for a dominant cause. We use the method of elimination to identify this cause. We partition all the causes of variation into families and then use process investigations or available data to rule out all families but one as the home of the dominant cause (see Chapter 9). We use elimination recursively to quickly narrow down the potential dominant causes to one or a few varying inputs. We describe a large number of methods to help search for the dominant cause in chapters 9 to 13.

When the team identifies and verifies a particular input as the dominant cause, they can consider the feasibility of one of the three cause-based approaches. If they rule out these approaches, they have three options:

- Reformulate the problem in terms of the dominant cause

- Reconsider the four non-caused-based approaches

- Search for a more specific dominant cause

If we decide to reformulate the problem in terms of the cause, we start the algorithm over with the goal of reducing variation in the identified input that is the dominant cause. Reformulation corresponds to "moving the problem upstream" or "searching for the root cause." We sometimes reformulate a problem several times. However, eventually we must select one of the variation reduction approaches.

Sometimes a team may reconsider the approaches with only partial knowledge of the dominant cause. They may have eliminated many possibilities but not be able to find a specific cause. With partial knowledge of a dominant cause, they may find one of the non-caused-based approaches more feasible. We describe the Choose Working Variation Reduction stage in detail in chapters 8 and 14.

At the next stage of the algorithm, Assess Feasibility and Plan Implementation of Approach, the team looks at the feasibility of the selected approach. They:

- Examine the process to see if it meets the conditions for the approach to be effective.
- Determine what further knowledge is required.
- Plan and conduct investigations to acquire the knowledge.
- Determine the solution, that is, what and how fixed inputs will be changed.
- Estimate the benefits and costs of the proposed changes.
- Look for possible negative side effects.

We describe assessing the feasibility of each approach in chapters 14 to 20.

If the selected approach is feasible, the team proceeds to validate and implement the solution. Otherwise, they must reconsider other variation reduction approaches.

We arrive finally at the Implement and Validate Solution and Hold Gains stage. Here the team assesses the baseline performance of the changed process to ensure that the project goal has been met. They must also examine other process outputs to make sure they have not created a new problem in order to solve the original one. Finally the team implements and locks the change into the process or its control plan. We recommend the team monitor the process output and audit the process change until they are certain the solution is effective and permanent. As well, the team should document what they have learned and identify future opportunities to reduce variation further. We discuss the Implement and Validate Solutions and Hold Gains stage of the algorithm in Chapter 21.

4.2 HOW TO USE THE ALGORITHM EFFECTIVELY

We have seen the proposed algorithm work well on a large number of projects and fail on others. We believe the key drivers to success can be divided into two groups related to process and management issues.

Process Issues

The process needs to be under reasonable control before starting a project using the formal algorithm. We once were asked to assess the likelihood of success of a project to reduce the amount of rework due to dirt defects on painted bumpers. In a quick walk-through of the painting process with a painting expert, we saw fiberglass bats fall off the ceiling onto the painting line, paint drips everywhere, operators sweeping in areas with unbaked painted parts, and so on. The process was a mess, riddled with poor practices, even to our untrained eyes. We suggested there was little value in using the algorithm with this uncontrolled process. Instead, we gave the simple message:

Fix the obvious!

We think that this is an important message both at the start and during any project to reduce variation. Many times, we have learned after starting a project that scheduled equipment maintenance has been abandoned, sometimes for years, or that operators are ignoring the control plan for the process, often because no one told them that such a document existed. A solution that results from any initial work needed to get a process under control may be thought of as fixing the obvious. In general, we expect to apply the algorithm to processes where:

- A control plan is being followed.
- Equipment is maintained.
- Gages are calibrated.
- Personnel are adequately trained.
- Housekeeping is addressed.
- Industry standard operating principles (see, for example, Todd et al., 1994) are followed.

Quality standards, such as ISO 9000 (Hoyle, 2001) and QS-9000 (AIAG, 1998) provide a mechanism to establish reasonable control. Bhote and Bhote (2000) call this activity *process certification*. We do not require that the process be under statistical control, that is, stable as defined in terms of a control chart, to implement the algorithm. See the discussion about classifying causes in the supplement to Chapter 2 for more explanation.

The algorithm is best suited to address chronic problems (long-standing adverse situations; Juran and Gryna 1980, p. 99) rather than problems that are sporadic. Sporadic problems, where the status quo is suddenly adversely affected, are difficult because they require immediate attention and quick solution. Sporadic problems often lead to containment of product and a corresponding large cost due to logistics, delay, and lost inventory. In this context, applying the proposed stage-by-stage algorithm, with its contemplative nature, is likely not an option. The algorithm has no mechanism for containment. However, many of the specific tools and methods discussed in this book as part of the algorithm are useful to find the cause of a sporadic problem and a solution.

The appearance of a sporadic problem is sometimes used to initiate a project whose goal is to look at both the new sporadic problem and the related long-term chronic problem. For example, a sudden large increase in brake rotor balance rejects from 25% to 50% prompted a process improvement project whose goal was to reduce balance rejects to less than the chronic rate. It is also important to realize that a recurring sporadic problem is best thought of as a chronic problem. Firefighting to address sporadic problems is not effective in the long run. Using the discipline of the proposed algorithm provides greater assurance of finding a permanent solution.

To apply the algorithm, we need to measure process inputs and outputs in a timely fashion. If we are unable to make such measurements, we may not gain the process knowledge required to move through the stages of the algorithm. In a project to reduce warranty costs, the team defined the focused problem in terms of the failure rate of wheel bearings within the warranty period of three years. Their goal was to reduce this failure rate from 3% to less than 0.3%. The key output is the time to failure of the bearing. Other than historical data stored in the warranty database, the team recognized that it would be difficult to apply the algorithm because it takes such a long time to measure the output for any new vehicle. To proceed, they decided to replace the time to failure in the field by a surrogate measure, the bearing failure time measured under extreme conditions in a laboratory.

In another situation, a shipping company set out to reduce the frequency of short and wrong stock shipments to its only customer. There was a lag of up to six months in the customer's reporting of shipping errors and little confidence in the accuracy of the reports. The long lag meant that these measurements were difficult to use for variation reduction. Before the algorithm could be applied, the customer-based measurements were replaced by local measurements based on the results of a daily preshipment audit. It was assumed that changes to improve the process expressed in terms of the audit results would also improve performance for the customer.

To apply the algorithm, we need a high- to medium-volume process that produces frequent parts or units. We assume that we can conduct process investigations relatively quickly and at low cost. If we cannot do so, then the algorithm is likely to fail or we will be very limited in our choice of approaches. In the bearing failure example, each measurement of failure time in the lab was very expensive, so that it was not feasible to carry out many investigations to find the cause of the variation in failure time. The team decided to make the process robust, the only feasible variation reduction approach.

The algorithm is designed to identify low-cost changes to the process or product that will meet the goal of the project. We consider the application of the algorithm a failure if the team cannot identify a low-cost solution. In some instances, the team has a solution in mind (for example, expensive new equipment) and does not consider all of the possible approaches at the Choose the Variation Reduction Approach stage. They do not adequately consider lower-cost changes such as an improvement to the control plan or a change of process setting to desensitize the process to variation in a dominant cause. On the other hand, there may be no cost-effective solution.

We urge you to remember that in most variation reduction problems, the process is currently producing a high proportion of excellent output. Because the process does an excellent job, most of the time, there likely will be a low-cost way to make it even better.

Management Issues

Management makes critical contributions to the successful application of the algorithm. Some specific management tasks are to:

- Choose projects and set reasonable goals.
- Select the process improvement team.
- Provide a supportive infrastructure (training, time, project reviews, and so on).
- Conduct or review cost/benefit analyses for any suggested process/product changes.
- Avoid local optimization in the choice of projects and proposed solution.
- Ensure that organizational learning occurs.

It is difficult for managers to make good decisions in terms of these points if they do not understand the algorithm and how it functions.

The algorithm is project-based, so we want to focus on projects with the potential to have a large impact on customer satisfaction, cost, or both. We know that there are costs associated with the algorithm and that the outcome is uncertain. We can use customer surveys, market research, and warranty data to highlight important customer concerns or desires. We need to link customer concerns to process outputs that can be measured at the manufacturing site; we can use a tool such as quality function deployment (Revelle et al., 1998). We can identify good cost reduction or productivity improvement projects by examining scrap and rework rates and using Pareto analysis (Juran et al., 1979) to rank possible projects. In many cases, other factors will also influence the decision. For instance, management may be planning to update or remove a production process. If such wholesale change is imminent, we should not waste resources addressing identified problems for that process.

When choosing a project (and also later when choosing a potential solution), managers must consider the issue of global versus local optimization. Local optimization may occur when we have a narrow focus and forget that the process of interest is likely part of a much larger system. The output of one process is an input to the next. Reducing variation in an output that has little effect on the downstream process is an example of local optimization. Goldratt (1992) provides an interesting discussion of the local optimization issue (for example, bottlenecks) in the context of cycle times and machine scheduling.

An example of local optimization occurred in the engine block leak example introduced in Chapter 1. The team solved the problem of center leaks by adding several chaplets (small steel inserts) to better support the core. This solution reversed a previous process change. The chaplets had been removed some months earlier to reduce cost. At that time, it was not realized that the cost savings from removing the chaplets would be overwhelmed by increased scrap costs. The truck pull problem described in Chapter 1 provides another example. The problem was initiated because variation in truck pull is noticeable to the customer. However, pull is a derived output characteristic that depends on other outputs such as camber and caster. See Chapter 6 for further discussion. Local optimization would have been a concern had management presented the problem as one to reduce variation in camber and caster. It turns out that camber variation has little influence on pull (relative to the effect of caster). The team could have spent considerable time and effort reducing variation in camber, but the (local) improvement would not have been noticeable to the customer.

The managers must specify the project goal, usually in monetary terms, and provide a schedule. Without a goal, the team will not be able to decide if an approach is feasible, or when the project is finished. The goal must be reasonable; the use of the algorithm will not (often) produce miracles. Resources must be allocated to fit the schedule. A good reference on project management is Lewis (2002).

Teams should include participants with expert knowledge of the selected process. As well, the team must include at least one individual familiar with the algorithm and the associated statistical methods. The core team should be relatively small; we recommend one to three people. Many others will be consulted as necessary to help with specific tasks.

Successful teams must have strong management support during the conduct of the project. Management must provide team members time away from other duties and help in obtaining necessary resources such as testing time. Management can provide training in the use of the algorithm if necessary, can facilitate access to experts, and can provide contacts with customers and suppliers as needed. Management should conduct periodic reviews of the project to keep the effort focused and to provide a mechanism to terminate the project if the likelihood of success appears too small or if costs are too high.

Management also has a strong role to play in weighing the likely costs and benefits of proposed solutions. The team can provide a business case to justify any suggested process/product change. Management approval is necessary to authorize and pay for such changes.

At the conclusion of a project, management must ensure that important lessons learned are disseminated throughout the organization. This may involve updating the corporate memory by changing design guidelines, and so on. A useful source for information on learning organizations is Senge (1990).

Key Points

- We present a variation reduction algorithm specifically designed to produce low-cost improvements to high- to medium-volume processes.

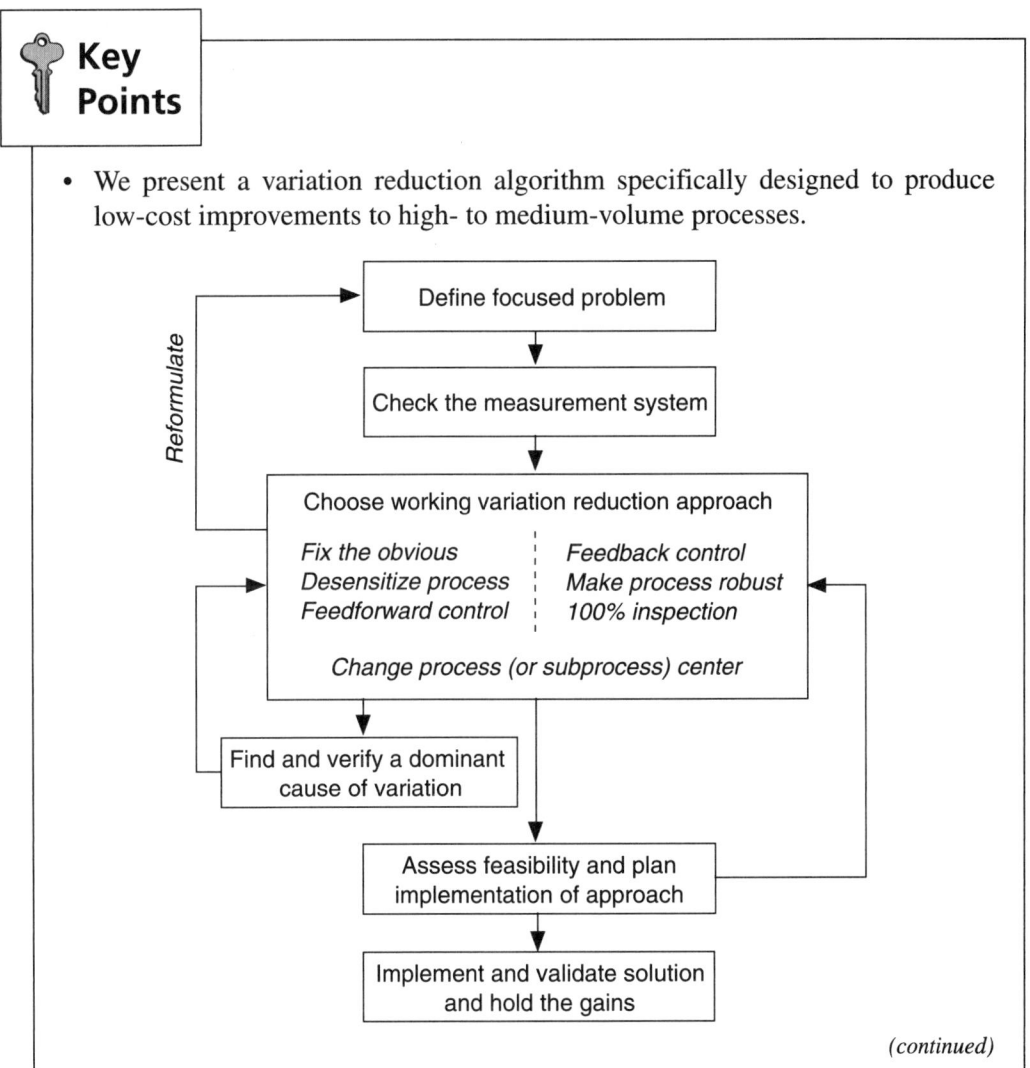

(continued)

- The Choose Working Variation Reduction Approach stage is designed to force the team to consider a wide range of possible solutions in terms of cost, likelihood of success, and prerequisite process knowledge.
- Key drivers to the successful application of the algorithm include the proper selection and management of projects, the provision of necessary resources including people's time, and training.

Endnote (see the Chapter 4 Supplement on the CD-ROM)

1. There are many other algorithms for reducing variation and improving process performance. The Six Sigma algorithm DMAIC (define, measure, analyze, improve, control) is widely used. We discuss a selection of these alternative algorithms and compare them to the algorithm proposed here.

5

Obtaining Process Knowledge Empirically

There is no substitute for knowledge.

—W. Edwards Deming, 1900–1996

New process knowledge is an essential ingredient to the variation reduction algorithm introduced in Chapter 4. Remember, the basis for the algorithm is the idea that we can find a cost-effective solution by better understanding how and why the process behaves as it does.

We use an empirical approach to acquire new process knowledge. Empirical means derived from experiment and observation. That is, we learn about process behavior by carrying out investigations (see Box, 1999). We follow a five-step process (we call this process a *framework* to avoid confusion with the process we are trying to improve) to help plan, execute, and draw conclusions from such investigations. We refer to the framework by QPDAC, an acronym that stands for Question, Plan, Data, Analysis, and Conclusion (Oldford and MacKay, 2001). The purpose of each step is:

Question: Develop a clear statement of what we are trying to learn.

Plan: Determine how we will carry out the investigation.

Data: Collect the data according to the plan.

Analysis: Analyze the data to answer the questions posed.

Conclusion: Draw conclusions about what has been learned.

The QPDAC framework highlights the important issues and forces us to think critically about the inevitable trade-offs necessary in designing and conducting an empirical investigation. At the end of the chapter, we give a checklist of issues that you should address in every such investigation. By so doing, you can be assured of solid conclusions at reasonable cost.

We expect to apply QPDAC several times in any application of the variation reduction algorithm. We need to use empirical investigations to increase process knowledge in most of the stages of the algorithm. Examples include:

- Quantifying the problem baseline (Chapter 6)
- Assessing the effectiveness of the measurement system (Chapter 7)
- Finding a dominant cause of variation (chapters 9–12)
- Verifying the dominant cause of variation (Chapter 13)
- Determining the feasibility of the variation reduction approaches (chapters 14–20)
- Validating that the proposed solution has addressed the original problem (Chapter 21)

5.1 QUESTION, PLAN, DATA, ANALYSIS, AND CONCLUSION (QPDAC) FRAMEWORK

In this section, we describe the QPDAC framework in more detail. To make the ideas concrete, we look at an investigation that was part of a project to reduce the variation in the diameter of machined aluminum V6 pistons. The manufacturer selected this project because the cost of internal scrap due to diameter variation was high. At the Define Focused Problem stage of the variation reduction algorithm, the team planned and carried out an investigation to establish a baseline measure of process performance. Maximum diameter (as the piston was rotated) was measured at a specified height above the skirt of the piston.

Question

In the Question step, we specify *what* we are trying to learn about the process. The goal is to pose one or more clear questions that we can address in the subsequent steps of QPDAC. Without clear questions, it is impossible to determine a good plan and draw appropriate conclusions.

In the Question step, we need to answer the following:

- To what group of units do we want to apply the conclusions?
- What input and output characteristics are needed to specify the question?
- What attributes of the process specify the question?
- What is the question of interest?

To introduce some terminology, we call a *unit* an individual realization (or product) of the process under investigation and the *target population* the group of units about which we want to draw conclusions. We sometimes specify the target population in terms of the *target process* that produces the units.

In the example, to establish a baseline performance measure, the team wanted to learn how the piston-making process would behave if it was left to operate normally. A unit was a piston and the target population was all pistons to be produced in the future under the current operating conditions (the target process).

We use the process language from Chapter 2 that defines input and output characteristics. In the example, we are only interested in the diameter of the piston in order to pose the question. In more complex investigations, we will specify a number of input and output characteristics to help define the questions.

To state clearly what we are trying to learn, we specify *attributes* of the target population. An attribute is a function of the characteristics over *all* the units in the target population. In the piston example, the team needed an attribute to quantify the output variation in the target population. They decided to use the standard deviation of the diameters in this set of pistons. Note that none of these pistons had yet been produced.

Attributes can be numbers such as averages, proportions and standard deviations, or pictures such as histograms or scatter plots. We define attributes in terms of one or more input and output characteristics.[1] In many applications of QPDAC, we formulate several questions, so we define several attributes.

We use the selected attributes to specify the question. In the piston example, to establish a baseline measure of process performance, the team asked the specific question:

What is the standard deviation of the diameters of pistons to be produced in the future if we leave the process to operate as it is currently?

After applying QPDAC, the team was able to provide an answer to this question. They then proceeded to look for new ways to operate the process that would reduce the variation. This search involved several applications of QPDAC. When they found a promising new method of operation, they asked and answered a question about another attribute of the target population:

What is the standard deviation of the diameters of pistons to be produced in the future if we operate the process under the new method?

By comparing the two attributes, the team gained valuable information helpful in making the decision about whether or not to change the method of operating the process. The cost of the change and potential side effects also entered into the decision. With knowledge of both attributes, the team had the process knowledge to make a decision to permanently change the operating method.

In the example, the target population contained a large unknown number of pistons, all to be produced in the future. This is the common situation where it is impossible to examine each unit in the target population. As a consequence, we will never know the target population attributes exactly. Our goal in the final four steps of QPDAC is to learn enough about the attributes of interest, subject to the constraints of time and cost, to make good decisions, in spite of the inherent uncertainty.

The outcome of the Question step is one or more clear questions about well-defined attributes of the relevant target population or process.

Plan

In the Plan step, we specify *how* we will answer the questions generated in the Question step. The result of this step is a plan to gather a *sample* of units, to measure a prescribed set of characteristics on these units, and to store the information collected. To get a detailed plan, we need to determine:

- What are the units and population available for the investigation?
- How will we select units to be included in the sample?
- What characteristics of the sampled units will we measure, deliberately change, or ignore?
- For those characteristics that we plan to measure, do we have confidence in the measurement systems?
- For those input characteristics that we will deliberately change, how will we make such changes?
- How will we deal with the logistical issues?

The *study population* is the collection of units from which we can choose the sample. In the V6 piston diameter example, the team chose the study population to be all pistons produced by the process in the next week. That is, they planned to collect the sample over the next week of production.

What were the consequences of this choice? First, the team would take at least a week to complete the investigation. Second, if the standard deviation of the diameters of pistons produced in the given week was different from the corresponding standard deviation in the long-term future (the target population), there will be an *error* in the conclusions. The team could have reduced the likelihood of this *study error* by extending the time of sampling to a month or even longer, but then they would pay a price in terms of time and cost.

We can only suspect that study error might be present. We cannot quantify the error without complete knowledge of all units in both the study and target populations. If we have this knowledge, then the investigation is pointless because we already have the answer to our question. You should always think about the relationship between the target population and a proposed study population in terms of a trade-off between possible study error and cost. Remember that the study population is the set of units from which we will get our sample. Even though we cannot quantify the study error, it is clear that some choices of study population are much better than others. In the V6 piston diameter example, the team would have been unwise to define the study population as the next 100 pistons to be produced, since it is likely that a dominant cause may take longer to vary over its normal range.

In most applications of QPDAC, it is not feasible to examine every unit in the study population because of cost and time constraints. Rather, we collect a sample of units using a *sampling protocol*. The sampling protocol specifies how we select the sample and how many units we choose. The goal of the sampling protocol is to produce a sample of units with attributes that match those of interest in the study population. We define *sample error* as the difference between the sample and the study population attribute. We cannot determine the sample error because we do not know the attribute in the study population.

We have numerous choices for a sampling protocol, including random sampling, haphazard sampling, systematic sampling (for example, sample every 10th unit), and convenience sampling (take what we can get).[2] When specifying a sampling protocol, we need to balance cost and convenience against possible sample error. For a given sampling method, larger sample sizes are more likely to yield a smaller sample error. However, it is more expensive to gather and deal with larger samples.

In the V6 piston manufacturing process, about 10,000 pistons were produced per day for five days in the week. The team decided to commit resources to collect a sample of 500 pistons. The issue of sample size is a difficult one, and the choice is usually driven by resource considerations. For convenience, the team decided to use a systematic sampling protocol in which they would pick every 100th piston. They expected such a sample to give a good representation of the week's production. That is, they thought the standard deviation of the diameters of the sampled pistons would be close to that of all the pistons produced over the week. In other words, they expected little sample error.

Next we decide what characteristics of the sampled units to measure, to ignore, or to change. We must measure or record any input or output characteristic used in the definition of the attribute of interest. However, it is often advantageous to measure additional characteristics. In the V6 piston diameter baseline investigation, for very small cost, the team recorded the time at which each piston in the sample was measured. They hoped that the pattern of variation of diameter over time would give valuable clues to be used later in the problem solving.

We typically ignore most characteristics of the sampled units; this is a conscious choice. In the V6 piston diameter example, the team decided to record the diameter, the time, and the day of measurement and to ignore all other input and output characteristics. They made this choice because of the question of interest. However, as a general rule, if there is little cost involved, we should record other inputs and outputs, especially if we have automated measurement and data collection systems available. We may be able to use these data later in the variation reduction algorithm.

In the Plan step, we specify how we will measure the selected output and input characteristics on the sampled units. The gages, operators, methods, materials, and environment all make up the *measurement system*. The difference between the measured and true value is called *measurement error*. Due to measurement errors, the attribute calculated using the measured values of the sampled units might differ from that using the true values. We need to worry about the contribution of errors from the measurement system to the overall error.

In some investigations, we deliberately change an input characteristic on one or more units in the sample to understand the effect on the output characteristic. If no input characteristics are deliberately changed, we call the plan *observational*. The plan to determine the baseline performance of the piston process was observational since the team did not deliberately change any inputs as the sample pistons were collected. Of course, many inputs changed from piston to piston in the sample. The key point is that the team let these changes occur naturally and did not deliberately manipulate any input.

If we deliberately change one or more input characteristics on the sampled units, we call the plan *experimental*. For example, as part of a project to reduce the proportion of steel stampings that rusted during shipping, a team specially oiled a number of stampings and shipped these parts in the same crates as stampings that were not oiled. The team hoped to show that the oiling significantly reduced the frequency of rusting. This was an experimental plan. The oiling procedure would not have happened without deliberate intervention. In Chapter 13, we look at the details of experimental plans, which are often called *designed experiments*.

We require an experimental plan for many questions, such as the previous example regarding the effect of oiling on rust. To investigate the effects of fixed inputs that do not normally vary, we must intervene and change them. Such interventions can cause difficulty in the production process. For other questions, we may have a choice between an experimental and observational plan. Where possible, we use observational plans because they have the major advantage of not disrupting the current operation of the process. Also, observational plans are usually cheaper to conduct than experimental plans.

A final task in the Plan step is to organize the logistics of the investigation. This can be nontrivial, especially for complex plans. We must consider who, how, where, and when as related to the investigation. For example, we must decide who will collect the sample of units, make the appropriate measurements, and record the data. We must take care to ensure that everyone involved knows what they are supposed to do and that everyone not directly involved knows what they are not supposed to do.

In the V6 piston diameter example, imagine the confusion generated if someone had decided to make an unexpected change (that is, change a fixed input) to the process during the week in which the team was collecting the sample. Remember the question:

What is the standard deviation of the diameters of pistons to be produced in the future if we leave the process to operate as it is currently?

If the fixed input had been changed in the middle of the sampling period, the team would have little confidence in the conclusions from their investigation. Worse, they may not have known that such a change was made.

We have also seen the opposite problem. A team took great care to explain that they were collecting data to establish a baseline and that no process changes should be made. The operators took the team at their word and ignored the adjustment procedure in the control plan until a supervisor noticed that a large number of out-of-specification parts were produced. The team had failed to explain that the current control plan was part of the process and was to be executed as usual.

If the plan is experimental (that is, involves deliberate changes to the process), we must make all interested parties aware of what is happening since we may put customers at risk. The changes may have unexpected consequences, and we must take special care if we plan to ship the product produced during the experiment.

Part of the logistics is to plan for data storage and processing. For small investigations, we can write down the measurements as they are made. Usually we want to store the data electronically for processing. We recommend a row/column spreadsheet format, where each row represents a different unit and each column represents a different characteristic, as described in Appendix A.

The output of the Plan step is, not surprisingly, a detailed plan for carrying out the investigation. In most applications of QPDAC, we consider all the substeps in the plan but not necessarily in the order that they are presented here. As well, we often iterate among the substeps. Sometimes, in the middle of the planning, we are forced back to the Problem step to clarify the question being asked.

In Figure 5.1, we summarize the connections among the target population and the data and show where errors can occur. Our task is to create a plan that is not too complex or expensive and yet controls the potential errors.[3]

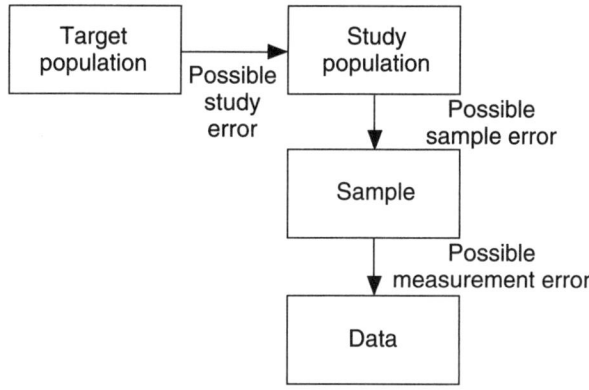

Figure 5.1 Target population to data.

Data

In the Data step, we execute the plan and collect the data. This is often the most time-consuming and costly part of the investigation. This is especially true for those investigations where we paid little attention to the plan, since there is a great opportunity for things to go wrong. In the Data step, we also record any deviations from the plan.

In the example, the team measured 469 pistons and recorded the diameter, day, and hour of measurement. The data were recorded in production order. We give only the last three digits of the diameters in microns. The diameter is measured in millimeters and the deviation from 101 millimeters is multiplied by 1000 before being recorded. The process did not operate on the last part of the shift on day four. The data from the piston baseline investigation are stored in the file *V6 piston diameter baseline*.

Analysis

The goal of the Analysis step is to use the data to answer the question(s) posed in the Question step.

For most of the investigations discussed in this book, we use simple numerical and graphical summaries. However, we also consider some more advanced statistical analysis techniques provided by MINITAB. We introduce these analysis methods and tools as needed in subsequent chapters.

The standard deviation of the sample data in the V6 piston diameter example is 3.32 microns. In Figure 5.2, we show a histogram and run chart of the sample piston diameters.

By plotting the data in several ways, we can detect unusual values (outliers). There is a concern, because many of the numerical attributes such as the average and standard deviation are sensitive to outliers. As a result, any outliers that greatly affect the calculated attributes should be identified.[4] When outliers are present, great care must be taken that the conclusions drawn from the investigation are truly representative of the target population.

58 Part One: Setting the Stage

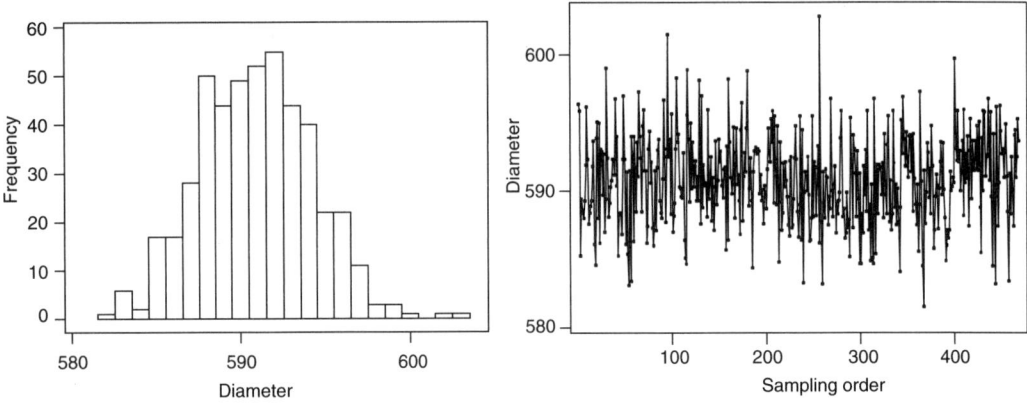

Figure 5.2 Histogram and run chart of V6 piston diameter.

Conclusion

In the Conclusion step, we answer the questions posed about the target population attributes of interest. We also discuss limitations of the answer due to possible errors, both as envisaged in the Plan step and due to deviations from the plan in the Data step. In thinking about errors, remember the three types: study error, sample error, and measurement error.

We use the results from the conclusion step to help us decide what to do next. We need to interpret the conclusion and the associated risk of error in the context of the variation reduction algorithm and the problem itself. To do so, we use basic understanding of the operation of the process, appropriate theory, and knowledge gained in earlier empirical investigations.

In the V6 piston diameter baseline investigation, the estimated process standard deviation was 3.32 microns, which served as a baseline measure of process performance against which the team eventually assessed the effectiveness of deliberate process changes. The team was confident that the estimated standard deviation was a good representation of the long-term variation in the current process, because their experience with the process suggested that in one week, most of the (important) varying inputs that could have changed would have changed. In addition, because of the well-designed plan, they expected that sample error would have little impact.

They specified the numerical goal for the project in term of the estimated standard deviation and decided to proceed to the next stage of the variation reduction algorithm.

5.2 EXAMPLES

Here are two more examples to help reinforce the ideas and language of QPDAC.

V6 Piston Diameter: Comparing Two Measurement Systems

In the process to make V6 pistons, there were gages for measuring the diameter at Operation 270, the last grinding operation, and at Operation 310, final inspection (see Figure 5.3).

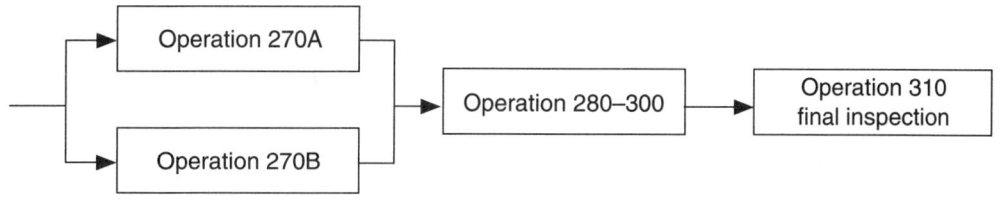

Figure 5.3 Part of a V6 piston machining process.

Operators used the measurement system at Operation 270 to control the two grinders that operate in parallel. The gage at final inspection operated in a controlled temperature environment. Every piston was inspected to ensure that the diameter and several other characteristics were within specification.

During an application of the variation reduction algorithm, the team decided to investigate the relationship between the two measurement systems to help understand the causes of diameter variation at Operation 310.

The team defined a unit to be the act of taking a measurement on a piston. The target population was the set of all such acts that would occur in the future under current conditions. The output characteristic was the measured diameter. One key input was the measurement system used.

To define an attribute useful for comparing the two measurement systems, think of measuring the same piston twice, once with each measurement system. Repeat this over all possible pistons and times (here we think of doing this only conceptually since the number of possible pistons and times is very large). Finally, plot the measured diameters from the Operation 270 gage against the corresponding measured diameter from the final gage. The scatter plot (the attribute of interest) might look like Figure 5.4. Note that the closer the plotted points fall to the 45° line shown on the plot, the better the agreement between the two systems. A point on the line corresponds to equal measured diameters by both systems. Of course, the scatter plot might be very different from the one that is shown.

The specific question was to estimate the scatter plot and determine some of its properties.

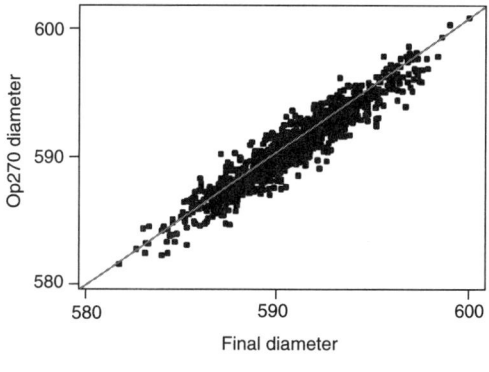

Figure 5.4 Hypothetical target population scatter plot comparing two measurement systems.

The team decided to carry out the investigation the following day. Every hour, the operator at Operation 270 measured four pistons, two from each of the parallel grinders. A designated team member marked these pistons so that they were uniquely identified, recorded the measured diameter, and set the four pistons aside. After six hours, he removed the pistons to the final gage room and let them to come to ambient temperature. During the next shift, the final gage operator measured the 24 pistons and recorded the diameters.

We use the language introduced in the previous section to examine this plan in detail. What is the study population? Measurements could be made on one day. As well, only the 24 selected pistons could be measured with both gages. There is a possibility of study error if the relationship between the two systems changed over time (day to day) or was dependent on the nature of the pistons being measured. To help control this source of study error, the team made a good decision to use pistons from both Operations 270A and 270B.

The sampling protocol specifies which units (acts of making a measurement) will be selected from the study population. Here, the team decided to use 24 pistons and measure each twice at specified times. We can criticize the step in which all 24 pistons were measured by the final gage over a short time. There may be sample error if the 24 pairs of diameters did not represent the relationship between the two gages on the day of the investigation. In terms of sample error, the team could have improved the plan by making the measurements with the final gage throughout the day.

For each piston, the team recorded the diameter from both gages, the time of measurement and the process stream (A or B) at Operation 270. They ignored all other process characteristics such as other dimensional and physical properties of the pistons, the ambient temperature at Operation 270, and so on. The team did not change any inputs, so this was an observational plan.

Since this was an investigation of the measurement systems, there was no need to consider the issue of measurement systems separately.

The logistics of who does what and when were well organized, since one person was assigned responsibility for executing the plan and storing the data in row/column format in a spreadsheet, convenient for subsequent analysis.

The team executed the plan without difficulty. They collected and stored the data in the file *V6 piston diameter gage comparison*.

The analysis consisted of simple numerical and graphical summaries of the data. We show a scatter plot of the pairs of measurements from the sampled units in Figure 5.5 and give the average diameter by gage over all 24 pistons in Table 5.1.

Table 5.1 Piston gage comparison results.

Gage	Average diameter
Operation 270	589.63
Final	591.14

The team concluded there was a strong relationship between diameters measured on the two gages. However, the final gage gave a measured diameter that was systematically higher by about 1.5 microns on average. The difference was small but unexpected. The

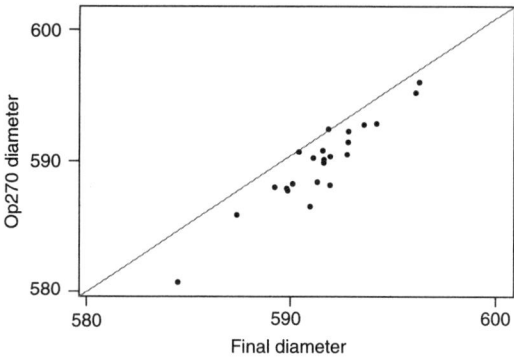

Figure 5.5 Scatter plot comparing sample diameters at the final and Operation 270 gages.

team later explained the difference when they discovered that steel, not aluminum, masters were used to remove the effect of temperature changes from the Operation 270 gage.

The team's only reservation was that the investigation had been limited to one day and there was some concern that the systematic difference between the two gages might become larger on a hotter day.

The team used the knowledge gained to recommend a new calibration procedure for the Operation 270 gage.

Painted Fascia Ghosting

In the painting of two color plastic fascias (the cover over a car bumper), the black prime coat was applied to the whole part. Next, some of the black area was masked with paper taped to the fascia, and then the color coat was added. When the masking was removed, there was occasionally a residual pattern, called ghosting, on the matte black surface under the tape. See the process map given in Figure 5.6.

Customers could detect ghosting since the black surface of the fascia was prominent. The plant could rework the surface to remove the ghosting with added cost and lost production. Management assigned the process engineer the task to eliminate the ghosting problem that occurred on about 3% of the fascias.

The engineer followed the Statistical Engineering algorithm. The process operators visually judged ghosting on a scale of 1 to 5 (1 = no ghosting, 5 = heavy ghosting). The plant reworked fascias with a ghosting score greater than 2 and reluctantly shipped those with a score of 2. The engineer carried out a small investigation to verify that the measurement system was repeatable; that is, the operators were consistent in their assessment of ghosting. Next, after several QPDAC applications, she was convinced that the cause of the problem

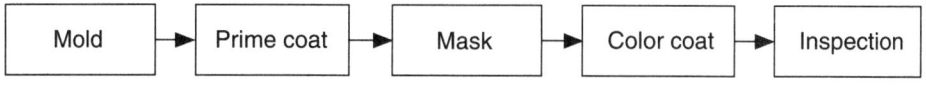

Figure 5.6 Fascia molding and painting process.

was some environmental factor such as ambient temperature or humidity that changed from day to day and could not be easily controlled. More ghosting problems seemed to occur on hot, humid days. She also knew that the ghosting appeared under the tape during the baking of the color coat and that the problem seemed more frequent with certain colors. Two other tape suppliers claimed that their products would not produce ghosting. She decided to investigate whether the other sources of tape were robust to the effects of the environmental factors and color. In the jargon of the variation reduction algorithm, she chose the Desensitize the Process approach. The engineer decided to carry out an experimental plan to compare the performance of the current tape, here denoted C, against the two other possibilities, A and B.

A unit was a fascia and the target population was all fascias to be made in the future. The output characteristic was the degree of ghosting and the key input characteristic was the type of tape used to mask. The three attributes of interest were the proportions of fascias with ghosting level 1 if the tape type was A, B, or C. The three questions of interest were:

What is the proportion of fascias in the future production with score 1 if tape A [or B or C] is used?

The engineer knew that she required an experimental plan since the tape type was a fixed input that could not change without intervention. Because of the high cost of scrap, she decided to use fascias that had been scrapped upstream of the masking operation for her study population. She also decided to use only those fascias produced on a hot, humid day and to paint these fascias with the color having the most frequent ghosting problem. Furthermore, she planned to mask three small areas, labeled I, II, and III on the primed surface on both sides of each fascia with the three different tapes. See Figure 5.7.

The study unit was one of the small areas on the scrap fascia. The study population was all such units available on the selected day. The study units and conditions under which they were processed were very different from the target population. There was likely to be study error because the conditions were selected to be extreme, so the degree of ghosting seen in the investigation was likely to be high. As well, the engineer knew that she was unlikely to get a large number of scrap fascias for her investigation, so she expected a relatively small sample size.

Accordingly, she decided to revisit the Question, which was changed to the following:

Under extreme conditions, what is the average ghosting score for tapes A, B, and C?

Figure 5.7 Schematic of the study units.

Note that these new questions involved averages, not proportions. The engineer thought that if one tape performed better on average in the extreme conditions, then this tape was likely to produce a higher proportion of fascias with score 1 in the future.

Going back to the Plan step, she decided to use all available scrap fascias (with a maximum of 15) produced on one day. She would chose the day based on a weather forecast. She planned to record the fascia number, the side, the position, the tape type, and the ghosting score for each unit in the sample. She had already established a reliable measurement system for ghosting.

Since this was an experimental plan, she had to decide how she was going to select which tape was applied to each unit. Her plan was to use all three tapes on both sides of each fascia available. Within a side, she would assign the tapes to the three positions at random.

The logistics were critical to the success of the plan. The engineer involved an operator to decide how to mask the small areas. She carefully labeled the tapes to avoid confusion. She prepared 15 schematics like Figure 5.7 that showed which tape went on each position. She planned to give these schematics to the operator one at a time to help him use the correct tape to mask each area in the sample. She would identify each fascia with a number and write it on the schematic. When the operator assessed ghosting, he could write the score on the sheet above each box.

She notified the process owners of the plan to ensure that the experimental fascias would be set aside after painting and to avoid interference with normal production procedures. A time to run the experimental parts could be set up once the day was selected.

The plan was executed without any hitches. Twelve fascias were processed and a total of 72 ghosting scores were recorded. The data were transferred from the schematics to the file *fascia ghosting robustness*. Table 5.2 is a summary of the data collected.

Table 5.2 Summary of ghosting scores.

	Ghosting score					
Tape	1	2	3	4	5	Average
A	16	7	1	0	0	1.375
B	5	9	7	2	1	2.375
C	7	9	6	2	0	2.125

Under the extreme conditions of the plan, tape A was clearly superior with a much lower average score. All but one area had a score of 1 or 2. The major limitation was study error; it was possible that tape A would not perform better than tapes B or C under normal operating conditions.

The engineer validated the performance of tape A under normal operating conditions with another application of QPDAC. Since the costs of the three brands of tape were similar, management accepted the engineer's recommendation to switch to tape A. In the longer term, the proportion of fascias with any detectable ghosting (score >1) fell to less than 0.3%.

5.3 SUMMARY

We summarize the terminology used within QPDAC as follows:

QPDAC Terminology

Term	Meaning
Attribute	Numerical or graphical summary of the characteristics over a collection of units
Experimental plan	Plan where some inputs are deliberately changed on the sample units
Measurement error	Difference between the measured and true value of a characteristic
Measurement system	Gages, people, methods, material, and environment involved in measuring a characteristic
Observational plan	Plan where all inputs vary naturally on the sample units
Sample	Units selected from the study population and measured in the Data step
Sample error	Difference between the attribute in the study population and sample
Sampling protocol	Procedure by which the sample is selected from the study population
Study error	Difference between the attribute in the study and target populations
Study population	Collection of units available for investigation
Target population	Collection of units produced by the target process that we want to draw conclusions about, usually all units produced now and in the future
Unit	A part or the act of making a measurement

None of the individual steps or substeps in QPDAC is difficult. However, pitfalls abound.

In the V6 piston diameter example, the team first established the baseline measure of process variation by measuring the diameters of 50 pistons produced consecutively. The estimated standard deviation was 2.27, suggesting a highly capable process. There seemed to be little reason to carry on with the project. However, everyone involved knew that the process needed improvement. The team had given little thought to the notions of the target and study population or the sampling protocol. Local management recognized the problem and the baseline investigation was redesigned as described in Section 5.1.

Here we provide a checklist for the major steps and substeps for any empirical investigation. Make sure you consider each substep as you plan and execute a process investigation.

A Checklist of Questions for Each Step of QPDAC

Question
What are the units?
To what population of units will the conclusions apply?
What characteristics are involved in the question?
What attributes of the target population make up the question?
What question(s) need to be answered? Be specific!

Plan
What units might be included in the investigation?
How well does this study population match the target population?
How will the sample of units be collected?
How well is the sample likely to match the study population?
What characteristics will be measured/ignored?
Are all the necessary measurement systems adequate?
What input characteristics (if any) will be deliberately changed?
How will changes be made?
Who will carry out the plan?
Who needs to know that the investigation is being conducted?
How will the data be stored?

Data
Was the plan executed as expected?
What deviations occurred?
Are the data stored as expected?

Analysis
What are the sample attributes? Draw a picture if possible!
Are there any unusual patterns (e.g., over time) of concern in the data?
Are there any unusual values (outliers)?

Conclusion
Has the posed question(s) been answered?
Are any of study, sample, or measurement errors likely to be large enough to affect future decisions?

Key Points

- We use the QPDAC (Question, Plan, Data, Analysis, and Conclusion) framework to carry out empirical investigations to increase our knowledge of the process.
- With careful use of the QPDAC framework, we can avoid or control study, sample, and measurement errors and generate the process knowledge required for the variation reduction algorithm.
- We expect to apply the QPDAC framework several times in any application of the variation reduction algorithm.

Endnotes (see the Chapter 5 Supplement on the CD-ROM)

1. We can use simple attributes such as averages, rates, standard deviations, histograms, and run charts to establish baselines in the Define Focused Problem stage of the variation reduction algorithm. In the supplement to this chapter we describe some other more complex attributes.
2. There are many different sampling protocols. The most famous, random sampling, plays a small role in the variation reduction algorithm given here. In the supplement, we look at the ramifications of choosing among different sampling protocols.
3. We show the nature of the possible errors in empirical investigations in Figure 5.1. Remember that we want to answer questions about attributes of the target population/process using measured values of characteristics in the sample. In the supplement, we describe study, sample, and measurement errors in more detail.
4. Outliers can have a large influence on the conclusion drawn from an investigation. In the supplement we explore the effect of outliers and describe methods for their identification.

 Exercises are included on the accompanying CD-ROM

PART II
Getting Started

If you know a thing only qualitatively, you know it no more than vaguely. If you know it quantitatively—grasping some numerical measure that distinguishes it from an infinite number of other possibilities—you are beginning to know it deeply.

—Carl Sagan, 1932–1996

In this second part of the book, we explore the first three stages of the variation reduction algorithm: Define Focused Problem, Check the Measurement System, and Choose Working Variation Reduction Approach. We require a quantitative goal for the problem and the full extent of variation of the output to help design and analyze investigations conducted in the search for a dominant cause or a solution, and to check that a proposed solution meets the problem goal. Next, we look at how to assess and improve measurement systems. A measurement system is necessary to support all process investigations and may itself be a dominant cause of variation. Finally, we discuss how to choose a working approach among the seven variation reduction approaches. In any application, this choice is important because it determines what we do next.

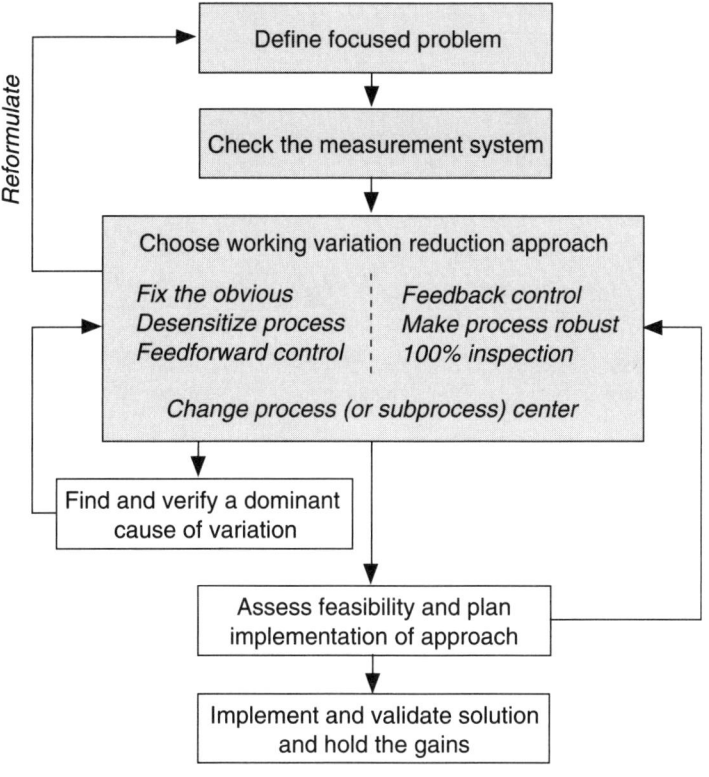

Statistical Engineering variation reduction algorithm.

6
Defining a Focused Problem

Our plans miscarry because they have no aim. When you don't know what harbor you're aiming for, no wind is the right wind.

—Lucius Annaeus Seneca the Younger, 4 B.C.–A.D. 65

There are two major tasks in the Define Focused Problem stage of the Statistical Engineering algorithm. First, the team must translate the project and its goal into a problem statement about excessive variation in a measurable output characteristic. The problem should be specific so that it is likely that there will be only one or two dominant causes. Second, the team must quantify the magnitude of the problem. We require this baseline measure to:

- Help set the goal for the problem.
- Help design investigations and interpret the results as we proceed through the algorithm.
- Allow validation that the problem has been solved.

The baseline is important in all subsequent stages of the algorithm.

6.1 FROM A PROJECT TO PROBLEMS

We start many applications of Statistical Engineering with a project defined in vague terms, and a goal specified in terms of dollars or customer concerns. In Chapter 1, we discussed the block leakers project, where the goal was to reduce the scrap rate of engine blocks that leaked after machining. The management of the foundry assigned a team to reduce cost and improve the relationship with the engine plant, their direct customer. The goal of the project was to reduce the scrap rate from more than 4% to less than 1%.

The team started by turning the project into three problems. The engine plant inspected every block for leaks using a pass/fail test. The team instituted a new test for failed blocks that determined the location of the leak. Using a sample of 100 leaking blocks, they found that 92% of the leakers fell into three classes based on the location of the leak, as shown in Table 6.1.

Table 6.1 Leak classification for 100 leaking blocks.

Class of leak	Percent
Center	32
Cylinder bore	26
Rear intake wall	34
Other	8

The team suspected that each class of leak would have its own dominant cause. Hence they defined three problems based on the three classes. In each case, they set the ambitious goal of eliminating the entire leak class. If they could achieve these goals completely, they would far surpass the project goal.

Some projects are defined in terms of processes that are replicated at different sites or in terms of a class of products. We may choose to focus the problem by concentrating on a single product or manufacturing line. In the truck alignment problem described in Chapter 1, several assembly plants built the same truck using the same components and assembly process. To concentrate resources, upper management assigned the project to a team in one plant. The idea was to apply the same solution to all plants.

In specifying a problem, we may need to define an output characteristic that can be measured locally. A team at an engine assembly plant was charged with reducing warranty claims due to excessive oil consumption. There were only a few claims, but the plant management initiated the project because of the potential damage to the long-term reputation of the engine. To define a specific problem, the team first spent considerable time and effort developing a dynamometer test that could reproduce in the lab the failure mode seen in the field. This effort was necessary because the field failures were so rare. The team specified the problem in terms of the dynamometer measurement. They were confident that if they could reduce oil consumption in the dynamometer test that they could eliminate the field failures.

We try to specify the problem in terms of a continuous output to improve the efficiency of the problem solving. If we use a discrete output, such as pass/fail, we require larger sample sizes for all of the subsequent investigations. For example, in the engine block porosity project described in Chapter 3, the project goal was to reduce the scrap rate from about 4% to below 1%. The team invented a new output, a porosity score based on the size, location, and number of holes on the surface. A block that was scrapped had a high score, but more importantly for the problem solving, every block could be assigned a porosity score that reflected the severity of the problem on a continuous scale. The team defined the problem goal in terms of reducing the variation in the porosity score.

We need to be careful defining the output. For instance, in a problem with seat cover appearance, the team measured shirring on a six-point scale using boundary samples. Scores of 1 to 4 were acceptable, and high scores came from *either* too much or too little shirring. With shirring score defined in this way, the team found it difficult to find a dominant cause of shirring variation, because both high and low values of the cause led to high scores. The choice of output forced the team into the Make Process Robust approach, which failed. See the Chapter 19 exercises for further details.

We need to be able to measure the output characteristic quickly. We sometimes aggravate or accelerate the usage conditions so that the problem occurs sooner. In the oil consumption example, the team created a dynamometer test with aggravated conditions in an attempt to quickly simulate use of the engine under extreme field conditions. There can be considerable difficulties linking the original management project goal to a goal for the output measured using an aggravated or accelerated test. We want to avoid study error where the cause or the solution of the problem under the aggravated conditions is not the same as under the normal conditions. To avoid study error, we need to check that the field failure mode can be replicated using the aggravated test.

We can sometimes use aggravation to deal with a problem with a binary output. For example, in a painting process, there were line defects on the roof of about 10% of the vehicles painted. The team used panels with an increased clear coat film build to make the defect occur more often. Using the panels under the aggravated conditions, they felt they could find clues about the dominant cause more quickly since the defect occurred more often. This advantage had to be balanced against the risk that the dominant cause of line defects on panels with increased clear coat film build (that is, under the aggravated conditions) was not the same as the dominant cause of lines on vehicle roofs with the normal clear coat film build.

To illustrate the transition from a project to problem(s), consider a project to reduce scrap in a process that produced piston rods. Based on past performance, the monthly scrap rate averaged 3.2%. Management set a goal of reducing the rate to less than 1.6%. We show a finished rod in Figure 6.1.

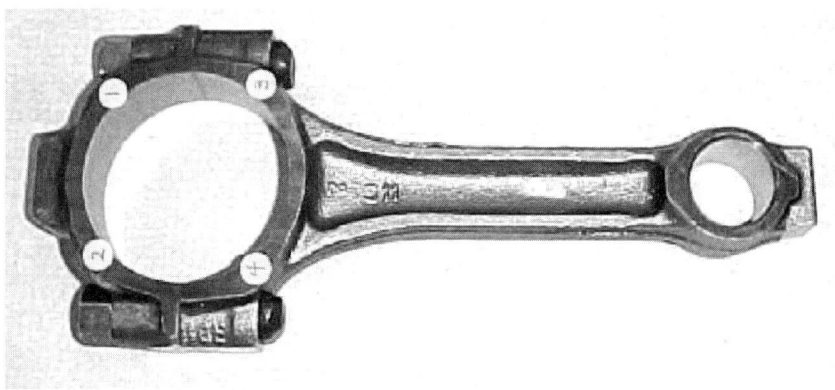

Figure 6.1 A connecting rod.

Looking at process records, the team found that scrap occurred at several processing steps. To focus the problem, they used Pareto analysis on the records from one month, as shown in Figure 6.2. They found that 62% of the scrap was identified at a grinding operation. Further investigation revealed that about 90% of the grinding scrap was due to undersized rods. After grinding, the thickness was measured on every rod at the four locations, marked (with small white circles and faint numbers) in Figure 6.1. A rod was scrapped if the thickness at any location was below the lower specification limit. Rods with thickness above the upper specification limit were reworked. The team focused their attention on reducing variation in rod thickness, a continuous output.

The team set the problem goal to produce all rods within the thickness specification. Achieving this goal would eliminate 90% of the scrap at the grinder, or 56% (90% of 62%) of the total scrap, and hence meet the project goal.

In summary, the key elements in focusing a project to one or more problems are:

- Identify and address the most important failure modes.

- Replace a binary or discrete output characteristic by a continuous one, if possible.

- Define the problem in terms of an output that can be measured locally and quickly.

- Choose the problem goal to meet the management project goal.

If multiple problems arise from a single project, we recommend that the team address the problems one at a time. We need to be careful that, in improving the process with respect to one output, we do not make it worse with respect to another, unless the gain outweighs the loss.

We find it helpful to think about what, where, when, and to what extent the problem exists. We can use a Pareto analysis to help focus the project to a problem that is both narrowly defined and a major contributor to the concern that generated the original project.

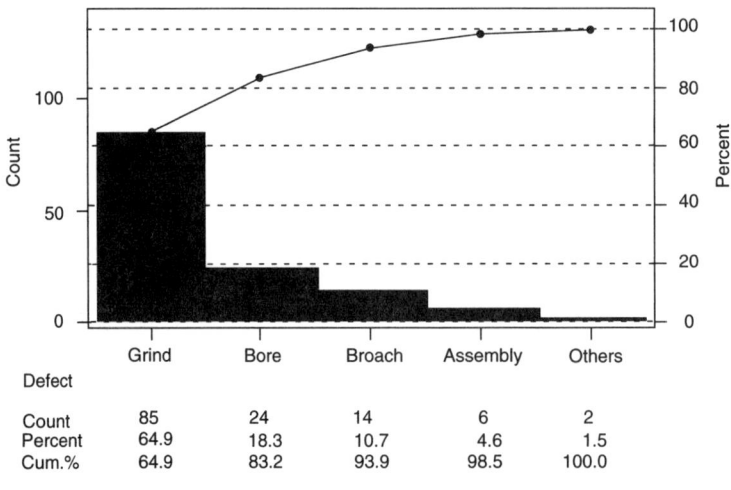

Figure 6.2   Pareto chart for rod scrap by operation.

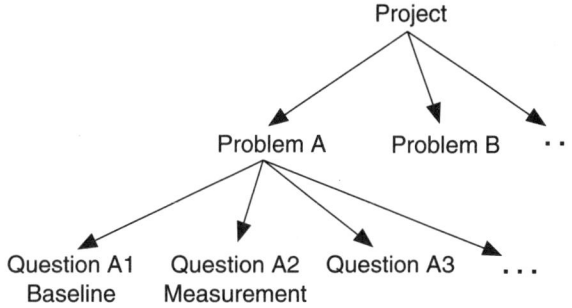

Figure 6.3 Hierarchy of project, problems, and questions.

Figure 6.3 shows the hierarchy we propose to define projects, problems, and questions. Company management initiates projects using concerns regarding quality and cost. The improvement team focuses the project into one or more problems. The team then attacks each problem using the Statistical Engineering variation reduction algorithm given in Chapter 4. In each application of the algorithm, the team will ask many questions about the process behavior. They address these questions using the QPDAC framework described in Chapter 5.

6.2 THE PROBLEM BASELINE

The *problem baseline,* also called the *baseline,* is a numerical or graphical summary of the current process performance. In the language of QPDAC, the baseline is an attribute of the process operating under current conditions. We select a baseline and estimate its value or appearance in the Define Focused Problem stage.

We specify the problem goal by stating how the baseline should be changed. At the Validate Solution and Hold the Gains stage of the variation reduction algorithm, we recalculate the baseline with the changes to the fixed inputs to demonstrate the improvement and to determine whether or not we have met the problem goal. If we have focused the problem, we should be able to translate the change in baseline to progress toward the overall project (management) goal.

We can choose among many attributes for a baseline. If the output that defines the problem is continuous, we may use the average, the standard deviation, a capability ratio such as P_{pk} (see Chapter 2 supplement), a histogram, or a run chart, among many others. For a binary output, we use a rate or a series of rates taken over time, often plotted on a run chart. We summarize the choices in Table 6.2.

We prefer not to use a capability ratio such as P_{pk} since it combines the average and the standard deviation. We use different approaches to change the process center and to reduce unit-to-unit variation, so we prefer to address two problems when we need to change the process center and reduce variation in the output.

Table 6.2 Problem types and corresponding baseline measures.

Problem type	Possible baseline measures
Too much scrap or rework	Proportion of scrap or rework, run chart of proportion scrapped
Process center off target	Average
Unit-to-unit variation too large with two-sided specifications	Standard deviation, histogram, P_{pk}
Unit-to-unit variation too large with one-sided specifications	Average, standard deviation, histogram, one-sided P_{pk}
Measurement	Measurement variation and bias

In the rod thickness example, the team selected a histogram with specifications limits as the baseline. See Figure 6.4. We discuss the investigation to produce this histogram in the next section. The team set the problem goal to reduce variation in thickness so that the histogram would fall entirely between the specification limits 10 to 60.

The data are found in the file *rod thickness baseline*. The height was recorded in thousandths of an inch, measured as the deviation from 0.900 inches. The summary is:

```
Variable       N       Mean    Median    TrMean    StDev   SE Mean
thickness     800     34.575   36.000    34.840    11.023   0.390

Variable    Minimum   Maximum        Q1        Q3
thickness    2.000    59.000     28.000    43.000
```

The standard deviation in the baseline investigation was 11.0. Since the process is roughly centered and the histogram is bell-shaped, the team needed to reduce the standard deviation by about 25% so that six times the standard deviation matches the specification range.

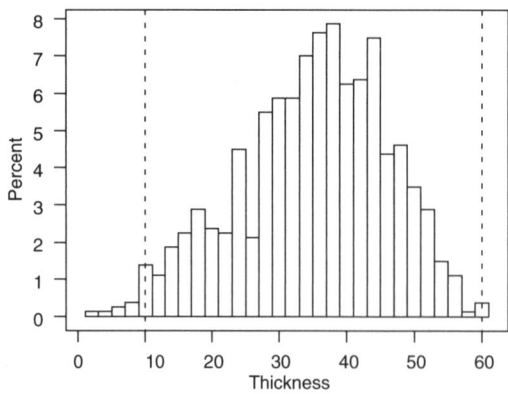

Figure 6.4 Baseline histogram of rod thickness.

There are many choices for a baseline measure. We once asked a process engineer how he could tell if a planned process change would lead to an improvement. After some thought, he told us that he would receive fewer phone calls per week from his customers. In a problem with a system to measure silicon concentration in cast iron, the baseline was the measurement system R&R (repeatability and reproducibility). The problem goal was to reduce R&R to less than 20% of the process variation. See Chapter 7 and its supplement for more information on assessing measurement systems.

As part of establishing the baseline for a continuous output, we determine the *full extent of variation* in the output. We use the full extent of variation in planning and interpreting the results in subsequent process interpretations. When looking at a histogram of the output, we define the full extent of variation as the range within which the vast majority of values lie. The range (minimum to maximum) defines the full extent of variation when the sample size is large (that is, the sample size is in the hundreds) and there are no outliers. For the rod thickness example, the full extent of variation is 2 to 59 thousandths of an inch.

More generally, for a histogram with a bell shape (as given in Figure 2.11), the full extent of variation corresponds to the range given by the average plus or minus three times the standard deviation. Defined in this way, the full extent of variation covers most of the output values.

Sometimes, we use the baseline investigation to generate clues about the dominant cause of the variation. For the rod line example, 40 rods were measured at the four positions each day for five days. We summarize the day-to-day performance using box plots in Figure 6.5.

We see most of the full extent of variation within each day. We can draw two important conclusions from this observation:

- The dominant cause of variation changes within days and we can rule out slowly varying causes that change from day to day or over a longer time frame.

- In subsequent investigations to search for a dominant cause, we can set the study population to be rods produced on a single day.

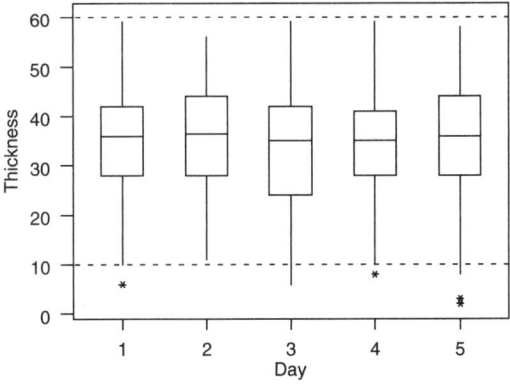

Figure 6.5 Rod thickness by day (dashed lines show specification limits).

6.3 PLANNING AND CONDUCTING THE BASELINE INVESTIGATION

To estimate the baseline, we need to carry out a process investigation. The purposes of the baseline investigation, in order of importance, are to:

- Estimate the baseline for the current process.
- Determine the full extent of variation in the output.
- Generate clues about possible dominant causes.

We need to keep the investigation simple, so we pay little attention to the third purpose.

We discuss common issues in baseline investigations using the QPDAC framework with the rod line scrap reduction problem as the example.

Question

The target population is all units produced by the process now and in the future (assuming no changes are made). When we focus the problem, we specify the output characteristic. In the rod line example, the output is the rod thickness measured at any of the four locations. The attribute of interest (we call this the baseline) is a measure of the process performance related to the goal. In the rod example, the team chose the histogram of the thickness in the target population as the baseline.

Plan

We need to specify a study population from which we will collect a sample of units. The big issue is the period of time over which the study population extends. In the example, the team decided that the study population was all rods ground within a week. They felt that a single week was sufficient to see the full extent of variation in the target population. There is always a trade-off between avoiding study error with a longer period and the cost and time it takes to complete the baseline investigation.

Remember, the goal of the baseline investigation is to understand the process performance as the process currently operates. As a result, process adjustments should be made according to the current control plan.

We choose a sampling protocol that spreads the sample units across the study population; in other words, we sample across the time period selected. In most cases, a combination of a systematic and haphazard method is used. If the output is measured on all units and the data are stored automatically, we can use the whole study population. In the rod example, 40 rods were selected haphazardly throughout each day for five days.

We specify the sample size to be large enough to avoid substantial sample error. As a rule of thumb, we suggest sample sizes of hundreds of units if the output characteristic is continuous and thousands if the output is binary. These sample sizes may seem large, but tables 6.3 and 6.4 demonstrate the difficulty of estimating a small proportion and the advantage of a continuous output.

In Table 6.3, we show roughly how well we can estimate (with 95% confidence) a standard deviation as a function of the sample size.[1]

Table 6.3 Relative precision for estimating a standard deviation as a function of sample size.

Sample size	Relative precision
50	±20%
100	±14%
200	±9%
500	±6%
1000	±5%

In the rod thickness example, we can estimate (with 95% confidence) the baseline standard deviation in the study population to within about ±5% with a sample of size 800.

For a binary output, we cannot give such a simple table since the precision of the estimate depends on both the sample size and the unknown defect rate. In Table 6.4, we give the relative precision (95% confidence) for estimating a proportion defective with a given sample of size.

Table 6.4 Relative precision (%) for estimating a proportion p with 95% confidence.

Proportion defective	Sample size			
	1000	2000	5000	10000
.05	±27%	±19%	±12%	±8%
.04	±30%	±22%	±14%	±10%
.03	±35%	±25%	±16%	±11%
.02	±43%	±31%	±19%	±14%
.01	±62%	±44%	±28%	±20%

In the example, the known baseline scrap rate is about 2%. Since the team examined only 200 rods in the investigation, the sample proportion of defectives is almost useless as a baseline measure since it is so imprecise. If instead they had used a baseline sample of 2000 rods they would have estimated the proportion to within about ±0.6% (2% × 0.31).

In the baseline investigation, we must measure the output and input characteristics that define the problem. We can decide to measure other characteristics, such as the time the unit is produced, to generate clues about a dominant cause. We do not need to assess the measurement system at this point because we want to include variation due to the measurement system in the baseline.

Data

The plan is implemented in the Data step. In the rod line example, the data were collected according to the plan without incident. The data are given in the file *rod thickness baseline*. We store the data in MINITAB using the row/column format convenient for statistical analysis as shown in Figure 6.6. Each row corresponds to a single measurement, and the columns give the corresponding day, batch, position, thickness, and rod number. See Appendix A for more on row/column data storage.

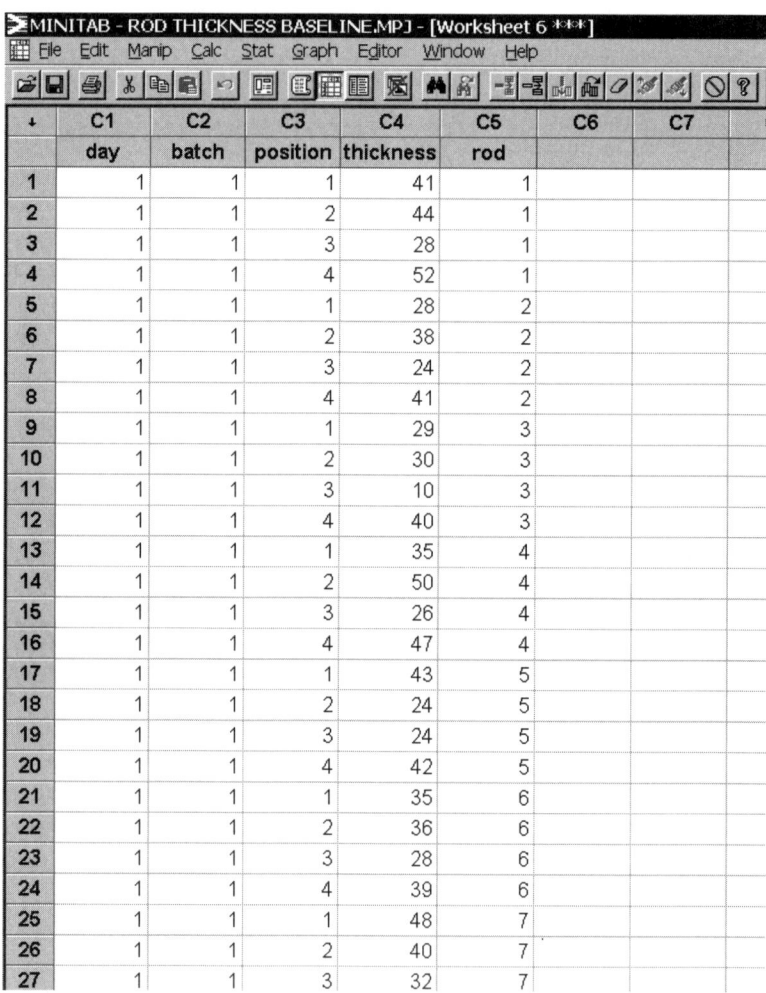

Figure 6.6 Row/column format storage of rod thickness baseline data.

Analysis

To quantify performance, we use the sample attribute corresponding to the population attribute selected in the Problem step. In the rod line example, we gave the sample histogram (Figure 6.4) and estimated the standard deviation as 11.0. We also look at other simple numerical and graphical summaries such as box plots and run charts to look for unusual values and patterns in the data.

For example, suppose we collect data as in the rod line investigation and we see a run chart as in Figure 6.7. Since there is an obvious trend, we would worry that the one-week study population was not long enough to capture the full extent of variation in the target population.

Similarly, if there were a large day effect, for example as illustrated by Figure 6.8, we would worry about study error. In this case, five days is not long enough to obtain a good

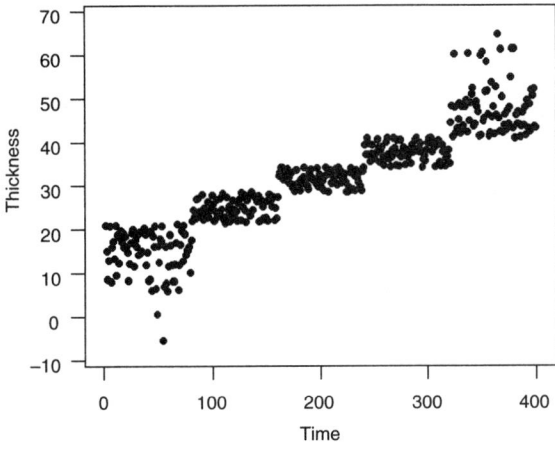

Figure 6.7 Artificial example showing a thickness trend.

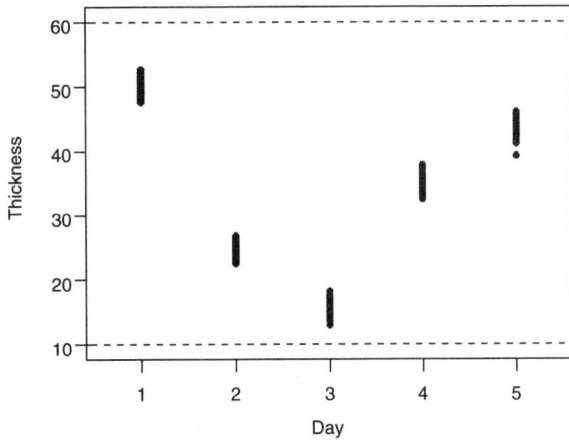

Figure 6.8 Artificial example showing a large thickness day effect.

estimate of the long-term process variation (target population). Here, since the variation within each day is so small, the effective sample size for estimating the overall variation is closer to five (the number of days) than 800 (the number of thickness measurements). With a large day effect, the additional measurements taken each day give information about the within day variation but do not provide much more information about the overall variation.

Conclusion

In drawing conclusions from the baseline investigation, we report the estimates of the process attributes of interest such as the standard deviation, histogram, and run chart. In addition, we note any material limitations due to possible study or sample error. As illustrated by figures 6.7 and 6.8, we may discover these limitations from the sample data. We may also have limitations due to concerns about the plan. If the sample size is small, we worry about sample error. In the rod example, the baseline is given by Figure 6.4. The full extent of variation is 2 to 59 thousandths of an inch.

Baseline Investigation Summary

Question

The purpose of the investigation is to:
- Estimate the baseline, an appropriately chosen attribute of the current process
- Determine the full extent of variation of the output characteristic

The team must select an appropriate baseline—for example, a histogram, a standard deviation, or a proportion.

Plan

- Choose a study population covering a period long enough to see the full extent of variation in the output.
- Determine what outputs and inputs to measure. The inputs should include the time of production.
- Select a sample well spread across the study population with respect to time and other (possibly) important inputs such as machine, position, and so on. The sample size should be hundreds of parts for continuous outputs and thousands of parts for binary outputs.

Data

Record the input and output values with one row for each output value measured (row/column format).

Analysis

- Summarize the data using the appropriate sample performance measure(s). For:
  - a continuous output, use the average, standard deviation, histogram, and run chart
  - a binary output, use the proportion defect and a run chart
- Check for patterns in the output over time (and possibly other inputs).
- Check for outliers.
- Estimate the full extent of variation in the output.

> **Conclusion**
> - State the problem and goal in terms of the estimated performance measure(s).
> - Determine the minimum time required to see the full extent of variation.
> - Consider possible study and sample error.

Comments

We can use existing data to establish a baseline. With the QPDAC framework we can examine the data and how they were collected to ensure that we have confidence that the calculated baseline reflects the true long-term process performance. For example, suppose the output is monitored using a control chart (Montgomery, 1996; Ryan, 1989). We can select an appropriate time period (that is, a study population) from the recent past and use the control chart data over that period. We do not require the process to be stable, as defined by the control chart, to estimate the problem baseline.[2]

Some projects will generate several problems, defined in terms of different outputs. We may be able to plan and carry out a single investigation that simultaneously establishes the baseline and full extent of variation for each problem. We need to be careful to avoid study and sample error for each of the outputs.

6.4 EXAMPLES

We give two further examples of focusing the problem and estimating the baseline.

Truck Pull

We described a project in Chapter 1 where management identified truck alignment as a key issue based on a Pareto analysis of warranty costs and customer quality surveys. They set a project goal to match the performance of a competitor based on these surveys. The same truck was built at several plants. To concentrate resources, the management established a team at one plant to work on the project. They assumed that any remedy found at that plant could be applied to the others.

From mathematical modeling of the truck geometry, the team knew that customers could detect pull, a torque on the steering wheel. Pull is a function of the front wheel alignment as well as many other characteristics such as tire design and pressure, the road camber, and condition. Within the assembly plant, pull was measured as a function of caster and camber only. We have

$$\text{pull} = 0.23*(\text{left caster} - \text{right caster}) + 0.13*(\text{left camber} - \text{right camber}) \quad (6.1)$$

The specification limits for pull are 0.23 ± 0.35 Newton-meters. The plant measures and records caster and camber on every truck. Any truck with pull outside the specification limits is repaired before release.

To focus the problem, the team started by relating pull as measured in the plant to warranty costs due to alignment problems. They divided the pull specification range into seven classes, each of width 0.10. Using historical data on the alignment characteristics and a warranty database, the team grouped trucks into the seven classes, and for each class calculated the average warranty cost due to alignment issues. Figure 6.9 shows that warranty costs are higher for the extreme classes, corresponding to pull values near the specification limits. This suggests that warranty costs can be reduced by reducing variation in pull. The team assumed reducing pull variation would improve customer satisfaction as measured by the quality surveys.

To establish a baseline, the team selected data from the previous two months. The data are given in *truck pull baseline*. They felt that this time period was long enough so that they would see the full extent of the variation in the process. The histogram for these data is shown in Figure 6.10, with the specification limits given by the dashed lines.

The team also constructed box plots of pull by day as shown in Figure 6.11. We see clear evidence of some day-to-day variation, with drifts in the pull center over time.

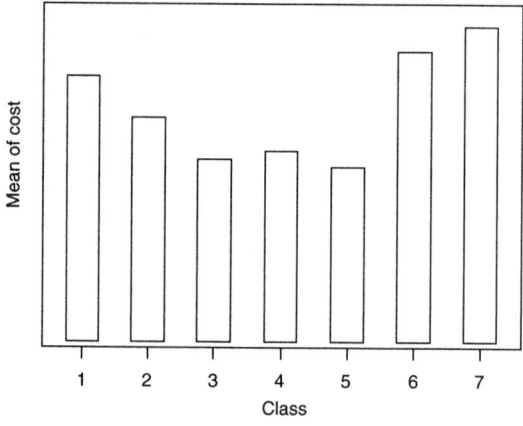

Figure 6.9   Average warranty cost versus truck pull class.

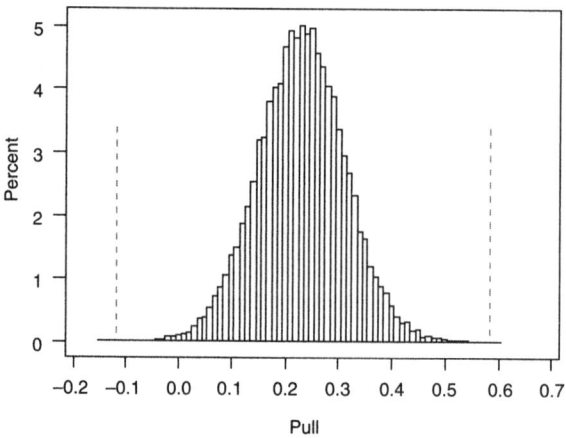

Figure 6.10   Truck pull baseline histogram.

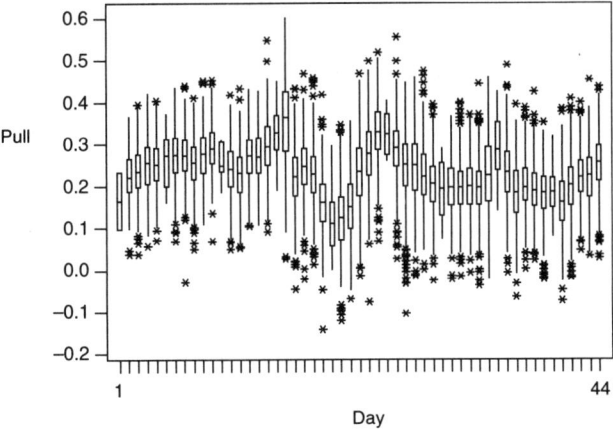

Figure 6.11 Box plots of truck pull by day.

The numerical summary of the pull data is:

```
Variable       N        Mean      Median     TrMean      StDev     SE Mean
pull         28258    0.23093    0.23084    0.23070    0.08234    0.00049

Variable    Minimum   Maximum        Q1         Q3
pull       -0.14060   0.60120    0.17618    0.28502
```

The team set the problem goal to reduce the standard deviation of pull from 0.082 to 0.050. If they could achieve this goal, almost all trucks would leave the plant with pull in classes 3, 4, and 5, and warranty costs would be reduced.

To further focus the problem, the team used Equation (6.1). Pull is a simple function of cross caster (difference between left and right caster) and cross camber. Looking at summaries of the cross caster and camber in the baseline data, we have

```
Variable          N        Mean      Median     TrMean      StDev     SE Mean
cross caster    28258   -0.99982   -0.99600   -0.99795    0.36724    0.00218
cross camber    28258    0.00748    0.01500    0.00991    0.20124    0.00120

Variable       Minimum   Maximum        Q1         Q3
cross caster   -2.73600   0.49800   -1.23900   -0.75700
cross camber   -1.01500   1.08400   -0.11800    0.14100
```

Using Equation (6.1) and ignoring the small correlations (the correlations between the right and left sides of both camber and caster are around –0.2, and the correlation between cross camber and cross caster is around 0.2), we have

$$\text{stdev}(pull) = \sqrt{(0.23)^2 \, stdev(cross\ caster)^2 + (0.13)^2 \, stdev(cross\ camber)^2} \quad (6.2)$$

Since 0.23*stdev(cross caster)* is much greater than 0.13*stdev(cross camber)*, cross caster is a dominant cause of pull variation. Based on this knowledge, the team refocused the problem based on caster. There is little benefit to reducing variation in camber. Using Equation (6.2), if they cut the variation in cross caster in half from 0.367 to 0.185, they could achieve the goal of reducing the pull standard deviation to 0.05. Numerical summaries for right and left caster are:

```
Variable      N       Mean    Median   TrMean    StDev    SE Mean
l-caster    28258    3.5190   3.5230   3.5204    0.2241    0.0013
r-caster    28258    4.5188   4.5210   4.5192    0.2427    0.0014

Variable    Minimum  Maximum     Q1       Q3
l-caster    2.4490   4.5480    3.3700   3.6720
r-caster    3.0440   5.9380    4.3600   4.6780
```

The team could accomplish the goal if they reduced the variation in both right and left caster by half. They decided to address right caster first, hoping that any process improvements could also be made to the left side.

We give a histogram of right caster in Figure 6.12. There is a large number of trucks in the data set, so we need to be careful when defining the full extent of variation. The team decided not to use the observed range since there were several outliers, as shown in the box plot in Figure 6.12. Rather, looking at the histogram and the numerical output, they defined the full extent of variation as 3.80° to 5.25° (4.52 ± 3*0.24).

In summary, the team focused the project as defined by the management to reducing variation in right caster from the baseline standard deviation of 0.24° to 0.12°. They connected the management and problem goal through an investigation of the warranty database. They could not connect the problem directly to the customer satisfaction surveys. They established the full extent of right caster variation. They were confident that there was little

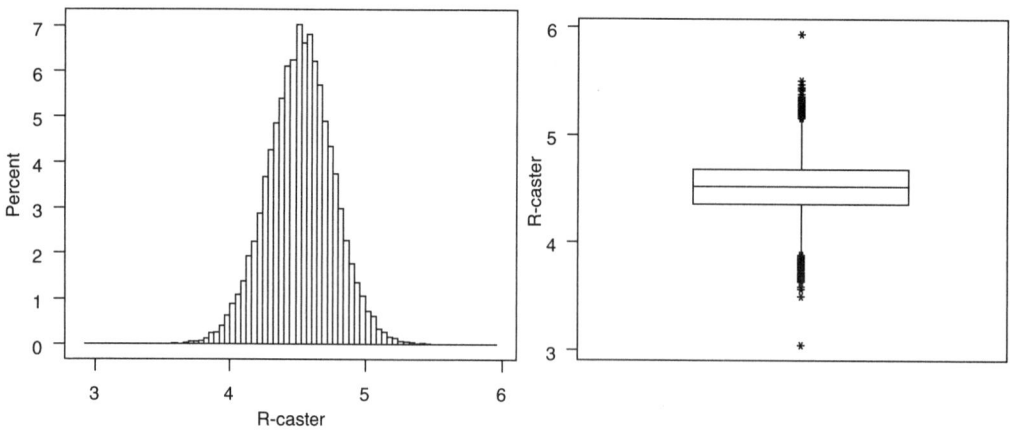

Figure 6.12 Histogram and box plot of right caster from baseline.

study error in their conclusions since they believed two months of data described the performance of the alignment process in the long term.

Pump Noise

A finished vehicle audit at an assembly plant repeatedly detected unacceptably noisy windshield washer pumps. The plant charged for the repairs and put great pressure on the pump manufacturer to solve the problem. The pump manufacturer established a project to eliminate complaints due to noisy pumps.

A team began by developing a noise measurement system that could be used in the manufacturing facility before the pumps were assembled into vehicles. Each member of the three-person team subjectively assessed 24 pumps in vehicles for noise using a five-point scale. The vehicles selected were both acceptable and unacceptable to the vehicle assembly plant. The average score for each pump was recorded. The 24 pumps were then removed and the noise was measured using the new in-plant system. After some adjustments, the team was able to achieve a strong correlation between the subjective human and the in-plant measurement systems (see Figure 6.13). The data from the final measurement investigation are given in *pump noise measurement*. Using the subjective measure, the team judged that a score of 4 or greater was unacceptable to the customer. Accordingly, they set a limit of 8 as the upper noise specification as determined by the in-plant system.

The team was confident that they could detect noisy motors in their facility using the new measurement system. Over a two-day period, they selected 100 motors haphazardly from the current production and measured the noise level. The data are given in *pump noise baseline*. The baseline histogram is given in Figure 6.14.

In the sample, 18% of the motors had a measured value exceeding the new specification limit. The problem goal was to reduce this percentage to 0. The full extent of variation in the baseline using the in-plant measurement system is 0 to 15.

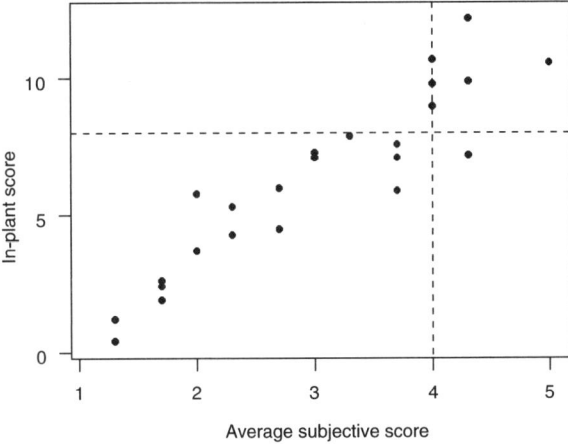

Figure 6.13 Correlation between in-plant and subjective pump noise measurement systems.

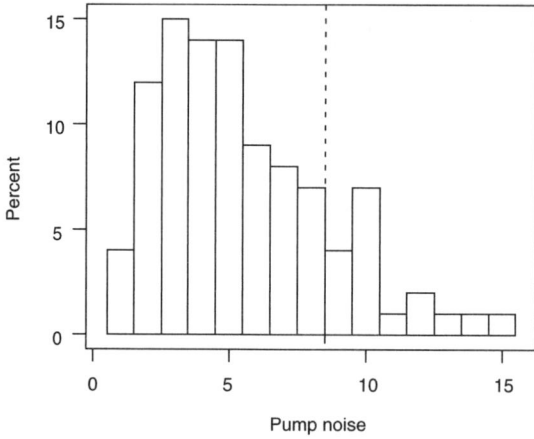

Figure 6.14 Baseline histogram for noisy pumps.

The team did not consider possible study or sample error. They assumed that the sample of 100 motors from a two-day period would accurately describe the long-term performance of the process. They were driven to proceed by the urgent nature of the problem.

6.5 COMPLETING THE DEFINE FOCUSED PROBLEM STAGE

Define a Focused Problem is the first stage of the Statistical Engineering algorithm. The purpose of the stage is to understand and quantify the nature of the problem in a way that makes sense in the production environment. When we start this stage, we have a project description and goal from management, and some prior experience with process. The key tasks necessary to complete this stage are:

- Define a focused problem or problems with an established link to the project goal.
- Estimate the problem baseline, that is, give a graphical or numerical summary of process performance.
- Set a problem goal, stated in terms of the baseline, that matches the project goal.
- Estimate the full extent of output variation in the current process.

The first task may be the most difficult. We must translate a vague management description, such as too much scrap or too many complaints, into specific problems expressed in terms of an output characteristic and the baseline. Since we assume that there is a single dominant cause of the problem, we need to be as specific as possible. We may have to revisit this stage later.

We try to describe the problem in terms of a continuous output. This may not be possible. In many applications, the team invents a new continuous output and a way to measure it to specify the problem. The pump noise problem is a good example.

We need to specify a baseline that we use to define the goal of the problem. We estimate this measure of current process performance using a baseline investigation or existing data. In either case, we must be careful that the time frame for the study population is long enough to give an accurate picture of the current process performance.

We use the full extent of variation to help plan and analyze future process investigations. In cases where there are outliers (unusual values) in the baseline investigation, we ignore them in determining the full extent of variation, unless the outliers define the problem.

The full extent of variation for a binary output is given by the two possible values of the output. The notation of the full extent of variation is not that helpful in planning investigations in this case, other than to suggest we need to examine both defective and nondefective units.

Key Points

- A project should be translated into one or more specific problems, each of which:
  – Has a single dominant cause
  – Can be quantified in terms of a measured output characteristic
- The goal of each problem should be directly linked to the project goal.
- We estimate a problem baseline to quantify the goal, to assess a proposed solution, and to help in the search for a dominant cause and solution.
- For problems with a continuous output, we determine the full extent of variation to help plan and analyze subsequent investigations.

Endnotes (see the Chapter 6 Supplement on the CD-ROM)

1. See the supplement to this chapter for a brief introduction to confidence intervals in the context of estimating the process standard deviation and other attributes. We explain how to use confidence intervals and how to determine them using MINITAB.
2. Does a process need to be stable in the sense of Statistical Process Control before we can establish a baseline performance measure? We consider this issue and firmly answer no!

Exercises are included on the accompanying CD-ROM

7

Checking the Measurement System

When you can measure what you are speaking about, and express it in numbers, you know something about it; but when you cannot measure it, when you cannot express it in numbers, your knowledge is of a meager and unsatisfactory kind. It may be the beginning of knowledge but you have scarcely, in your thoughts, advanced to the state of science, whatever the matter may be.

—William Thompson (Lord Kelvin), 1824–1907

Check the Measurement System is the second stage of the Statistical Engineering algorithm. The purpose is to ensure that we have an effective measurement system for the output characteristic that defines the problem. We assess the measurement system for two reasons:

- The measurement system may be home to the dominant cause of variation.

- We use the measurement system to produce data in subsequent investigations in the variation reduction algorithm.

The measurement system provides a window to view the process. If the window is foggy, we are not able to see clearly what is going on. We must improve the measurement system if we judge it to be inadequate. Sometimes, we can meet the problem goal just by improving the measurement system.

There are many types of output characteristics and measurement systems. For example, the output of the system may be as complex as a force versus time curve, or as simple as a score on the scale from 1 to 5. In this chapter, we describe how to assess a nondestructive measurement system for a continuous characteristic. In the supplement, we discuss variants on the plans to assess measurement systems for binary characteristics[1] and for destructive measurement systems.[2]

7.1 THE MEASUREMENT SYSTEM AND ITS ATTRIBUTES

A measurement system is not just a device or gage. For any characteristic, it includes the:

- Gages, masters, and other materials
- People using the gages
- Procedures for use, including calibration and mastering
- Environment in which the gage is used

In a project to reduce camshaft scrap and rework, the team focused on journal diameter. At the final 100% inspection, the diameter of each of the four journals was measured at two locations. We show the front and rear locations for journal 1 in Figure 7.1.

The inspection gage had eight heads, one for each location. The operator placed the camshaft into the gage. The diameter was defined as the maximum value as the part was rotated through 360°. The specifications were ±12.5 microns, measured from the target value. Camshafts that did not meet specifications were sent to a rework area where operators used a similar gage to aid in repair or to confirm that a part should be scrapped. The two gages were mastered daily using a certified steel cylinder that could be placed in the two heads corresponding to each journal. There was also a yearly calibration procedure.

In this measurement system, there are two gages (or 16 if you count each head separately), a single master, several operators, a prescribed method, and a changing environment in which temperature, for example, was uncontrolled.

We do not get the true value of the characteristic when we measure a part. We call the difference between the true and measured values the *measurement error*. We cannot determine the measurement error since the true value of the characteristic is unknown unless we measure a part with certified known value.

We think of the act of measuring as a process. There is a direct analogy between a measurement process and a production process. Measurement systems produce measurements (the units) rather than parts. In this light, assessing the measurement system is similar to establishing the problem baseline (see Chapter 6).

We need to specify a target population. We plan to use the measurement system to determine process behavior and, perhaps, to monitor and control the process in the future. The target population is all the acts of measuring that we plan to make. For each such act,

Figure 7.1 Front and rear camshaft journal locations.

the output characteristic is the measurement error. We define two important attributes of the measurement system for this target population:

- *Measurement variation* (also called *precision,* or better *imprecision*): the standard deviation of the measurement errors

- *Measurement bias* (also called *accuracy,* or better *inaccuracy*): the average measurement error

We can picture measurement variation and bias as in Figure 7.2. The idealized histogram shows the measurement errors for all measurements in the target population. This is the set of measurement errors if we repeatedly measured many parts over a long time period.

There are many other attributes of the measurement system that may be of interest. For example, do measurement bias and variation depend on the true value of the part being measured? Or do measurement bias and variation change over time? We can define and estimate other attributes to look at these issues related to linearity and stability of the system (Automotive Industry Action Group [AIAG], *Measurement System Manual,* 1995).[3]

For many gages, the supplier provides a standard deviation to indicate the capability of the gage. However, these performance measures are determined under narrow conditions and almost always underestimate the measurement variation as we have defined it over a broad target population with varying parts, operators, environment, and so on.

For many problems, we do not assess measurement bias, especially when the problem baseline is quantified in terms of a standard deviation. If we remove bias from the measurement system and change nothing else, we shift the center of the output characteristic but do not change the variation. If there are two measurement systems within the same process, then we may look at the *relative measurement bias* of the two systems. See Section 7.3.

We start by discussing how to assess measurement variation.

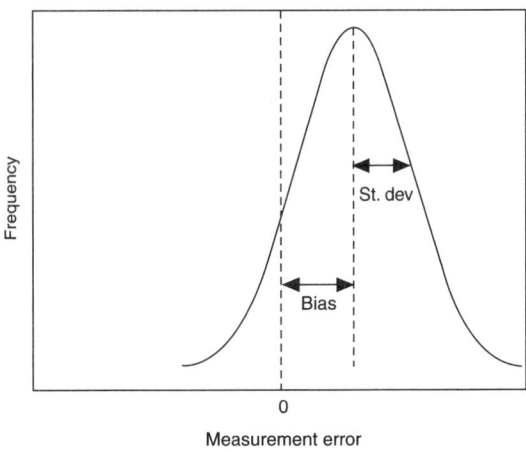

Figure 7.2 Representation of measurement variation (standard deviation) and bias.

7.2 ESTIMATING MEASUREMENT VARIATION

Measurement variation is the standard deviation of the measurement errors over a broadly defined target population. Surprisingly, we can estimate the measurement variation without knowing the true value of the characteristic for any part, that is, without knowing the measurement errors. Suppose that we measure a characteristic of the *same* part several times. For example, suppose we measure the diameter of the same camshaft journal at the same location five times with results

$$1.4, 1.1, 1.3, 0.2, 0.1$$

Since we are repeatedly measuring the same part, the variation in the measured values is due solely to the variation in the measurement errors. We are assuming that the measuring process does not change the true diameter, that is, that the measurement process is not destructive. If the five measurements were taken over conditions matching the target population, we can use the standard deviation of the five values to estimate the measurement variation.

We now describe an investigation to estimate the variation of the measurement system. We use the camshaft journal measurement system as the example restricting our attention to the final inspection gage and the front location on journal 1. In reality, the team looked at all eight locations on both gages.

Question

The purposes of the investigation were to:

- Estimate the standard deviation of the measurement errors over a wide range of parts, a variety of operators, changing environmental conditions, and a long period of time
- Compare the measurement variation to the variation due to the rest of the process

Plan

The first step in the Plan is to define a study population. In other words, what measurements can we possibly take in the investigation? We need to decide:

- Which parts to measure
- Which operators to include
- What time frame to use

We recommend using three parts with values spread across the full extent of output variation seen in the baseline investigation. Here, the team used three camshafts with initially measured diameter values –12.2, 0.9, and 12.8, as shown in Figure 7.3.

If several operators use the measurement system, then we recommend including at least two operators in the study population. If the gage is automated so that it is known that there is no operator impact, then we can use a single operator. In the camshaft example, there were three operators, one from each shift.

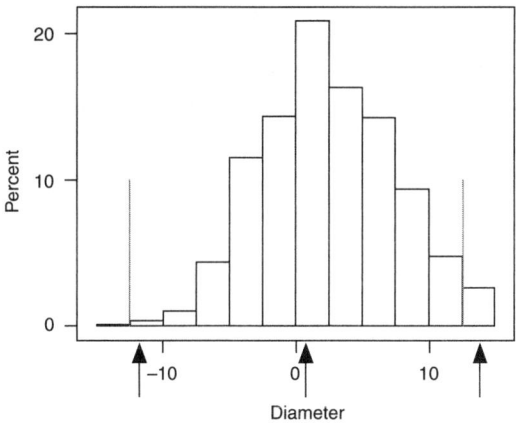

Figure 7.3 Baseline performance for journal 1 front diameter. Arrows show diameter values chosen for measurement investigation.

We can get some guidance from the baseline investigation to help select an appropriate time frame for the study population. Ideally, we assess the measurement system over a period long enough to see the full extent of variation given in the baseline. This ensures that the dominant cause (if it is in the measurement family) has time to act over the course of the investigation. In the example, the team selected a one-week period for the study population. In our experience, most teams make the mistake of selecting a time frame that is far too short because they want to finish the measurement assessment stage quickly.

We need to specify a sampling protocol to determine exactly what measurements will be taken. We plan to measure each part a number of times. We recommend selecting at least two time points within the chosen study period and having each operator measure each part at least twice at each time point.

In the camshaft example, two days were chosen, one week apart. On both days, the team planned to have each operator measure each camshaft three times. They would give the parts to the operators in random order for each determination. A team member would record the results. With this plan, each of the three parts is measured 18 times (2 days by 3 operators by 3 determinations). The sample has a total of 54 measurements.

Other than day, operator, part number, and time, the team decided to record no other characteristics. If they had suspected that the attributes of the measurement system would change due to some environmental factor such as temperature, they could record the temperature at each determination. If the measurement system is the home of the dominant cause of the overall variation, the team can use these data to look for clues about why the measurement variation is so large.

The plan should be executed under normal conditions. For example, the operators taking the measurements should use their usual method, and gages and parts should not be specially cleaned.

It is a good idea to randomize the order of measurement within each time period so that the operators cannot remember an earlier result for a particular part characteristic. This point is especially important if there is a subjective element to the measurement system.

There are many other possible plans.[4]

94   Part Two: Getting Started

Figure 7.4   Camshaft diameter measurement investigation data in row/column format.

Data

In the camshaft journal diameter example, the team executed the plan. The 54 measured values and the corresponding time, operator, and part number were stored in the file *camshaft journal diameter measurement*. Figure 7.4 shows some of the data stored in the suggested row/column format (see Appendix A) in a MINITAB spreadsheet.

Analysis

We estimate the measurement variation by calculating the standard deviation of the measurements made on each part. The variation in these measurements is due solely to variation in the measurement errors. For the camshaft example the average and standard deviation for each part are:

```
Descriptive Statistics: diameter by part

Variable   part    N     Mean    StDev
diameter    1     18   -10.90    0.638
            2     18     1.20    0.796
            3     18    12.68    0.820
```

We combine the three standard deviations with the formula

$$sd(measurement) = \sqrt{\frac{0.638^2 + 0.796^2 + 0.820^2}{3}} = 0.756$$

to produce the estimate of the measurement variation. We can get this value directly from a one-way analysis of variance (known far and wide as ANOVA) using MINITAB as described in Appendix D.

We can also look at how the measurement variation changes over time, over parts, and over operators. We find this analysis useful to look for outliers (see the supplement to Chapter 5) and to generate clues about a dominant cause if the measurement variation is too large.

In Figure 7.5, we plot the measured values versus the part number (left panel) and, more revealingly, the deviation from the part average versus part number (right panel). We see that the variation due to the measurement system is roughly the same for the three parts and that the measurement system can distinguish among these parts. There are no outliers.

In Figure 7.6, we show the results for the two weeks used in the example investigation. We see no obvious changes in the measurement variation over time. If there were clear differences over time, we would be concerned about study error and would recommend repeating the plan over a longer time frame.

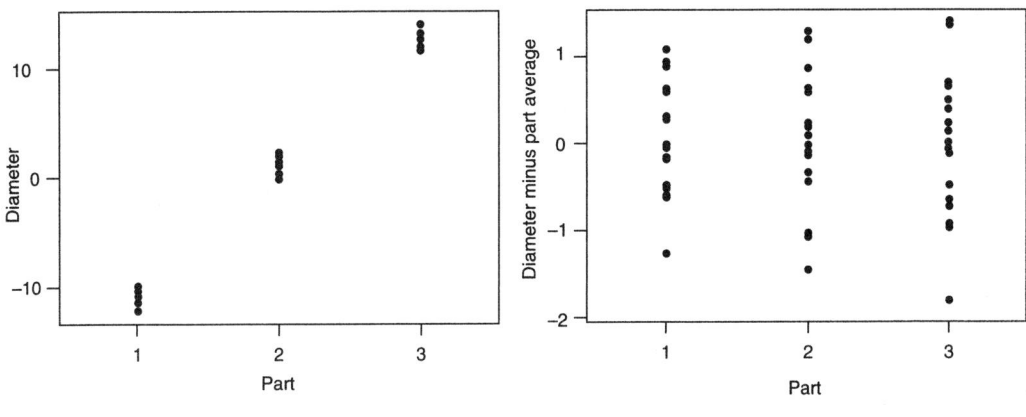

Figure 7.5 Diameter and diameter minus part average by part number.

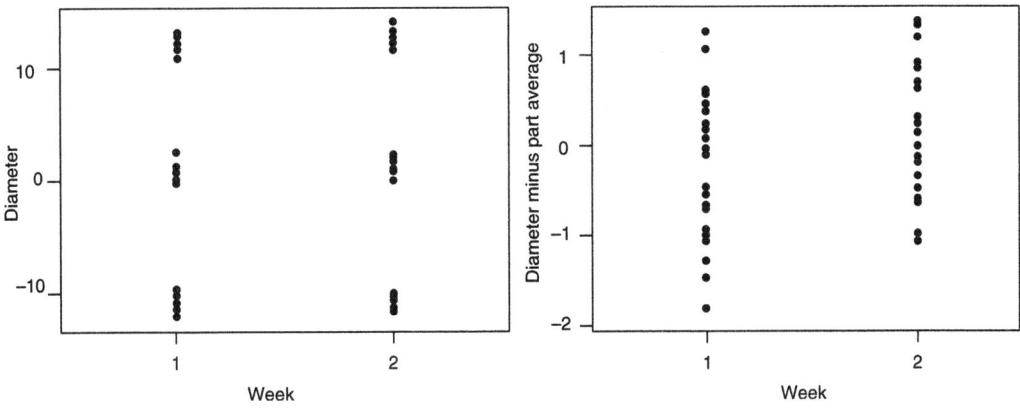

Figure 7.6 Diameter and diameter minus part average by week.

Conclusion

We estimate the measurement variation to be 0.756 microns. We cannot interpret this standard deviation in terms of how close a measured value is likely to be to the true value because we do not know if the measurement system has bias. Perhaps the easiest interpretation is to say that if we measure the same part at two different times with different operators, then the difference in the two measurements is likely to fall in the range[5]

$$\pm 2 \times \sqrt{2} \times (\text{estimated measurement variation}).$$

In the camshaft example, this range is ±2.14 microns. Since we looked at only 54 measurements, there is possible sample error in the estimate of the order ±20% (see Table 6.2).

To quantify the effect of measurement variation, recall from the model described in Chapter 2 that we can partition the overall variation into two pieces—one due to the variation in the true values of the characteristic and the other due to the variation from the measurement system. That is, we have

$$sd(total) = \sqrt{sd(\text{due to } process)^2 + sd(\text{due to } measurement)^2} \quad (7.1)$$

From the baseline investigation in the Define Focused Problem stage (discussed in Chapter 6), we have an estimate of *sd(total)* that we used as a baseline to define the problem. We labeled this estimate *stdev(total)*. From the measurement investigation, we estimate *sd(measurement)*. Then we can use Equation (7.1) to judge if the measurement system is a major contributor to the overall variation.

In the camshaft example, the estimated total standard deviation is 6.055 (determined from a baseline investigation) and the estimated standard deviation from the measurement system is 0.756. Using Equation (7.1) with the estimates, we can solve for the contribution from the rest of the process, that is,

$$stdev(\text{due to } process) = \sqrt{stdev(total)^2 - stdev(\text{due to } measurement)^2} \quad (7.2)$$

In the example, *stdev(due to process)* is $\sqrt{6.055^2 - 0.756^2} = 6.008$. We see that the measurement system has very little impact on the overall variation.

The effectiveness of the measurement system depends on the relative sizes of the variation due to the process and measurement. We summarize the measurement effectiveness using the *discrimination ratio* D given by Equation (7.3). Larger values of this ratio are better since we are better able to distinguish among parts using the measurement system.[6]

$$D = \frac{stdev(\text{due to } process)}{stdev(\text{due to } measurement)} \tag{7.3}$$

If D is less than about 2, the measurement system is home of a dominant cause of variation. We should reformulate the problem in terms of the measurement system. In this case, improving the measurement system may solve the original problem.

If D exceeds 3, then we know that the measurement system is not the home of a dominant cause and we can proceed with the next stage of the algorithm and choose a working variation reduction approach.

If D falls between 2 and 3, the measurement system is not a dominant cause, but the measurement system should be improved (see Section 7.5). This recommendation may be ignored depending on the nature of the problem and the difficulty and cost of improving the measurement system. If the discrimination ratio is between 2 and 3, we may have difficulty interpreting the data in future process investigations.

Measurement Variation Assessment Investigation Summary

Question

For all acts of measurement with the measurement system in the future:
- What is the measurement variation, i.e., the standard deviation of measurement errors?
- Is the measurement variation sufficiently small?

Plan

- Ensure that the measurements are made under normal operating conditions.
- Select:
    - Three parts that cover the full extent of variation in the output
    - Two or three time periods that cover a period over which you expect to see the full extent of the variation in the output, if feasible
    - Two or three operators, if multiple operators are normally used
    - Two or three gages, if multiple gages are normally used
- Make three measurements under each combination of part, operator, time period, and gage.

Data

Record the measured output and corresponding operator, gage, time, and part number with one measurement per row.

Analysis

- Calculate the average and standard deviation of the measurements by part number.
- Combine the within-part standard deviations to estimate the measurement variation.

$$stdev(measurement) = \sqrt{\frac{stdev(part\ 1)^2 + stdev(part\ 2)^2 + stdev(part\ 3)^2}{3}}$$

- Plot the measurements and the deviation from part average by the part number, time period, operator, and gage. Look for unusual patterns and outliers.

Conclusion

- Calculate the discrimination ratio,

$$D = \frac{stdev(\text{due to } process)}{stdev(\text{due to } measurement)}$$

- If:

  $-D < 2$, the measurement system is the home of a dominant cause. Reformulate the problem in terms of measurement variation.

  $-2 < D < 3$, the measurement system is not the home of a dominant cause but should be improved.

  $-D > 3$, the measurement system is not the home of a dominant cause and is adequate.

- If the measurement system is not adequate, look for clues pointing to the dominant cause of measurement variation.

In the camshaft example, there were eight diameters measured on each part. We may not be able to find three parts with the full extent of variation on each characteristic. Finding and using different parts for each diameter was too complicated, so the team selected parts based on journal 1 only.

We may be able to use the results from recent gage repeatability and reproducibility (R&R) investigations to estimate the measurement variation.[7] However, we must assess the risk of study error since typical R&R investigations are conducted over a short time frame. In many cases, the estimate from R&R substantially underestimates the measurement variation.

7.3 ESTIMATING MEASUREMENT BIAS

To estimate measurement bias we must measure units with known values. We use *known* here in a relative sense. As a rule of thumb, the variation and bias of the measurement system used to determine the known values should be at least 10 times less than the variation and bias of the system under investigation.

The units with known values may be certified standards such as gage blocks or they may be actual parts that have been measured on another measurement system that has low variation and bias. For the camshaft journal diameter system and others that have specialized fixtures, we recommend the use of certified parts to avoid possible study error.

We follow the same plan as we used to estimate the measurement variation. In the camshaft journal diameter example, the original problem was excessive scrap and rework. The rework operators had noticed that some of the parts rejected by the final inspection gage were acceptable when measured by the rework gage. The team decided to assess the bias of both measurement systems. Here we consider only the final inspection system and, as earlier, the front location on journal 1.

There were no certified parts available. Instead, the team used the three camshafts that were described in Section 7.2. For each part, the journal diameter was measured five times on the in-house coordinate measuring machine (CMM). The data are given in Table 7.1.

Table 7.1 Journal diameter CMM data.

Part	Repeated measurements					Summaries	
	1	2	3	4	5	Average	Standard deviation
1	−8.62	−8.35	−8.61	−8.26	−8.68	−8.504	0.19
2	4.24	4.00	4.17	3.93	4.47	4.162	0.21
3	14.64	14.75	14.62	14.20	14.56	14.554	0.21

The team took the average CMM values to be the true values. Note that the variation of the CMM was about 0.20 microns. By averaging five readings, this variation was reduced to 0.09 ($0.2/\sqrt{5}$; see "Use Combined Measurements" in Section 7.5). The team assumed that the bias of the CMM was negligible.

To estimate the bias in the final gage, we calculate the measurement error for each of the 54 diameters measured in the initial investigation. Recall that

$$\text{measurement error} = \text{measured value} - \text{true value}$$

The data file *camshaft journal diameter measurement2* gives the measurement errors for this investigation. We use the average of the observed measurement errors as the estimate of bias.

From the numerical summary that follows, we estimate the measurement bias to be −2.44 microns. The final gage gives consistently smaller measured diameters than the CMM.

```
Variable        N       Mean     Median     TrMean      StDev    SE Mean
meas. error    54     -2.437     -2.434     -2.432      0.866      0.118

Variable        Minimum    Maximum         Q1         Q3
meas. error      -4.455     -0.511     -2.954     -1.780
```

We plot the measurement errors against the part number and week in Figure 7.7. The plot shows that the measured values at the final gage were consistently smaller for all parts and there was no change from one week to the next.

The team considered two possible limitations to this conclusion. First, was there something peculiar about the three parts used or the time during which the investigation was conducted? Since the bias persisted across all three parts (see Figure 7.7) and did not change much over the week, the team was confident that there was a systematic difference between the final gage and the CMM. Second, was the bias due to the CMM? The team, with all members from production, had more faith in their own system that was specially designed to measure journal diameters than the CMM that was an all-purpose gage. Recall that there were eight gage heads to measure the two positions on the four journals in the inspection system. We have given the results for one position on journal 1. The estimated bias for the other seven heads was very small, less than 0.5 microns. Since the same CMM program was used to certify all four journals, the team concluded that the observed bias on journal 1 was not due to the CMM.

Given that the diameter specifications were ±12.5 microns and the process performance used up more than the full tolerance (Figure 7.3), the team knew that the bias was large enough to cause inspection errors. They arranged maintenance on the gage, and a subsequent bias investigation showed that the problem had been corrected. They implemented a once-a-shift check of the system using a reference part (one of the parts certified by the in-house CMM) to indicate if the bias problem recurred.

Note that the plan to investigate measurement bias is identical to that for estimating measurement variation except that we use parts or standards with known values. We can simultaneously estimate both bias and variation from such a plan.

When we have two or more measurement systems for the same characteristic, measurement variation and bias can be the source of acrimony and confusion among the users. We have seen battles between customers and suppliers, each insisting that the value produced by their own measurement system is correct. In one instance, we saw a process in which a transmission part was inspected on four different systems, the last being at the customer. The customer

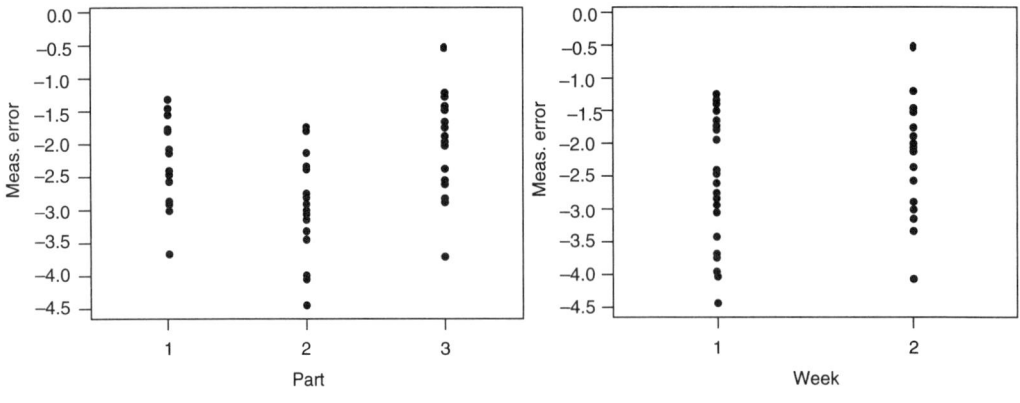

Figure 7.7 Measurement error stratified by part number and week.

occasionally found parts out of specification even though they had been through three upstream inspections. Needless to say there was a lot of finger pointing and high inspection costs.

The *relative bias* of two measurement systems is the difference in bias. We can estimate the relative bias without using parts with known values. We need to measure a number of parts on both systems and compare the results. There are many possible plans. We discussed an investigation to estimate the relative bias of two measurement systems for V6 piston diameter in Chapter 5.

We find it difficult to give a general rule to decide if bias is too large. We need to assess the bias in terms of the original problem, the overall variation, and the consequences.

In the example, given the baseline variation, the team knew that the negative bias in the final inspection gage was large enough that they were shipping oversized parts. The bias also helped to explain why the repair gage accepted parts that were undersized at final inspection.

7.4 IMPROVING A MEASUREMENT SYSTEM

If the variation or bias of the measurement system is too large, we must improve the system to solve the problem or to proceed to the next stage of the variation reduction algorithm. There are several possible strategies.

Fix the Obvious

The first rule, as always, is to fix the obvious. In the camshaft example, the team decided that maintenance of the gage was necessary because the bias was large enough to explain false rejects. They also knew how to implement and use the reference part procedure to detect recurrence of the bias.

In general, we can adjust for bias when it is consistent across parts and time by subtracting an offset (equal to the estimated measurement bias) from the measured value. We need to assume linearity (constant bias for all parts) for this offset strategy to be effective. If the properties of the measurement system change too quickly over time, we may calibrate the system more frequently. Or we may discover that maintenance of the system is long overdue.

Use Another Measurement System

In some cases we may be able to use a better measurement system for the duration of the problem solving. For example, we may be able to use an offline measurement system that has smaller variation but a cycle time that is too large for use in regular production. Another idea to reduce variation in a measurement system with a large subjective component is to use a single operator to make all of the measurements in subsequent investigations.

Alternately, we may invent a new measurement system. In a bottling operation, management raised a concern over crooked and misplaced labels. The label was to be placed on the neck of the bottle a set distance below the bead. In establishing the problem baseline, a height gage was used that measured the distance between the bottom of the bottle and the bottom of the label. As the bottle was rotated, the system measured the maximum and minimum distance and calculated the average height and crookedness (maximum height – minimum height). In the measurement system assessment, the discrimination ratio for average height was 3.1. Although the measurement system was not the dominant cause of the variation, the team believed they would have difficulty in subsequent investigations because of the measurement variation. They decided to use a feeler gage to measure the distance between the label and the bottle bead. The discrimination ratio of the feeler gage system was 6.5. The team felt comfortable proceeding to the next stage of the variation reduction algorithm using the feeler gage.

Use Combined Measurements

We can sometimes combine a number of measurements to reduce bias or variation. For example, if there is a significant bias in a weighing process, to determine the weight of an object we may first weigh the object together with a container and then weigh the container on its own. The weight of the object is the difference. This method eliminates the measurement bias as long as bias does not depend on the true weight. There is an increase in measurement variation.

We can reduce variation by a factor of $1/\sqrt{2} = 0.7$ if we average the values from two determinations on each part. More generally we reduce variation by a factor of $1/\sqrt{n}$ if we average n measurements on each part. Averaging has no effect on bias. For this method to be effective, we assume that the repeated measurement of a part by the same operator over the short term captures the long-term measurement variation in the measurement system. This method was used in the camshaft journal diameter example to certify parts using the in-house CMM. The diameter of each camshaft journal was measured five times. To help ensure that the five measurements reflect the long-term measurement variation, the camshaft was removed from the CMM and refixtured between measurements.

Reformulate the Problem and Use the Statistical Engineering Algorithm

We may find a way to reduce measurement variation using the Statistical Engineering algorithm. The results from the measurement investigation can serve as a problem baseline. As in other applications of the algorithm, to determine the appropriate (measurement) variation reduction approach, we need to examine how the measurement system behaves. We provide examples of applying the various approaches to measurement systems later in the book.

If we record the values of other inputs such as temperature, time, and so forth, in the measurement system investigation, we can use these data to search for clues to the dominant cause of the measurement variation.[8] Again, note the analogy to the baseline investigation. See further examples in chapters 9 through 12.

7.5 COMPLETING THE CHECK THE MEASUREMENT SYSTEM STAGE

The purpose of the Check the Measurement System stage is to ensure the measurement system is not a dominant cause of the variation and that the measurement variation and bias are sufficiently small to support further process investigations. To complete this stage, the key tasks are:

- Estimate the measurement variation and possibly bias.
- Calculate the discrimination ratio, D, as given by Equation (7.3).
- If the measurement variation or bias is too large, reformulate the problem in terms of the measurement system.
- Fix obvious problems in the measurement system.

If we substantially increase the discrimination ratio, is it necessary to repeat the baseline investigation using the altered measurement system?

If $D < 2$, then we reformulate the problem in terms of the measurement system and use the results from the measurement system investigation as a baseline for the reformulated problem. We will reassess the original baseline when we validate the solution in the last stage of the Statistical Engineering algorithm. If $D > 2$, the measurement system is not a dominant cause of variation. Thus, reassessing the baseline is unnecessary because any change to the measurement system will have little impact on the full extent of variation.

Key Points

- We must assess the measurement system for the output characteristic that defines the problem because it may be the home of the dominant cause of variation and because it will be used in future process investigations.
- The key attributes of the measurement system are bias and variation. Bias is the average and variation is the standard deviation of all measurement errors made in the future.
- To estimate the measurement variation, we define a study population over a wide range of conditions, including a variety of parts, operators, gages, and times.
- To estimate bias, we apply the same plan proposed for estimating measurement variation using parts with known values.
- To assess the adequacy of the measurement system, we look at the impact of the bias and compare the measurement variation to the standard deviation of the true values of the characteristic.
- If necessary, for the purpose of problem solving, we may use a different measurement system than that used to define the baseline performance.

Endnotes (see the Chapter 7 Supplement on the CD-ROM)

1. Binary measurement systems that classify units into one of two classes are relatively common. In the supplement we discuss how to assess such systems.
2. We have provided plans and analysis methods to estimate variation and bias for continuous characteristics that are not affected by taking a measurement. In the supplement, we look at similar methods for destructive measurement systems.
3. We consider the AIAG definitions of repeatability, reproducibility, linearity, and stability. We compare these attributes to measurement variation and bias and discuss their usefulness in the variation reduction algorithm.
4. We discuss gage R&R and Isoplots™ (measurement system scatter plots) as tools for estimating the measurement variation. We compare these assessment procedures to that proposed in Section 7.2.
5. It can be difficult to interpret the estimate of measurement variation. We provide an explanation in terms of how large a difference you might see between two measurements on the same unit.
6. We have suggested a criterion for assessing the variation of a measurement system and making decisions on how to proceed. We look more deeply at this criterion and compare it to other standards such as to gage R&R < 30%. For example, we look at the effect of measurement variation on power and sample size and the effect of measurement variation when comparing measured values to specification limits.
7. See note 4.
8. There are many other attributes of the measurement system apart from bias and variation. These become important when we wish to improve a system that is not adequate. We can stratify results by operators, gages, or time. We can use the data from the basic plan for estimating variation to generate clues if we want to reduce the variation in the measurement system.

 Exercises are included on the accompanying CD-ROM

8
Choosing a Working Variation Reduction Approach

Change is not made without inconvenience, even from worse to better.
—Samuel Johnson, 1709–1784

In this stage of the Statistical Engineering algorithm, we select a working approach from the seven possibilities described in Chapter 3. You may be tempted to skim over this stage, since our directions are somewhat vague. However, without a working approach, it is not clear what to do next. If you do take further action, you are implicitly adopting a working approach. We believe that explicit consideration of the approach at this point will lead to a better choice and produce better results sooner.

At this stage of the algorithm, we may not know enough to choose the approach that will eventually be implemented. Based on what we do know, using both engineering knowledge and previous process investigations, we select the working approach and then gather the required information to assess its feasibility. So at this point, we are trying to pick the most feasible approach with incomplete knowledge. We may have to return to this stage of the algorithm a number of times as we obtain more process knowledge. We fix obvious faults in the process as we uncover them.

When a team arrives at the Choose Working Variation Reduction Approach stage, they will have the following:

- The problem expressed in terms of an output characteristic and a baseline measure of process performance

- A goal expressed in terms of the baseline measure tied to the project goal

- Some knowledge of the process and its behavior (for example, a process map, a control plan, results of a baseline investigation and other past investigations, applicable science and engineering knowledge, and so forth)

- Some knowledge of the constraints (for example, economic reality, span of control, political and cultural constraints)

- Confidence in the measurement system unless the problem is defined in terms of this system

106 *Part Two: Getting Started*

We return to the question of choosing a working approach in Chapter 14, where we specifically discuss the consequences of having searched for, and hopefully found, a dominant cause. In this case, the team will have acquired considerable new knowledge of the process behavior.

We have the following options at this point in the algorithm:

- Search for a dominant cause of variation. Identifying a dominant cause is a prerequisite for reformulation or the cause-based approaches:

    –Fixing the obvious

    –Desensitizing the process to variation in a dominant cause

    –Feedforward control

- Assess the feasibility of one of the non-caused-based approaches:

    –Feedback control

    –Making the process robust

    –100% inspection

    –Moving the process center

We consider each approach based on the requirements for that approach and our current knowledge of the process. We also consider the likely implementation costs, classified into three broad categories:

- Costs related to determining if the approach can be effective
- Costs of the change to the process to implement the approach
- Ongoing costs to maintain the approach

We incur the first type of cost when we search for a dominant cause, when we try to find process settings that are robust or less sensitive to variation in a dominant cause, when we look for an adjuster to change the process center, and so on. We incur initial costs such as changes to the product or process design, changes to the control plan, training of personnel, and so forth when we first implement an approach. Finally, we incur ongoing costs such as the measurement and adjustment costs needed for feedback or feedforward control and inspection.

At this first iteration of the Choose Working Variation Reduction Approach stage, we are unlikely to have sufficient process knowledge to make a good assessment of the implementation costs. We consider implementation costs for each approach in detail in chapters 14 through 20.

The best option is always problem specific and we find it difficult to give precise directions. Instead, we pose a series of questions that the team should ask.

8.1 CAN WE FIND A DOMINANT CAUSE OF VARIATION?

We need to identify a dominant cause of the variation in order to implement one of the cause-based variation reduction approaches. We strongly recommend that the team search for a dominant cause and then decide which of the three to adopt as a working approach. We have seen many problem-solving efforts flounder because the team tried to change the process without first identifying the dominant cause of the variation. If we know a dominant cause, we have a much better chance of finding a cost-effective process change that will meet the problem goal.

Despite our strong preference for trying to find a dominant cause, in some problems the team may know enough to answer no to this question.

Manifold Blocked Port

In the casting process that makes exhaust manifolds, described in Chapter 2, management adopted a project to eliminate a defect called "blocked ports." The foundry customer detected this rare defect only a few times per year. At each occurrence, there was a loss of goodwill and an expensive containment effort that typically found no other defective manifolds.

The project team knew that observing the process to find the cause would likely be fruitless since there were very few blocked ports. Changing fixed inputs to see their effect was equally hopeless. They decided to look at the feasibility of automated 100% inspection to check every manifold for blocked ports.

Electric Box Durability

The management of a firm that built metal boxes for exterior electrical components decided to increase the durability of the boxes to satisfy its customers, large utility companies. The management had received numerous complaints about premature rusting on boxes in service.

The team considered looking for the cause of the variation in durability. They could trace field failures to only a few processing conditions, such as the lot of paint. They could not determine the values of most varying inputs for any particular box in the field. The team also suspected that the dominant cause of variation was environmental and outside of their control. Finally, they had little confidence in the measurement of the failure time as determined by the utility companies.

The team could measure durability in-house by painting standard panels and using an accelerated test in a salt spray chamber. They decided not to search for a dominant cause of durability variation and instead considered process changes that could increase average durability in the salt spray chamber. That is, they adopted Move Process Center as the working approach.

Comments

One other reason for not looking for a dominant cause is historical. If many teams have failed to find the cause in previous projects, it may be inefficient to keep searching.

Having recommended looking for the cause in almost all cases, we believe the team should consider the four non-caused-based approaches before reaching a decision to search for a dominant cause. Based on current knowledge and potential costs, the team may see that one of these approaches is feasible.

8.2 CAN WE MEET THE GOAL BY SHIFTING THE PROCESS CENTER WITHOUT REDUCING VARIATION?

In other words, should we adopt the Move Process Center approach? For example, if the baseline histogram matches one of those shown in Figure 8.1, and the goal is to reduce the proportion of parts out of specification (specification limits shown by the dashed lines), then we can consider this approach.

To make the approach effective, we need to find a change to some fixed input that will shift the center of the process without high cost or large negative side effects. We also must be able to maintain the change in the long term.

Sand Core Strength

In Chapter 1, we introduced a problem of excessive breaking of sand cores in a foundry operation. The baseline histogram of core strengths is reproduced in Figure 8.2.

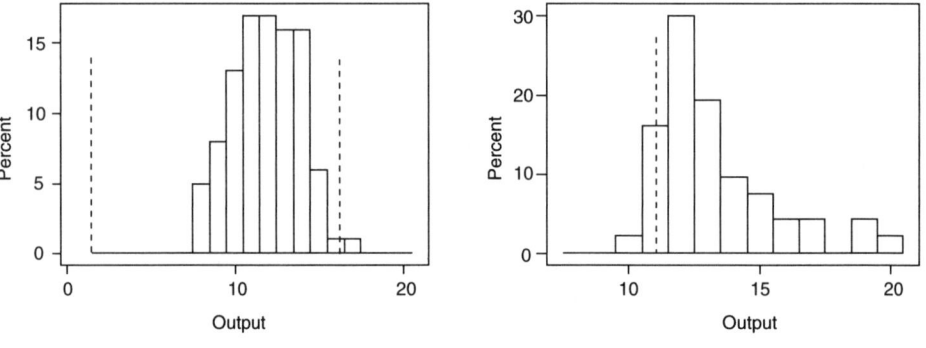

Figure 8.1 Two cases for shifting the process center.

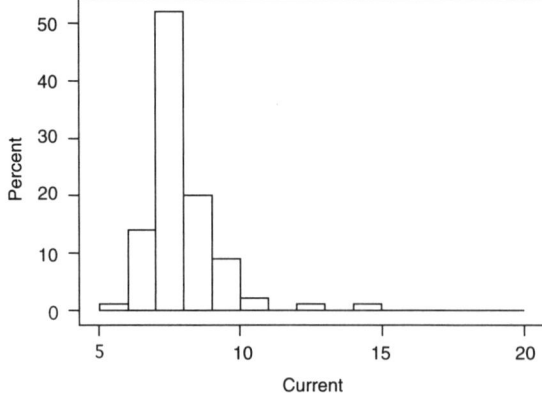

Figure 8.2 Baseline histogram of sand core strength.

The team knew they could increase average core strength by increasing the resin concentration in the core molding process. They could easily assess the cost of the change and maintain it in the future. The team did not know how much extra resin was needed and how well the stronger cores would behave in the casting process. They worried that it would not be possible to shake out the core sand if the cores were too strong. Residual core sand stuck in the casting is a serious defect.

Because of the available knowledge, the team adopted Move Process Center as their working approach. They planned an experimental investigation to quantify the effects of increasing the resin concentration on both core strength and casting quality.

Wheel Bearing Failure Time

A team was assigned a project to reduce warranty costs and customer complaints due to failure of a wheel bearing in a light truck model. The failures occurred within the first three years of use. Using the warranty database and a record of customer complaints, the team focused the problem to the failure of a seal under harsh driving conditions. They developed an accelerated test in the laboratory that could reproduce the field failure mode within a few hundred hours of testing. The results of five tests to failure for the current design are shown in Figure 8.3.

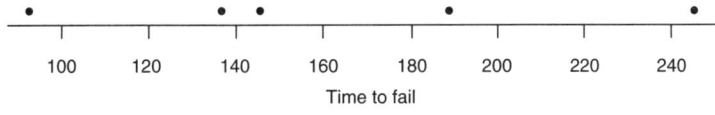

Figure 8.3 Wheel bearing seal failure times.

The team decided to look for a design change that would increase the failure time on the accelerated test to at least 200 hours for all seals. That is, they adopted Move Process Center as their working approach. They rejected the approaches based on finding the cause of the failure (what is different about the seals that explains the different failure times seen in Figure 8.3) because it was too expensive to collect the necessary data. Based on their knowledge of the failure, the team made a design change that they then tested by repeating the five-piece investigation. They accepted the risk that the proposed design change might prove ineffective in the field, where conditions were different from the accelerated test.

Camshaft Lobe Runout

We discussed the problem of base circle (BC) runout of a camshaft lobe in Chapter 1. The baseline histogram is shown in Figure 8.4.

The goal of the project was to produce a greater proportion of lobes with smaller runout. Recall that runout must be greater than zero and is a measure of variation itself. Defining the problem in terms of another output characteristic that is not a measure of variation would have been preferred. The team did not expect to directly discover a change to

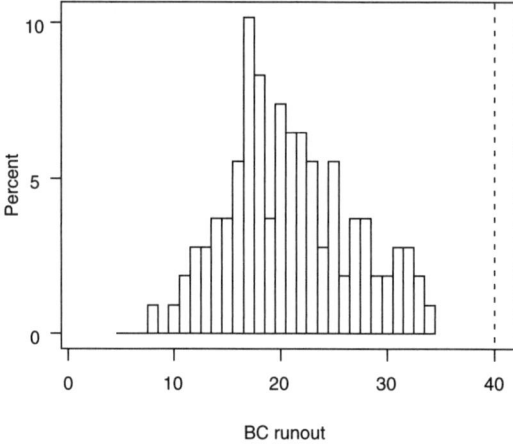

Figure 8.4 Baseline histogram of camshaft lobe BC runout (dashed line shows specification limit).

the process that would shift the histogram to the left. That is, they initially decided not to investigate one or more process changes that might meet the goal. This had been tried in the past without any substantial progress. Instead they decided to look for the dominant cause of lobe-to-lobe variation. Once the cause was found, they could reconsider which approach to select. They were willing to accept the investigation costs with no certainty that they could find the cause or that they would adopt one of the cause-based approaches.

8.3 CAN WE REDUCE VARIATION BY CHANGING ONE OR MORE FIXED INPUTS WITHOUT KNOWLEDGE OF A DOMINANT CAUSE?

If we answer yes, we adopt the Robustness approach. For this approach to work, we need to find a new setting for some fixed input or inputs that reduces the effect of the unknown dominant cause of variation. This approach is often adopted out of desperation after a search for the dominant cause has been fruitless. Without knowledge of the dominant cause, it is difficult to know what fixed inputs to consider.

Transmission Shaft Diameter

In a grinding process for an output shaft, a transmission component, the team set out to reduce shaft diameter variation. They established the baseline performance using process control data (every 50th part was measured) from the previous month. The team assessed the measurement system and found that it was not a major contributor to the overall variation. Having just returned from a course in designed experiments (taught by one of the authors), they decided to run an experiment with 16 runs in which six fixed inputs were varied. The inputs included feed rate, speed, coolant concentration, and so forth. Each experimental run consisted of grinding five shafts. That is, the team selected robustness as their working approach.

For each run the team calculated the logarithm of the standard deviation of the five diameters and then analyzed these 16 performance measures. They found the combination of levels of the fixed inputs that produced the smallest value of the performance measure.

The team then decided to confirm the improved settings of the fixed inputs by running these levels for one week under otherwise normal production conditions. They found no improvement in the process variation and abandoned the new settings because there was an increase in cost.

The team was disappointed in designed experiments as an improvement tool since they had spent a great deal of effort for no reward. What went wrong? The major problem was that the experiment set out to reduce short-term diameter variation (variation within five consecutive shafts), and it turned out that this was a small component of the baseline variation. The team adopted the working approach without sufficient knowledge. In this case, the project was doomed by this decision.

Speedometer Cable Shrinkage

In a famous positive example of making a process robust, Quinlan (1985) reported on a project to reduce both the average and variation in postextrusion shrinkage of a casing for a speedometer cable. The team did not find the cause of the shrinkage variation. Instead, they chose 15 fixed inputs and selected one new level for each (for each selected fixed input, there were two levels, existing and new). They then ran an experimental investigation with 16 runs, a highly fractionated design. In each run, approximately 3000 feet of casing was produced and four samples were selected haphazardly. The shrinkage was measured for each sample.

Here the team adopted the Move Process Center and Robustness approaches simultaneously, since reducing variation or shifting the process center could improve the process.

For each run, the team calculated a performance measure using the four shrinkage values. They then analyzed the performance measures to isolate the best combination of levels to shift the average shrinkage towards zero and reduce the variation. The new levels were confirmed and the process was much improved.

Comments

As shown by the speedometer cable shrinkage example, the robustness approach can be successful, but we feel it is better to first look for the dominant cause. The more you know about the dominant cause of variation, the greater the chance you will select fixed inputs to change that will mitigate the variation in the dominant cause.

In the transmission shaft example, the team would have been much better off if they had recognized that the dominant cause of variation acted over a longer time frame than five consecutive parts. They would not have planned the experiment as they did with this extra knowledge.

8.4 DOES THE PROCESS OUTPUT EXHIBIT A STRONG PATTERN OVER TIME?

If the answer is yes, we can predict future output values from present and recent values. If the team also knows of a cost-effective quick way to adjust the process center to compensate for the deviation between the prediction and the target (or feels they could find one), feedback control is feasible. A feedback controller does not require knowledge of a dominant cause.

Fascia Film Build

In the fascia film build example, introduced in Chapter 3, a team wanted to reduce variation in the paint flow rate. A baseline investigation showed the time pattern in the flow rate given in Figure 8.5.

The baseline data provided evidence that feedback may be feasible. The flow rate variation was much smaller over the short term than over the long term. Since the team also knew they could use a valve to adjust the flow rate, they adopted feedback control as the working approach. They planned to assess the benefit by simulating the effect of the feedback control scheme on the baseline data. At the same time they could assess the costs of the adjustments.

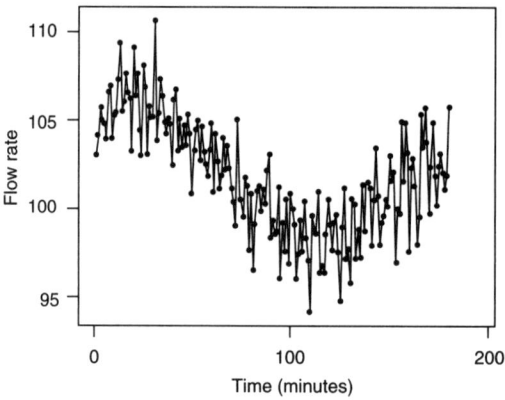

Figure 8.5   Variation in flow rate over time.

Parking Brake Tightness

In a vehicle assembly process, there was a problem with excess variation in parking brake tightness that led to considerable rework. A baseline investigation showed there were runs of high and low tightness values. The dominant cause acted slowly over time. Based on this knowledge, the team suspected that a dominant cause of tightness variation was the length of the front cable. The cables were delivered in batches and the average length differed substantially from batch to batch.

The team did not further explore front cable length; the cable arrived as part of an assembly, so it could not be measured in the plant. Because the dominant cause changed from batch to batch, they decided to consider feedback control.

The team knew that changing the depth of an adjustment nut would change parking brake tightness. To quantify the effect of this adjuster they planned an investigation where they would try different adjustment depths for a number of vehicles. In addition, the team still needed to decide on an adjustment rule, that is, when an adjustment should be made and by how much.

Comments

In the fascia film build and parking brake tightness examples, we saw a time pattern in the run chart of the output from the baseline investigation. We recommend a team use a sampling protocol in the baseline investigation that lets them see any systematic pattern in the output over time.

Patterns in the variation over time can have many forms. Some processes are setup dependent; that is, once a setup is complete, there is little variation from part to part. Many machining and stamping processes exhibit this behavior. If the team discovers such a pattern, and the output of the process can be measured immediately after setup and there is a low cost and quick way to adjust or redo the setup, then feedback control could be selected as the working approach.

In cases where feedback control is feasible, the output varies systematically over time. This implies that the dominant cause also exhibits the same time pattern. With this clue, it may be more economical to try to find the dominant cause than to adopt feedback control as the working approach. If the dominant cause is found, the team can address the cause directly, for example by applying feedback control to the cause. In the parking brake tightness example, feedback control was applied to the output despite partial knowledge of the dominant cause. This knowledge was used to help design the feedback control scheme.

8.5 SUMMARY

The thought process involved in choosing a working approach is summarized in Figure 8.6. Questions given at the same vertical height are addressed in either order or simultaneously depending on the problem context.

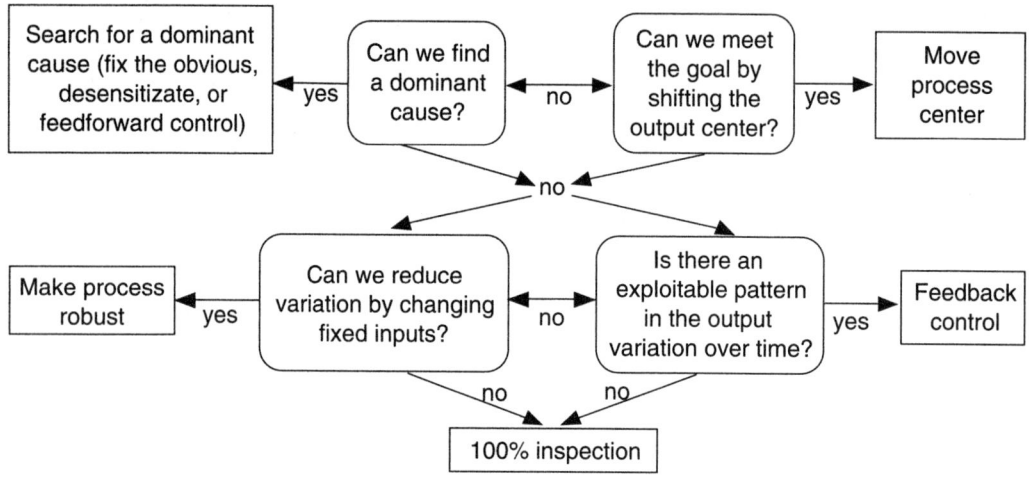

Figure 8.6 Flowchart to help choose a working approach before a dominant cause is known.

 **Key Points**

- We recommend the team select a working approach from the seven possible variation reduction approaches to guide further efforts:

  –Fixing the obvious using knowledge of a dominant cause
  –Desensitizing the process to variation in a dominant cause
  –Feedforward control based on a dominant cause
  –Feedback control
  –Making the process robust
  –100% inspection
  –Moving the process center

- The following questions can help to select the working approach:

  –Can we find a dominant cause of the unit-to-unit variation?
  –Can we meet the goal by shifting the process center without reducing variation?
  –Can we reduce variation by changing one or more fixed inputs without knowledge of a dominant cause?
  –Does the process output exhibit a strong pattern in the variation over time?

- We strongly recommend the team search for a dominant cause of variation, unless there is clear evidence that one of the non-cause-based approaches is likely to be feasible.

- After choosing a working approach, we conduct further process investigations to determine whether the selected approach is feasible.

- In making a choice of working approach, we try to assess potential implementation costs.

PART III

Finding a Dominant Cause of Variation

How often have I said to you that when you have eliminated the impossible, whatever remains, however improbable, must be the truth?

—Sir Arthur Conan Doyle (as Sherlock Holmes), 1859–1930

In many applications, identifying a dominant cause of variation leads to cost-effective process improvement. We recommend finding the dominant cause using the method of elimination and the idea of families of causes. With the use of observational plans and leveraging (comparing extremes), the search for a dominant cause of variation can be inexpensive and nondisruptive to the production process. We discuss the tools and methods used to support the search for a dominant cause. We introduce planned experiments to verify a suspect cause is dominant.

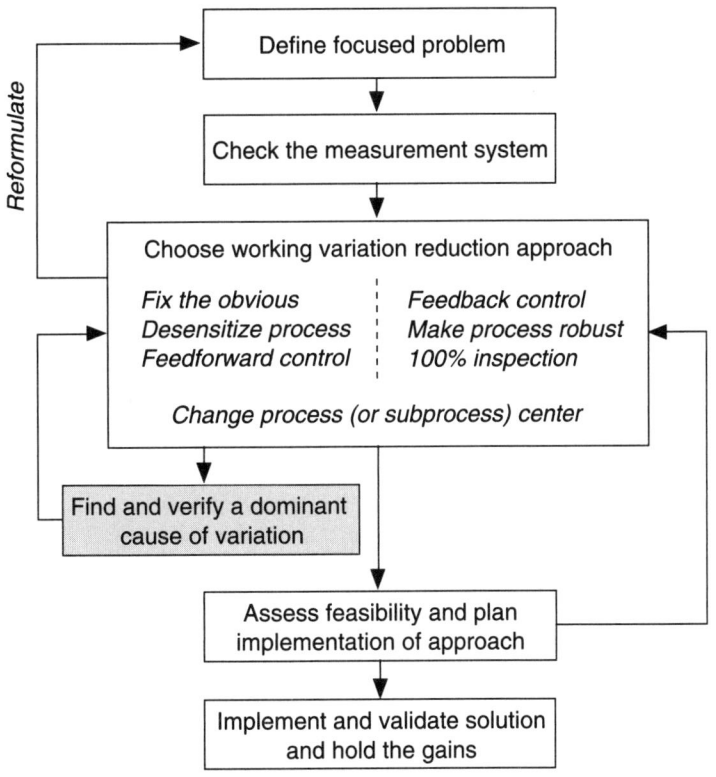

Statistical Engineering variation reduction algorithm.

9

Finding a Dominant Cause Using the Method of Elimination

We shall learn to be imaginative about what is causing product variability by going to the manufacturing area and figuratively talking to the product.

—Dorian Shainin, 1914–2000

In most problems, the team will decide to try to find a dominant cause before assessing the feasibility of a specific variation reduction approach. We recommend a search strategy based on the method of elimination to find a dominant cause. That is, we concentrate on ruling out possibilities rather than looking directly for the dominant cause. In this chapter, we discuss the principles behind this strategy; in chapters 10 through 13, we provide investigation plans, analysis tools, and numerous examples to demonstrate how it can be implemented.

The method of elimination can be explained using the familiar game 20 Questions. In one version of this game, we choose one of the 130,000 entries in the Canadian Oxford Dictionary (2002). We then give you the dictionary and ask you to determine the selected word using a series of yes/no questions. A poor strategy is to start asking about specific words. Unless you are lucky, you are quite likely to be exhausted before you get to the correct word. A much better strategy is to divide the dictionary in half and ask if the unknown word is in the first half. Whatever answer you get, you will have eliminated half the words in the dictionary with a single question. If you divide the remaining words in half at each iteration, you can find the unknown word with at most 17 questions (2^{17} is just greater than 130,000).

We apply the same idea to search for a dominant cause. We divide the set of all causes into families and then conduct an investigation to rule out all but one family. We repeat the exercise on the remaining family until a single dominant cause remains.

9.1 FAMILIES OF CAUSES OF VARIATION

In any process, there are many varying inputs that could have a large effect on the output. We partition the set of all such inputs into two or more families with common features such as the time frame or location in which they act. Then we use available data, investigations, and knowledge of the process to rule out all but one family as home of the dominant cause.

V6 Piston Diameter

Consider again the problem of reducing variation in V6 piston diameters introduced in Chapter 1. Figure 9.1 is a process map.

To start we divide the causes (varying inputs) into two families: measurement and the rest of the process. The measurement family includes all causes associated with the final gage and its operation and the rest of the process family includes all other causes. More specifically, the measurement system family includes those inputs that can change if we repeatedly measure the same piston: the operator, the position of the piston in the fixture, and the time since calibration of the gage, for example. The rest of the process family includes raw materials, the path through the machining process, the mold number, and so on. The dominant cause of diameter variation may be in either family.

To document progress in the search, we portray the families in a *diagnostic tree*. The top box describes the problem and the families of causes are listed below. The first part of the diagnostic tree for the V6 piston diameter problem is shown in Figure 9.2. As we proceed through the search, defining and eliminating families of causes, the diagnostic tree will grow.

To eliminate one family of causes, we carry out an investigation, as described in Chapter 7, to learn how much of the baseline variation can be attributed to the measurement system. In this investigation, two operators measured three different pistons three times each over two different days. The variation in the measurement system was relatively small since the discrimination ratio D was 8.3. We eliminate the measurement system family and concentrate on the rest of the process family as the home of the dominant cause of variation. We cross off the eliminated family on the diagnostic tree.

We always start with the two families: the measurement system and the rest of the process. We have explicitly included this partition of the causes in the Check Measurement System stage of the Statistical Engineering algorithm.

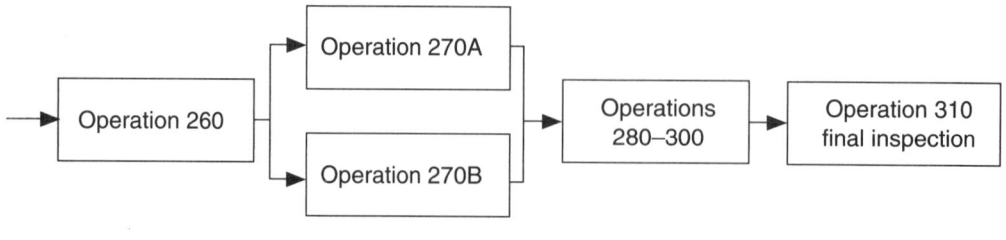

Figure 9.1 The piston machining process map.

Chapter Nine: Finding a Dominant Cause Using Elimination

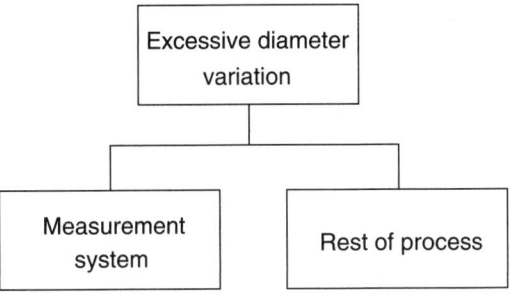

Figure 9.2 Diagnostic tree for piston diameter variation.

Bottle Label Height

In a bottling operation, management raised a concern over crooked and misplaced labels. The ideal label was a set distance below the bead on the neck of the bottle. The output characteristics were the average label height and the difference of the maximum and minimum heights as the bottle was rotated. The team assessed the measurement system that used a feeler gage and found that the measurement variation was small relative to the baseline for both characteristics.

Next, the team decided to proceed with the search for a dominant cause by partitioning the remaining causes into the bottle-to-bottle and the time-to-time families. The bottle-to-bottle family includes inputs that change quickly from one bottle to the next in the labeling process. The time-to-time family contains causes that vary more slowly and change substantially only from one hour to the next. To assess the relative effect of the two families, the team measured the label height for five consecutive bottles selected each hour for two shifts, a total of 16 hours.

We show the diagnostic tree for average height in Figure 9.3. Since there are two outputs in this problem, the tree may be different for each.

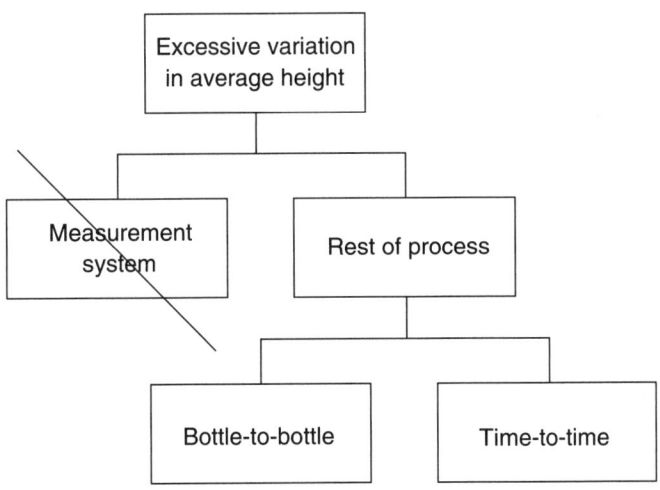

Figure 9.3 Diagnostic tree for average label height.

Engine Oil Consumption

A team was assigned the task of reducing warranty costs due to oil consumption in a truck engine. The engine was built at two different plants that had common equipment, common suppliers, and similar manufacturing processes. From the warranty database, the team noted that while the two plants had produced roughly the same number of engines, over 90% of the more than 1500 claims were associated with engines built in one of the two plants.

Here we divide the causes into two families: the within-plant and plant-to-plant families (see Figure 9.4). The within-plant family contains causes that vary in the same way for each plant, such as characteristics of components from a common supplier. The plant-to-plant family contains causes that have different values in the two plants, such as characteristics of components from different suppliers or differences in operating procedures.

Using the warranty data, the team ruled out the within-plant family. The dominant cause lived in the plant-to-plant family. To continue the search for a dominant cause, the team focused on the few varying inputs that were different in the two plants.

In this example, the team did not check the measurement system. The discrepancy in warranty claims was so great that they assumed that there was a real difference between the plants. The team made huge progress using the data available from the warranty system because the plant-to-plant family had relatively few causes.

Surprisingly, another team working on the same problem from an engineering perspective had decided (incorrectly) that changing the design of the PCV valve could solve the problem. The second team had implicitly adopted a robustness approach. By changing the PCV valve, they hoped to reduce the effect of the unknown dominant cause of oil consumption. Since there was a common supplier, no characteristic of the PCV valve could be a dominant cause because there was no systematic difference in the valves from plant to plant.

There are many ways to partition causes into families and different investigations and tools to eliminate all but one family. We look at numerous examples in the next three chapters.

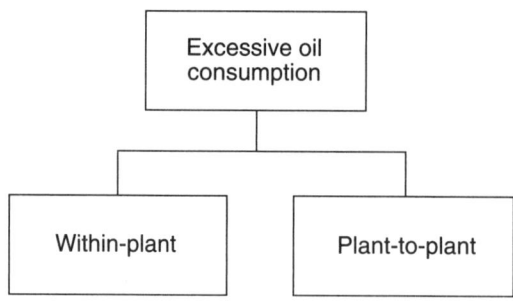

Figure 9.4 Partition of the causes of oil consumption.

9.2 FINDING A DOMINANT CAUSE USING THE METHOD OF ELIMINATION

To find a dominant cause, we apply the method of elimination repeatedly. There are three steps:

1. Divide the remaining suspect causes into two or more families.
2. Plan and carry out an investigation to eliminate all but one family as the home of the dominant cause.
3. Repeat steps 1 and 2 until only one suspect for the dominant cause remains.

The search for a dominant cause is complete if we find a cause or family that is specific enough to allow us to choose a variation reduction approach. In the engine oil consumption example, the dominant cause was in the plant-to-plant family. However, the team could not address this cause directly unless they were willing to shut down one plant! The team continued searching for a more specific cause in the identified family.

In some cases, we may stop the search when there are only a few possible causes left because it is more efficient to identify the dominant cause using an experimental plan (see Chapter 13).

V6 Piston Diameter

Consider again the V6 piston diameter example, discussed earlier in this chapter, where the problem was excess variation in piston diameter. The first investigation eliminated the measurement system family. Next, the team looked at two location-based families as shown on Figure 9.5.

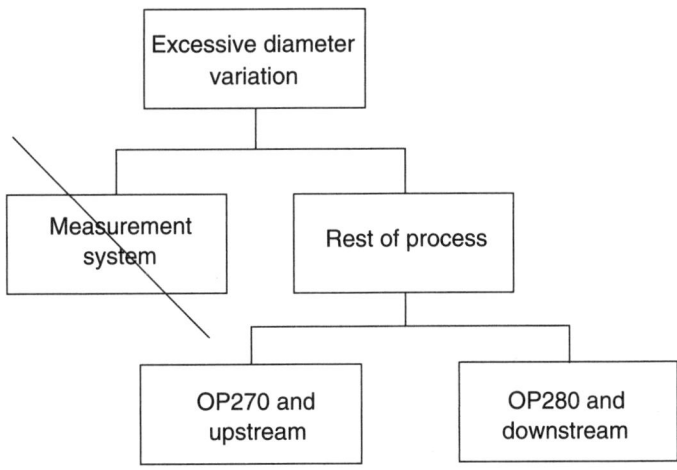

Figure 9.5 Diagnostic tree for V6 piston diameter variation.

The team planned an investigation in which they measured the diameter of 96 pistons after Operation 270 and at the final gage. This investigation eliminated the operations downstream from Operation 270 as the home of the dominant cause.

The team concluded that the diameter measured just after Operation 270 was the dominant cause of variation in the final diameter. This finding did not surprise the team, who understood the functions of Operation 280 through 310. However, by verifying that the dominant cause lived in Operation 270 or upstream, they had made substantial progress. The order in which the pistons are measured at the final gage was not the same as the order in which they were machined, whereas the order of production was preserved much better in the machining operations up to Operation 270. The preservation of order made it much easier to track pistons through the process and thus made further investigation easier.

The team was unable to directly address the diameter variation at Operation 270. They decided to reformulate the problem and to search for a dominant cause of diameter variation as measured after Operation 270. At the same time, they compared the measurement system at Operation 270 to the final gage. While the Operation 270 measurement system was not the home of a dominant cause of variation, the team discovered the relative bias between the Operation 270 and final gages as described in Section 5.2. They fixed this problem by changing an offset in the Operation 270 gage. They also knew that this obvious fix would not reduce diameter variation at either gage.

Next, the team decided to look at three families of causes of Operation 270 diameter variation as documented on Figure 9.6.

The team planned another investigation with 96 pistons. For each piston, the diameter was measured before and after Operation 270. Each parallel stream at Operation 270

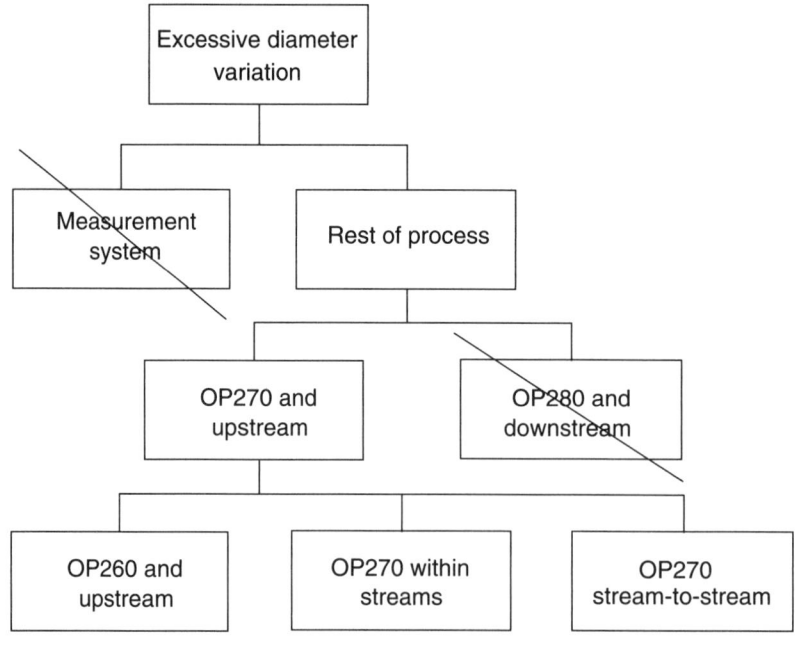

Figure 9.6 Diagnostic tree for V6 piston diameter problem after Operation 270 investigation.

processed half the pistons. Based on the observed data, the team concluded that the Operation 270 stream-to-stream family was the home of the dominant cause. They eliminated the upstream family. They concentrated their efforts on identifying what was different between the two streams that could explain the observed difference in process behavior.

At Operation 270, the grinders were of the same design, and there was a common source of unfinished pistons and a common control plan. However, there were different operators. The team interviewed the operators and found that each operator ran his machine differently; each was convinced that the method they used was superior to that in the written control plan.

Now the team suspected the dominant cause was a difference in the method of process control in the two streams. The final step was to verify this suspicion. We give the completed diagnostic tree in Figure 9.7.

We use the diagnostic tree to document the search for the dominant cause. We can add additional information to the diagram to support the logic of eliminating families. Remember that the tree is built up as we iteratively apply the method of elimination. We do not start by constructing the whole diagram.

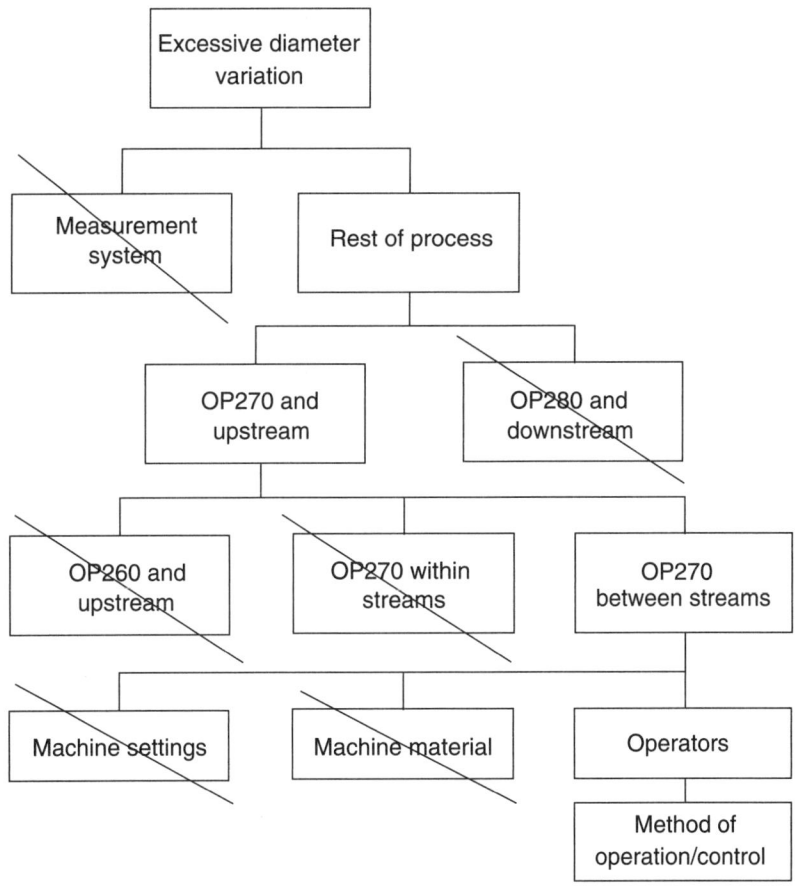

Figure 9.7 Final diagnostic tree for excessive variation in V6 piston diameter.

9.3 IMPLEMENTING THE METHOD OF ELIMINATION

In this section, we give some general suggestions for implementing the method of elimination that we illustrate with the examples in subsequent chapters.

Using the method of elimination, we do not start by listing possible causes and classifying them into families.[1] Instead, we start by considering broad categories that define the families. This is a major difference from other problem-solving algorithms. As we obtain more process knowledge, we subdivide the remaining families to further narrow down the suspects for a dominant cause. Our goal is to end up with a single cause or, in the worst case, a short list of possibilities. The final step is to verify (and select in the case of a short list) that we have indeed found a dominant cause.

For most problems, there are a large number of possible dominant causes and many ways to divide them into families. How do we start and proceed? We make the following suggestions.

Check the Measurement System First

To start, we always divide causes into those associated with the measurement system and those associated with the rest of the process. We can use a measurement system investigation as described in Chapter 7 to eliminate one of these two families.

Use the QPDAC Framework

The search for a dominant cause involves many process investigations. To help plan, execute, and draw conclusions from such empirical investigations, we use the five-step QPDAC framework as described in Chapter 5.

The investigations used to support the search for a dominant cause share many features. Using the QPDAC terminology, the *target population* is all units produced by the process now and in the future. The target population stays the same unless we reformulate the problem. The attributes of interest are the components of the output variation attributable to the different families or individual causes.

In the Plan step, we design the investigation to avoid large study and sample errors. We are unconcerned with measurement error since we have previously assessed the measurement system and determined that it is adequate. We use the results from the baseline investigation to help make the choice of study population. We choose this population so that the dominant cause has the opportunity to act within this population. We then select the sampling protocol so that we can estimate the variation attributable to each possible family. As always, we balance the statistical considerations against the constraints of cost and time.

Consider the problem of excess variation in average label height introduced earlier in this chapter. In the investigation to compare the bottle-to-bottle and time-to-time families of causes, the team decided to choose a study population defined by all bottles produced over two shifts. They made this choice because in the baseline investigation, where they had looked at historical records covering production from over a month, they had seen close to the full extent of variation in average label height within each shift. If, instead, the baseline investigation had shown that the dominant cause acted more slowly, for example, week to week, they would have needed a longer time frame for the study population. To obtain information about the bottle-to-bottle family, the sampling protocol must involve selecting some bottles made consecutively. To help avoid sample error, the team selected five consecutive bottles every hour throughout the study population.

In the Find and Verify a Dominant Cause stage of the algorithm, we assume the baseline investigation captured the performance of the current and future process with little study or sample error. With this assumption, if we see the full extent of variation in the output in any subsequent investigation, we are confident the dominant cause has acted. See further discussion of the full extent of variation later in this chapter.

Think About Families of Causes and the Possible Types of Investigations

In chapters 10 through 12, we present a variety of process investigations useful in the search for a dominant cause. Each type of investigation aims to compare different families of causes. We may decide to compare:

- Location-based families (stratification)
- Time-based families
- A combination of time- and location-based families (multivari investigation)
- Upstream and downstream families (variation transmission investigation)
- Assembly and component families (component swap)
- The effects of individual inputs (group comparison, input-output investigation)

The best choice of investigation depends on the problem, the process, and current process knowledge. At the start of the search, we want to compare large families of causes with the hope of eliminating many causes quickly. Later in the search, we are more willing to look at individual causes.

Use Available Data

There are often data available that have been collected for other purposes such as routine process control, scrap accounting, or warranty cost assessment. We also have the data from the baseline investigation. We suggest the team examine the available data and how it was collected by running through the steps of QPDAC. If the team has confidence that there is little study and sample error, they can use the data to rule out families of causes. Here, we start with the data and then think about the families.

The use of the warranty data in the engine oil consumption problem is a good example. For another example, in Chapter 3, we briefly discussed a problem where there was a high frequency of rejected engines at the valve train test. Every engine was tested and the results were recorded in a database. There were three test stands that operated in parallel and engines were assigned to a stand haphazardly. The stands checked a number of characteristics and an engine was rejected if one or more characteristics was out of specification. Rejected engines were torn down and repaired. There were many "no fault

Table 9.1 First-time reject rate by test stand.

Test stand	First-time reject rate
1	12.8%
2	4.2%
3	2.8%

found" rejects. The test stand operators repeatedly measured a rejected engine without teardown because they knew it would often pass with a second or third test.

A team was assigned the goal of reducing the reject rate. The baseline first-time reject rate was about 6.6%. Before conducting an assessment of the measurement system, the team used the available data to stratify the first-time reject rate by test stand over a one-week period. The results are given in Table 9.1.

The dominant cause of the high reject rate lived in the measurement system family. The team proceeded by looking at causes that could explain the stand-to-stand differences.

Use Knowledge of the Process

The method of elimination relies on process knowledge both in determining the plan for an investigation and interpreting the results. For example, in the V6 piston diameter example discussed earlier, the team suspected that the dominant cause acted upstream of Operation 280. They chose to split the process after Operation 270 so that they could rule out all the downstream operations in a single investigation.

Knowing what can be measured and where, how easy it is to trace parts through the system, when production order is preserved, and so on, is essential in selecting families and developing investigation plans.

Knowledge of the process can help the team interpret the data and determine what families are eliminated. For example, in a problem to reduce scratches on a painted fascia, the team made a concentration diagram using 100 damaged parts to show where on the part the damage occurred. Combining the new knowledge about the location of most of the scratches and existing knowledge of the handling system, the team was able to rule out large parts of this system as the source of the scratches.

Observe the Full Extent of Variation

We assume that there is a single dominant cause[2] that explains most of the output variation seen in the baseline investigation. When we carry out an investigation to eliminate one or more families, we need to see most of this baseline variation in the observed output values. Otherwise, we may decide the dominant cause is in the wrong family.

Consider the V6 piston diameter example with some hypothetical data. Suppose we start by considering the two families shown in Figure 9.8. Recall from the baseline investigation that the full extent of diameter variation was 582 to 602 microns.

Chapter Nine: Finding a Dominant Cause Using Elimination 127

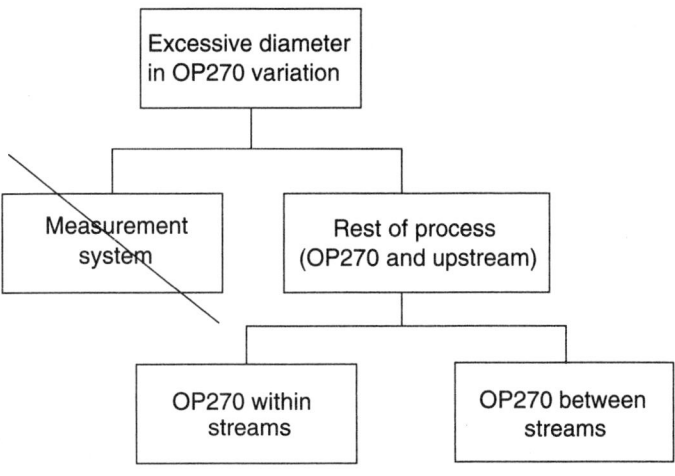

Figure 9.8 Diagnostic tree for hypothetical search.

To eliminate one of the families, we track 30 consecutive pistons from each stream at Operation 270 and measure the diameters at the final gage. What can we conclude if the results are as given by Figure 9.9?

If we are careless, we conclude that there is a large difference stream to stream and rule out the within-stream family. However, in the 60 pistons, the diameter varies only from 588 to 596 microns, about half the full extent of variation in the baseline investigation. The dominant cause has not acted fully during this investigation, and we may be incorrect in ruling out either family. We need to revisit the plan for the investigation. Perhaps we should have collected 30 pistons from each stream spread out over a day rather than 30 consecutive pistons to see the full extent of variation.

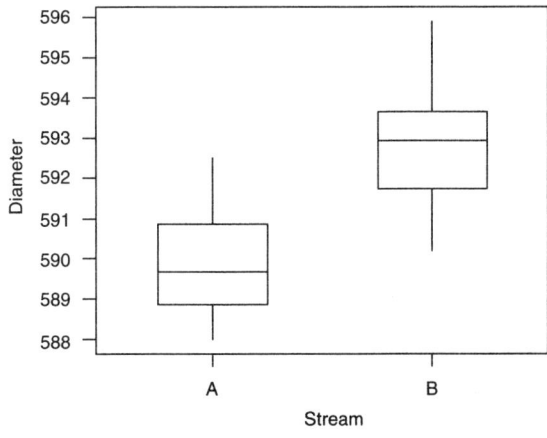

Figure 9.9 Box plot of diameter by Operation 270 stream from hypothetical investigation.

128 Part Three: Finding a Dominant Cause of Variation

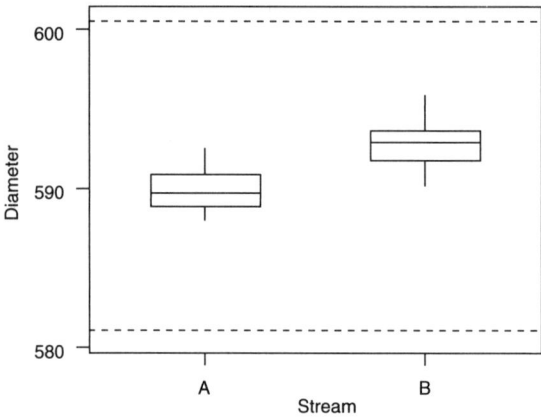

Figure 9.10 Box plot of diameter by Operation 270 stream from hypothetical investigation showing full extent of baseline variation.

To ensure correct interpretation, we recommend showing the full extent of variation in the baseline investigation on all graphical displays of data used in the method of elimination. So rather than use Figure 9.9, we present the results as in Figure 9.10. With Figure 9.10, we cannot conclude that the stream-to-stream family is home of a dominant cause.

The message here is that we need to consider the steps in QPDAC carefully even for these simple studies to rule out families of causes. We need to specify the study population and sampling protocol so that we can be confident that we will see the full extent of variation in the output.

Use Leverage Where Possible

Leverage means that we select units with output values at both ends of the full extent of variation. Assuming there is a single dominant cause, the values of this cause will be very different among these extreme units. This guarantees that we will see the full extent of variation in the output.

Consider again the problem of excessive variation in V6 piston diameters. The full extent of variation in the baseline was 582 to 602 microns. After checking the measurement system, suppose we select 10 pistons with diameter larger than 600 and 10 with diameter less than 584 microns. If the dominant cause acts in the stream-to-stream family at Operation 270, we expect almost all of the large pistons will come from one stream and almost all the small pistons from the other stream. By selecting pistons that are very different, we get leverage in comparing families of causes.

In the V6 piston diameter example, the team could not conduct such an investigation because they could not determine the stream of the finished pistons.

Leverage was used in the problem of noisy windshield washer pumps introduced in Chapter 6. The team established a reliable in-plant measurement system for noise and a baseline. They then selected five noisy and five quiet pumps relative to the full extent of variation. Next they disassembled and reassembled each pump and noticed that there were dramatic changes in noise levels. The noisy pumps became much quieter and the quiet pumps somewhat noisier when the pump housing was rotated relative to the motor housing in the reassembly. The dominant cause of noise was in the assembly family.

If we exploit leverage, we can use small sample sizes in many investigations. We need to ensure that the extreme output values are due to the dominant cause. If the extreme values are due to a different failure mode, leveraging may mislead us. By defining a focused problem and avoiding outliers, we hope to avoid this occurrence.

Take the Simplest Path

There are many ways to divide causes into families. For a given partition, you may or may not be able to think of a way to investigate the process that will allow you to rule out all but one family. In the next chapters, we describe numerous plans for investigations and the corresponding analysis tools. We recommend that you consider each of these plans in the context of your problem and where you are on the diagnostic tree. Then pick a simple plan that can help make progress.

In theory, you can look at many families simultaneously using a complex plan. This seems like a good idea because you can greatly narrow the search for the dominant cause with a single investigation. However, we strongly recommend using a small number of large families to keep the investigation simple. We know that this recommendation generates more iterations in the search for the dominant cause. We also know from bitter experience that complex plans often fall apart due to logistical difficulties and production pressures.

Most of the plans for ruling out families of causes are observational. We strongly recommend observational plans, where we "listen to the parts," over experimental plans that require an intervention in the process. Experimental plans are difficult to manage and expensive to execute in the production environment.

Be Patient

To use the method of elimination, we need to plan, execute, and analyze the results of a number of simple investigations. To be successful, we need to avoid the temptation to jump to a specific cause too soon. As consultants, we have met strong resistance from process managers when we suggested that we plan to conduct several investigations to find the cause of variation. The usual reaction is to ask that we get it over with quickly using a single investigation. In management review meetings, we have seen problem-solving teams struggle with the question,

"Well, have you found the cause (or solution) yet?"

The answer, "No, but we have eliminated a lot of possibilities," can be hard to defend if the problem is urgent and substantial resources have been committed to its solution. On the other hand, trying but failing to find the cause with a single complicated investigation is even harder to defend.

Reconsider the Variation Reduction Approach

Sometimes the method of elimination fails. We may have identified a particular family as home of the dominant cause, but we cannot see how to split this family further. We are left with two choices. We can begin testing the remaining individual suspects using the methods described in chapters 12 and 13. Alternately, we can abandon the search and adopt a variation reduction approach that does not require the identification of a dominant cause. We still use any knowledge gained about the family containing the dominant cause.

Key Points

- To find the dominant cause of variation, partition the possible causes into families and use the results of an investigation to eliminate all but one family as the home of the dominant cause. Iterate to further subdivide the remaining families until a dominant cause is found.
- "Let the process or product do the talking." Observational data can provide strong clues about a dominant cause of variation.
- Document the search for a dominant cause using a diagnostic tree.
- Implementation suggestions:
    –Check the measurement system first
    –Use the QPDAC framework to help plan the investigations
    –Use available data
    –Use knowledge of the process
    –Observe the full extent of variation
    –Use leverage where possible
    –Take the simplest path
    –Be patient
    –Reconsider the variation reduction approach

Endnotes (see the Chapter 9 Supplement on the CD-ROM)

1. The traditional strategy to find the cause of a problem starts by listing all the possible causes in a cause and effect diagram. We compare this strategy to the use of families and the method of elimination in the chapter supplement.
2. The method of elimination relies on there being one or two dominant causes. In the chapter supplement, we discuss what happens when this assumption does not hold.

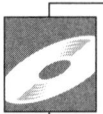
Exercises are included on the accompanying CD-ROM

10

Investigations to Compare Two Families of Variation

The world is full of obvious things which nobody by any chance ever observes.

—Sir Arthur Conan Doyle (as Sherlock Holmes), 1859–1930

In this chapter, we describe some plans and the corresponding analysis to compare two families of causes. We concentrate on a single iteration of the method of elimination where the goal is to eliminate one of the two families. For each example, we assume that the team has successfully completed the Define Focused Problem and Check Measurement System stages of the variation reduction algorithm and has decided to search for a dominant cause of variation.

We use plans that meet the criteria discussed in Chapter 9:

- Keep it simple.
- Use available data.
- Observe the full extent of variation in the output.
- Exploit leverage if possible.

10.1 STRATIFICATION

With stratification (Kume, 1985), we divide parts into distinct groups based on their source. For example, we may stratify by machine, mold, line, gage, plant, location within the part, supplier, operator, and so on.

Using stratification, we can assess whether the dominant cause impacts each group in the same way or each group in different ways. A dominant cause that impacts different groups differently is said to act in the group-to-group family. If such a cause is present, we will see large differences among the group averages and low variation within each group.

Casting Scrap

The baseline scrap rate for a casting defect was 2.4%. The castings were molded four at a time with cavities labeled *A, B, C,* and *D*. The cavity label was molded into the casting.

The team decided to stratify the causes into the cavity-to-cavity family (also called the *among-cavity* or *between-cavity* family in other references) and the within-cavity family. We have:

- Cavity-to-cavity family: causes that affect different cavities differently
- Within-cavity family: causes that affect each cavity in the same way

The team classified 200 scrap castings by cavity as shown in Table 10.1. Because the proportion of scrap castings from each cavity was about the same, the team ruled out the cavity-to-cavity family. The dominant cause affected all cavities in the same way.

In Table 10.1, we compared the number of scrap castings from each cavity directly because there was equal volume from all four cavities. If the cavity volumes were not the same, we would have compared the percent scrapped from each cavity.

Table 10.1 Scrap stratified by cavity.

Cavity	Number of castings
A	58
B	51
C	43
D	48
Total	200

Rod Thickness

In Chapter 6, we looked at an example where the goal was to reduce thickness variation in a connecting rod at a grinding operation. The team established a baseline and found the full extent of thickness variation was 2 to 59 thousandths of an inch (recorded as a deviation from a particular value).

In the baseline investigation, the team measured thickness of each sampled rod at four positions. The data are given in the file *rod thickness baseline*. With these data, we can compare the family of causes that produce different thickness within each position to the family of causes that act from position to position.

In Figure 10.1, we plot thickness by position. Since we are using the baseline data, we are guaranteed to see the full extent of variation for the output, given by the dashed horizontal lines. Position 3 has a smaller average thickness than the other positions. There is a cause in the position-to-position family that produces this difference. However, we can see that this cause is not dominant by imagining the plot with the average thickness

Chapter Ten: Investigations to Compare Two Families of Variation

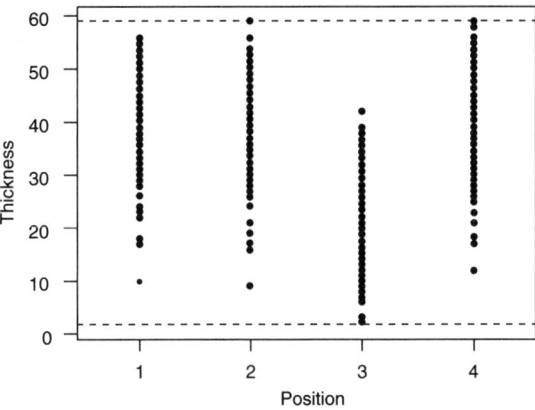

Figure 10.1 Rod thickness stratified by position (dashed horizontal lines give the full extent of variation).

aligned for each position.[1] The variation within each position is a substantial part of the overall variation. We cannot eliminate all the out-of-specification rods without reducing the within-position variation. Of course, we will implement an obvious fix to better align the thickness center at Position 3 if available.

Camshaft Lobe Runout

There are 12 lobes per camshaft in the problem with excess base circle (BC) runout described in Chapter 1. In the baseline investigation, the team measured BC runout on all lobes for nine different camshafts selected each day for 12 days. We can stratify the causes into the lobe-to-lobe and within-lobe families.

- Lobe-to-lobe: causes that affect different lobes differently
- Within-lobes: causes that affect each lobe in the same way

We use these data, given in *camshaft lobe runout baseline*, to try to eliminate one family of causes. Here again we are guaranteed the full extent of variation since the data came from the baseline investigation.

We can see the variation due to the within-lobe and lobe-to-lobe families in Figure 10.2. The center lobes (numbers 5–8) have much lower averages and standard deviations than do the end lobes.

We cannot eliminate either family based on the data. However, we get some valuable clues. The dominant cause is an input that acts in both families or involves inputs from both families.[2] We redefine the families. Since it is difficult to think of good labels, we proceed based on a diagnostic tree as given in Figure 10.3.

In the subsequent search, the team looked for causes that could explain the difference between the middle and end lobes and eliminated any that could not.

134 Part Three: Finding a Dominant Cause of Variation

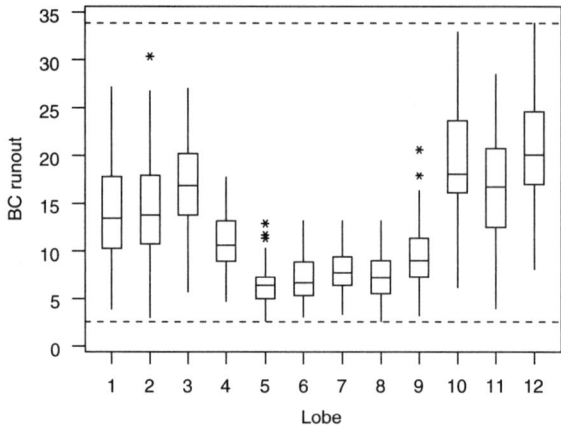

Figure 10.2 Camshaft journal runout by lobe (dashed horizontal lines give the full extent of variation).

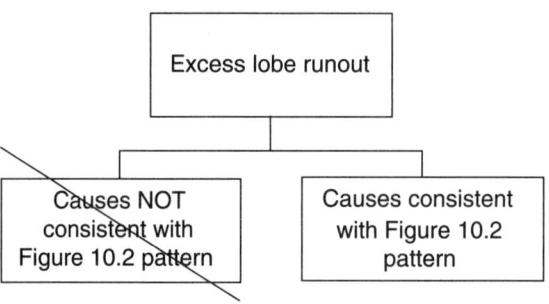

Figure 10.3 Diagnostic tree for camshaft lobe runout.

Comments

We can often use stratification with existing data for any multistream process if we can identify the stream associated with each part. Some examples are parallel processing steps, multiple suppliers, gages, product sources, operators, and so on. We saw two examples in Chapter 9: first, when oil usage warranty claims were stratified by the plant where the engines were built, and second when rejected engines were stratified by the test stand.

For a continuous characteristic, a box plot (using either the box summary or individual values depending on the data volume) is a convenient way to present the data to display the variation within groups and group to group. Adding dashed horizontal lines shows if we have seen the full extent of variation. The easiest way to assess the contributions of the two families is to shift the boxes (or individual values for each group) in your mind so that they all have the same centerline. The variation that remains is due to the within-group family. If this variation is large relative to the full extent of variation, we eliminate the group-to-group family as the home of the dominant cause.[3]

10.2 COMPARING TWO TIME-BASED FAMILIES

For many problems, we can separate causes based on the time frame within which they vary. Some causes, such as properties of raw materials and tool wear, change slowly, while others, such as position in a fixture and contamination by dirt, can change quickly from part to part. We use the term *part-to-part* to label the family of causes that vary from one part to the next in time sequence. An investigation to compare two time-based families is a simple special case of a multivari—see Chapter 11.

In Figure 10.4, we show a schematic of a sampling plan where we measure three consecutive parts once per shift. We plan to sample parts until we see the full extent of variation established in the baseline investigation.

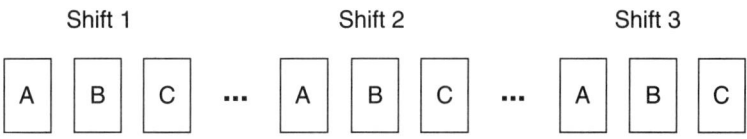

Figure 10.4 Part-to-part and shift-to-shift families.

Camshaft Lobe BC Runout

The plan of the BC runout baseline investigation was to collect nine camshafts per day for 12 days. To compare the effects of the within-day and day-to-day families, the team plotted the runout values by day as given in Figure 10.5. Since the pattern across days was similar, they were tempted to eliminate the day-to-day family.

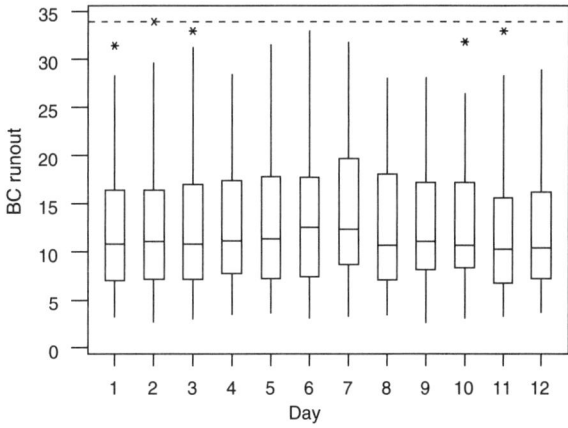

Figure 10.5 Box plots of BC runout by day (dashed horizontal lines give the full extent of variation).

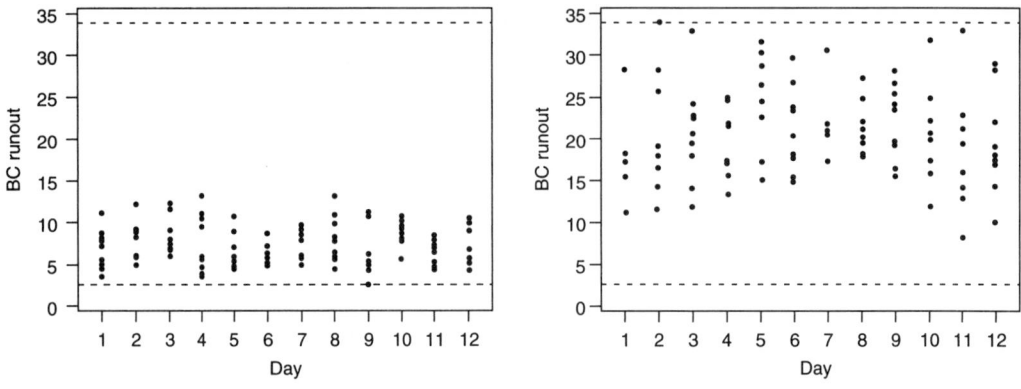

Figure 10.6 Box plots of camshaft journal runout by day for Lobe 8 (left) and Lobe 12 (right) (dashed horizontal lines give the full extent of variation).

However, recall that the dominant cause must explain the differences between end and middle lobes as discussed in Section 10.1. To see if the lobe-to-lobe differences depended on the day, the team examined BC runout by day for each lobe separately. Figure 10.6 shows the plots for lobes 8 and 12. In the baseline, Lobe 12 exhibited the most runout variation, while Lobe 8 gave the least.

Figure 10.6 suggests there was no interaction between causes that act lobe-to-lobe and day-to-day since for each lobe, the pattern in runout is similar across days. The team concluded the dominant cause acted within each day. Using this conclusion, they could plan subsequent investigations over shorter time periods (i.e., one or two days) and expect to see the full extent of variation.

Engine Block Porosity

We introduced a problem of reducing scrap due to porosity in cast-iron engine blocks (Figure 10.7) in Chapter 3. Porosity occurs because gas bubbles in the molten iron do not escape before the casting hardens. The scrap rate was about 4%, with half detected in the foundry (the porosity was visible on the surface) and the rest found after the initial machining of the bank faces of the block.

The team created a new measurement system to determine porosity on a continuous scale. The system measured the total hole size across the bank faces based on a classification system of hole sizes in 1/64 square inches. The team checked the measurement system and judged that it was adequate. With this measurement system, there was not a direct translation to the decision to scrap a block since that decision depended critically on the location of the porosity, not just the total size.

Although a new output was defined, the team did not establish a new baseline. The full extent of variation in the new porosity measure could have been easily determined by measuring porosity using the new continuous scale on a large number of blocks. The team decided not to take the time to establish the baseline for the new measure of porosity. This decision introduced the risk that they would make the wrong decision about the identity of a dominant cause.

Chapter Ten: Investigations to Compare Two Families of Variation   137

Figure 10.7   Plastic scale model of the engine block.

The team decided to compare block-to-block and time-to-time families as shown in Figure 10.8. The block-to-block family includes causes that change quickly from one block to the next as they are poured. The time-to-time family includes causes that change more slowly.

The team marked five consecutive blocks at the start of every hour for eight hours. They located the 40 blocks after the bank faces were machined and measured the porosity. Of the 40 blocks, 3 would have been rejected due to porosity had they been subject to the regular production inspection. This exceeded the historical reject rate, so the team was confident that the dominant cause was acting during the investigation. The data are stored in the file *engine block porosity multivari*.

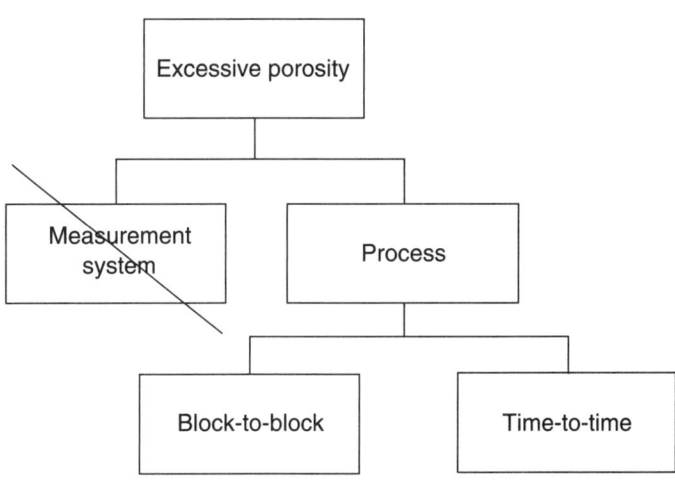

Figure 10.8   Diagnostic tree for excess porosity.

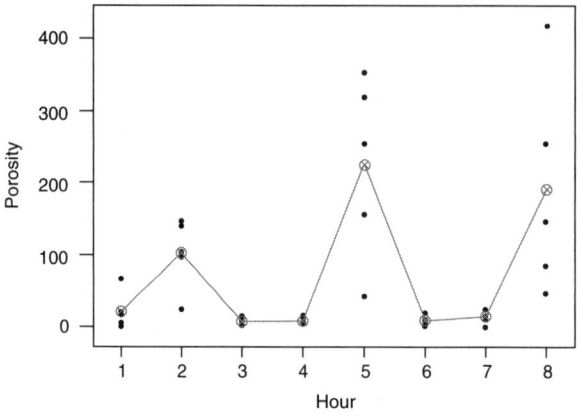

Figure 10.9 Multivari chart of block porosity versus hour.

We created the plot in Figure 10.9 using the multivari routine in MINITAB (see Appendix C). The dots on the plot correspond to the individual porosity values. The five block porosity averages are joined from hour to hour. We see large differences in the average porosity over time relative to the variation of the porosity within most hours. Based on this observation, we eliminate all causes, such as mold dimensions, that act from block to block. The average porosity (and the variation) is highest in hours 5 and 8. The team noted that these were times when the process was neglected due to lunch and preparations for the end of the shift. Combining this process knowledge with the patterns seen in Figure 10.9, the team looked for causes that behaved differently during these special times.[4]

10.3 COMPARING UPSTREAM AND DOWNSTREAM FAMILIES

We can divide causes into families based on where in the process they act. If we represent the process as a series of operations, we can split the causes in terms of whether they act upstream or downstream from a particular point.

When comparing upstream and downstream families we use a *variation transmission* investigation. We want to determine whether variation observed at an intermediate operation is transmitted through the downstream operation or if the downstream operations add substantial variation. We consider variation transmission investigations involving more than two process steps in Chapter 11.

Camshaft Lobe BC Runout

The team had discovered that for Lobe 12, the dominant cause of BC runout variation acted within days. From the baseline investigation they knew that the full extent of runout variation

Chapter Ten: Investigations to Compare Two Families of Variation 139

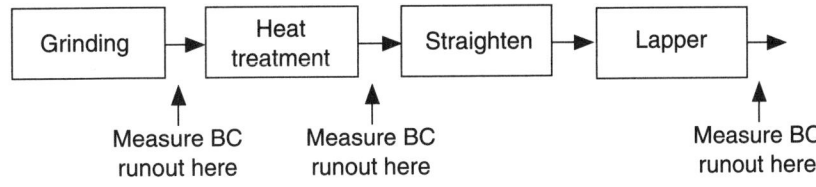

Figure 10.10 Camshaft process map.

was 2.6 to 33.9. Using the process map (Figure 10.10), they first split the remaining causes into two families:

- Causes that act downstream of heat treatment
- Causes that act in heat treatment or upstream

The team selected 32 parts over the course of one day. There were four camshafts from each of the eight lobe grinders. One camshaft from each grinder was processed on each heat treatment spindle. BC runout was measured after the heat treatment and after the final step of the process on each lobe. The data for Lobe 12 are in the file *camshaft lobe runout variation transmission*.

Figure 10.11 shows a plot of the Lobe 12 final BC runout versus the BC runout after heat treatment. Similar plots were produced for the other lobes. Across all the lobes, the final runout variation matched the baseline variation. Using Figure 10.11, we can separate the two families:

- Downstream family—causes that act downstream of heat treatment
- Upstream family—causes that act in the heat treatment step or upstream

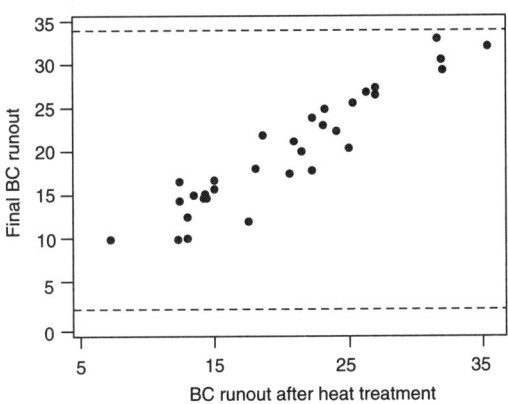

Figure 10.11 Lobe 12 BC runout after heat treatment and final (dashed lines give full extent of variation).

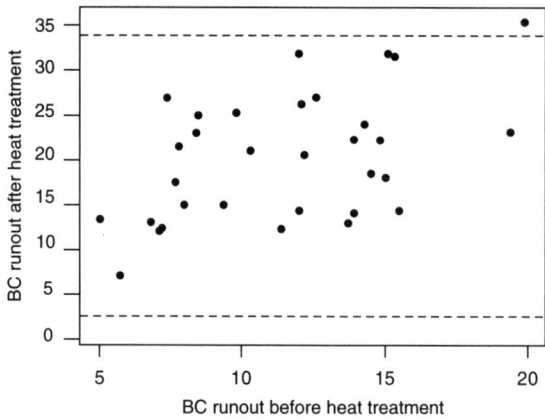

Figure 10.12 Lobe 12 BC runout before and after heat treatment (dashed lines give full extent of variation).

The plot shows a strong relationship between runout after heat treatment and at the end of the process. If we hold fixed the runout after heat treatment, there is little variation in final runout. We eliminate the downstream family that includes the straightening and lapping operations as the home of the dominant cause. The variation in runout is coming from the upstream family.

In the investigation, the team also measured runout for all lobes before the heat treatment step. Figure 10.12 shows very little relationship between the BC runout before and after heat treatment. If we hold fixed the runout before heat treatment, we see most of the full extent of variation after heat treatment. The team concluded that the dominant cause acts in the heat treatment. They eliminated all causes that act upstream of the heat treatment.

Cylinder Head Rail Damage

In the casting of cylinder heads, the problem was excessive scrap due to damaged head rails. The damage occurred somewhere in the shakeout and cleaning operations. The heads were 100% inspected.

To eliminate some causes, the team recorded the location of the damage for 100 scrap rails on a concentration diagram, a schematic of the part. See Figure 10.13.

The team noticed that the damage did not occur at random on the rails but was concentrated at particular locations. The nonrandom pattern ruled out several of the steps in the cleaning process as the home of a dominant cause. As a result of this simple investigation, the process owners canceled the purchase of new equipment ordered to deal with the damaged rails. The team concluded the dominant cause did not act in the processing step where the new equipment would have been installed.

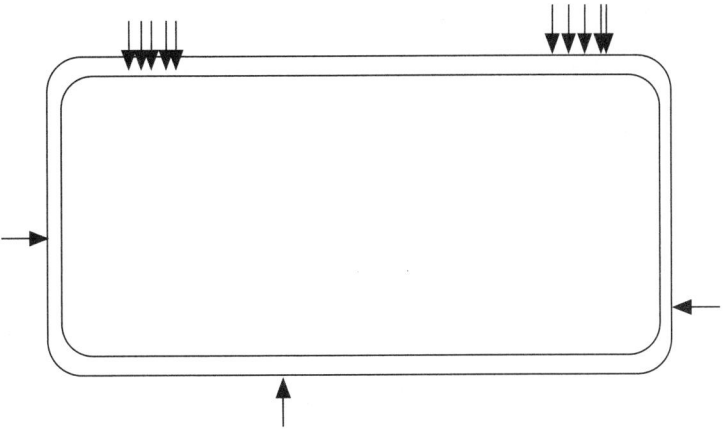

Figure 10.13 Concentration diagram showing location of rail damage.

Comments

We can use plans similar to that used in the camshaft lobe runout example to partition the causes upstream and downstream from an intermediate operation whenever we can measure the output after the intermediate operation.

In the chapter supplement, we look at other plans and analysis tools for comparing upstream and downstream process step families that have more limited application.[5]

10.4 COMPARING ASSEMBLY AND COMPONENT FAMILIES

For an assembled product, we can partition causes into those that reside in the assembly operation and those that live with the components. To rule out one of the families, we observe whether the output characteristic of a unit changes substantially when it is reassembled using the same components. This is only possible if we can disassemble and reassemble without damaging or changing the components. This type of investigation works well for rare defects.

To exploit leverage, we start with two assemblies with opposite and extreme performance in terms of the output.[6] In other words, the two selected units show the full extent of variation. Next we disassemble and reassemble the units at least three times, each time measuring the output. If the output does not change substantially for either unit, we eliminate the assembly operation as the home of the dominant cause.

Power Window Buzz

A team was charged with resolving a "buzz" problem with a power window regulator. Buzz was measured on a seven-point scale. In the baseline investigation, the team found regulators with scores ranging from 1 to 7. The regulator was assembled from four major components. The team split the causes into two families:

- Components: causes acting within the four components
- Assembly: causes acting in the process that puts the four components together

To eliminate one of these families, the team found two regulators, one that was extremely noisy with a score of 7 and a quiet one with a score of 2. They disassembled and reassembled the two regulators three times and measured the noise score each time. They found that the noise score did not change in either regulator. The team eliminated the assembly family and proceeded on the basis that the dominant cause acted in the components. We discuss this example in more detail in Chapter 11.

Door Closing Effort

High door closing effort was a relatively frequent complaint in a new vehicle owner survey. A team worked on the rear doors where there were more complaints. They measured door closing effort using a velocity meter that determined the minimum velocity (meters per second) necessary to close and latch the door. A baseline investigation showed there was considerable door-to-door variation in the velocity and that doors with high velocity were difficult to close.

Using the baseline data, the team selected two cars with extreme velocity values, one high and one low. The task of complete disassembly and reassembly of the door was difficult, so instead the team decided to remove and replace only the striker. For each car, this was repeated five times using production operators and tooling. The data are shown in Figure 10.14. The initial values are shown by the horizontal lines and indicate the full extent of velocity variation.

There was a large cause acting in both the assembly and components families. Within each car, there was considerable variation as the striker was removed and replaced but not the full extent of variation. The door-to-door average velocity differed considerably due to differences in the components of the two door systems. The team kept both families and worked on each separately. Since the assembly family consisted of only the position of the striker, they first took the striker assembly variation as one large cause and designed a new fixture to control this variation. Using the new fixture, there was a significant reduction in both the velocity average and standard deviation, but not enough to meet the problem goal. They also searched for the dominant cause of the remaining variation in the components family.

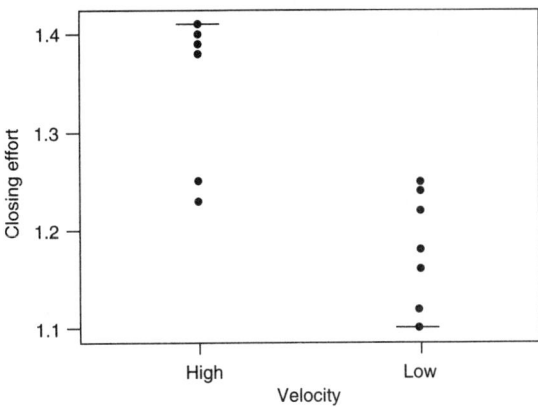

Figure 10.14 Closing effort by velocity (initial values from the baseline given by horizontal lines).

Comments

Figure 10.15 shows three possible plots of the results of a disassembly-reassembly investigation.

In the left panel of Figure 10.15, the dominant cause acts in the assembly family and in the right panel it acts in the component family. In the middle panel, there is an interaction between causes in the assembly and component families. For example, the presence of a burr on a component may make it difficult to assemble the unit so that it performs consistently. If the burr is absent, it is easy to assemble the product to get consistent performance.

There are several cautions when using this plan to compare assembly and component families. It is best to use the original assembly process, tools, and people to avoid study error. Otherwise there is a risk that any conclusions will not be relevant for the actual process. Also, we must ensure that parts are not damaged or changed in the disassembly-reassembly.

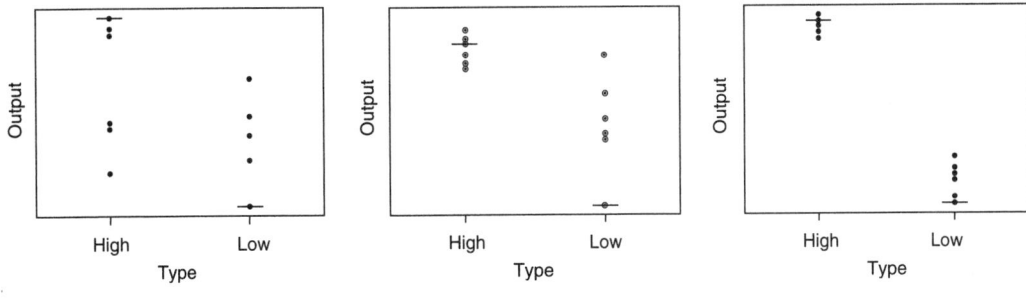

Figure 10.15 Hypothetical results for disassembly-reassembly investigation (initial value from the baseline given by horizontal lines).

> **Comparing Assembly and Component Families Summary**
>
> **Question**
>
> In the current process, does the dominant cause act in the assembly or component family?
>
> **Plan**
>
> - Select two assemblies with opposite and extreme output values relative to the full extent of variation.
> - Disassemble and reassemble each part three or more times. On each occasion measure the output.
>
> **Data**
>
> Record the original output values and each new measurement together with the corresponding part number, one measurement per row.
>
> **Analysis**
>
> - Plot the output by part. Use a special symbol to denote the original output values.
> - Use the plot to compare the assembly variation within each part to the difference between the two original output values.
>
> **Conclusion**
>
> If the assembly variation within each of the two parts is relatively:
>
> - Large, the dominant cause acts in the assembly family
> - Small, the dominant cause acts in the component family
>
> Otherwise the dominant cause involves inputs in both the assembly and component families.

10.5 COMMENTS

Binary Output

With a binary output characteristic and a low rate of defectives, we cannot look at the part-to-part family. In the block leaker problem described in chapters 1 and 6, the team made a false start. They sampled five consecutive blocks for 20 hours. They tested each block for leaks and found two leakers in separate hourly periods. From these data they could not eliminate either the block-to-block or hour-to-hour family because of the low failure rate. With a high-volume process, we can separate the hour-to-hour and day-to-day families by plotting the hourly rate of defectives by day.

Traceability and Order Preservation

In general, to stratify a process by time or process step we must be able to trace the parts measured to a specific time period or processing step. We must also ensure that the sampling plan matches the actual production order. If we call parts consecutive, then they should be processed consecutively as far as possible.

In the engine block porosity problem, the team had to specially mark the molds so that they could find the five consecutively molded blocks after the bank face machining. Once the blocks were poured, they were not necessarily finished and machined in pouring order. In a process that produced brake lines, the problem was a crack in a formed flange that led to leakage when the line was coupled to a valve. The crack was found at final inspection. The output characteristic was the size of the crack. The team considered an investigation to compare the part-to-part and hour-to-hour families. They abandoned this plan because the sequence of parts changed so many times in the process. Consecutive parts at final inspection were not processed consecutively through earlier process steps. The team could not isolate causes that acted in the part-to-part family because of this loss of order.

Formal Analysis

In this chapter we have described simple investigations supported by graphical displays such as scatter plots, box plots, multivari charts, and defect concentration diagrams. Since we are searching for a dominant cause, we can almost always draw the conclusion to eliminate a family without formal analysis procedures. Rarely do we need more formal methods such as the analysis of variance (ANOVA)[7] or regression analysis (see Chapter 12).

What If We Do Not Observe the Full Extent of Variation?

We must observe close to the full extent of variation in the output to draw conclusions from any investigation used in the search for a dominant cause. Otherwise, the dominant cause may not change during the investigation and we may only see the effects of nondominant causes. In QPDAC language, we have a study or sample error.

For example, suppose we are comparing part-to-part and time-to-time families. We measure five consecutive parts produced at the start of each hour and stop sampling after one day, as in the block porosity problem. We plot the data with two different vertical scales in Figure 10.16.

From the left panel of Figure 10.16, we might mistakenly conclude that the dominant cause lives in the hour-to-hour family. This conclusion is premature when we examine the right panel of Figure 10.16, where the dotted lines show the full extent of variation. Whatever the dominant cause and wherever it resides, we have not yet seen its full effect. We need to sample for a longer period or in a different way.

In the engine block porosity example discussed earlier, the team failed to establish a baseline in terms of their new porosity measure. As a result, they did not know the full extent of porosity variation. They fortunately were able to link the results of the multivari with the baseline, established in terms of reject rate. They then assumed the full extent of porosity variation was captured in the multivari investigation.

We can guarantee the full extent of variation by either using baseline data or using the idea of leverage where we compare units that are extreme with respect to their output values. We use leverage, for example, in investigations to compare the assembly and component families.

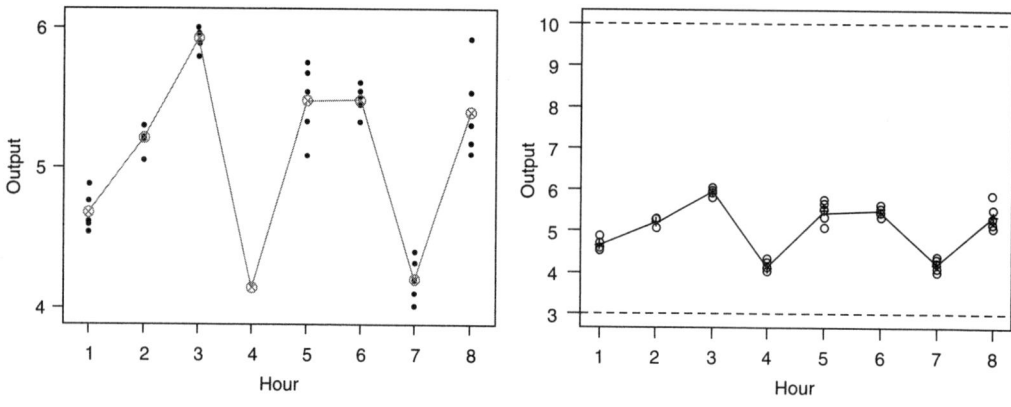

Figure 10.16 Two multivari charts of the same data with less than full extent of variation (dashed horizontal lines give the full extent of variation from the baseline).

Some Investigations Show Changes in Variation

Figures 10.2 and 10.9 showed that output variation changed by lobe and by time, respectively. Such patterns provide valuable clues about the dominant cause. For instance, in Figure 10.9, the difference in porosity variation of the output at hours 5 and 8 may be due to:

- A difference in the average or range of a dominant cause at different time intervals
- A different relationship between the cause and output in different time intervals
- A combination of these two reasons[8]

In both the examples, the change in variation was accompanied by a change in the average output level. To examine a situation where only the output variation changes, consider a hypothetical example. Suppose in the V6 piston diameter example discussed in Chapter 9, we collect 30 pistons from each stream over one shift so that we see the full extent of variation. We plot the results in Figure 10.17.

The average diameter is roughly the same for the two streams, so stream-to-stream differences are not a dominant cause of variation. However, stream B shows substantially greater diameter variation than does stream A. This pattern of unequal variation tells us a lot about the dominant cause. The dominant cause must act within the stream B family. We have eliminated all causes in the stream-to-stream and within-stream-A families. Note that this conclusion is more specific than the usual interpretation from stratification where we are able to rule out either the within-stream or stream-to-stream families.

What kinds of causes can explain the behavior seen in Figure 10.17?[9] Two possibilities are that there was a worn fixture in stream B or that the stream B operator made more frequent adjustments.

Chapter Ten: Investigations to Compare Two Families of Variation 147

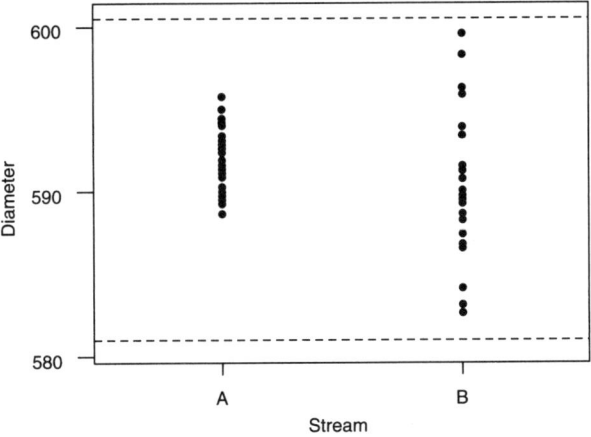

Figure 10.17 Box plot of diameter by Operation 270 stream from hypothetical investigation (dashed horizontal lines give the full extent of variation from the baseline).

 **Key Points**

- Simple investigations and graphical analysis tools can provide valuable clues about the home of a dominant cause of variation.
- We often find it useful to divide the remaining possible causes into two families of causes such as:

  –Short-time and long-time
  –Group-to-group and within-group
  –Upstream and downstream
  –Assembly and components

  and then carry out a simple investigation to eliminate one family.
- To avoid study and sample errors, we must observe the full extent of variation in the investigation.

Endnotes (see the Chapter 10 Supplement on the CD-ROM)

1. If we have two families and stratified data, we can use analysis of variance (ANOVA) to partition the overall standard deviation into two pieces, one associated with each family. We look at the details in the supplement.

2. By understanding the possible reasons for an observed pattern in the results of an investigation, we can eliminate many suspects. In the supplement, we look more closely at possible explanations for the changing average and variation pattern seen in figures 10.2, 10.9, and 10.17.
3. We identify a dominant cause primarily using graphical displays. We do not use a formal hypothesis test. In the chapter supplement we elaborate on why hypothesis tests are not appropriate in the search for a dominant cause.
4. See note 2.
5. We used a simple plan to compare upstream and downstream families. We examine operations swap and randomized sequencing, two more restrictive plans for separating these families.
6. By selecting units with extreme performance, we can compare the assembly and component families with a very limited number of disassembly-reassembly cycles. We examine this application of leverage in more detail in the supplement.
7. See note 1.
8. See note 2.
9. See note 2.

 **Exercises are included on the accompanying CD-ROM**

11

Investigations to Compare Three or More Families of Variation

Things which matter most must never be at the mercy of things which matter least.

—Johann Wolfgang von Goethe, 1749–1832

In Chapter 10, we compared two families of causes using simple plans and analysis tools. In this chapter, we consider investigations for simultaneously comparing three or more families to identify the home of a dominant cause of variation. We also consider the repeated application of investigations based on two families.

11.1 MULTIVARI INVESTIGATIONS: COMPARING TIME- AND LOCATION-BASED FAMILIES

For many processes, we can construct families of causes based on time and location. Suppose the goal of a project is to reduce variation in the weight of a molded part. The process has two four-cavity molds that operate in parallel, as shown in Figure 11.1.

We consider the following families:

Family	**Description**
Mold-to-mold	Causes that explain differences in weight from one mold to the other
Cavity-to-cavity	Causes that explain differences in weight among the cavities
Part-to-part	Causes that explain differences in weight among consecutive parts
Hour-to-hour	Causes that explain differences in weight from one hour to the next

150 Part Three: Finding a Dominant Cause of Variation

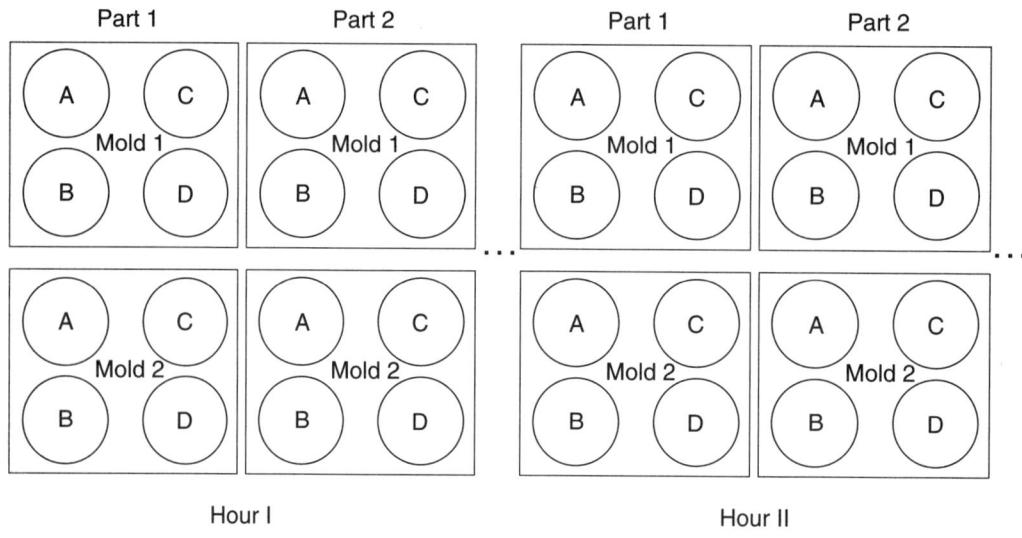

Figure 11.1 Cavity-to-cavity, mold-to-mold, part-to-part, and hour-to-hour families.

To isolate the effects of each family, we sample from the process in a systematic manner. We include parts in the sample that are produced consecutively and over hours from all molds and cavities. For example, we might select and measure five consecutive parts from each mold and cavity (40 parts in total) every hour for several hours.

With the proposed sampling plan, we can also detect interactions between the families. For example, we can see if mold-to-mold differences change from hour to hour.

We call this a multivari investigation (Seder 1950a, 1950b, 1990). See also Zaciewski and Nemeth (1995) and Snee (2001). We design the sampling protocol to see the variation due to the causes within each family of interest. We must be able to trace the parts according to the order and location in which they are produced. In the example, each part was labeled by cavity and mold. We may have more difficulty finding parts that were molded consecutively at the end of the process.

We use multivari charts (modified run charts) to display the results of a multivari investigation. Using MINITAB, we can choose to display the effects of up to four families on the multivari chart at the same time. Since charts with many families are difficult to interpret, we prefer to use a series of charts, each with fewer families. We can use analysis of variance (ANOVA) to quantify the effects of each family. ANOVA is useful when we cannot easily determine the dominant family from the multivari charts.[1]

We expect the causes in some families to act systematically. For example, if the dominant cause acts in the cavity-to-cavity family, we will see large differences in the average output across the cavities. In other families such as the part-to-part, we expect the causes to act haphazardly. We would be surprised to see large differences between the output averages (averaged over all cavities, times, and molds) of the first and second part sampled in different hours.[2] The effects of causes that change from part to part will likely produce haphazard variation. We need to plot the data carefully to avoid masking this type of variation.

We provide three examples of multivari investigations to illustrate the sampling plan, the presentation of the data, and the interpretation of the results.

Cylinder Head Scrap

Iron cylinder heads were cast in a green sand process using a four-cavity mold. In the jargon of the foundry, each cavity is called a pattern. The project was to reduce scrap at the machining operation due to a defect called *shift*. Shift occurs when the top and bottom parts of the sand mold are not properly aligned. Excessive shift causes problems with part location in the subsequent machining operation.

To measure shift, there are locator pads on the top and the bottom of the casting. Shift is determined by measuring the relative distance between the two pads in the x (side shift) and y (end shift) directions. The team recorded both measurements in inches for each part. Here we look at side shift only. The target value is 0.

The team carried out the baseline investigation by haphazardly selecting and measuring 20 castings per day from each of five days' production. The data are given in the file *cylinder head scrap baseline*. Since it was difficult to trace parts through the process, they recorded only the day of production.

We plot side shift stratified by day in Figure 11.2. The average side shift was –0.009 inches with a standard deviation of 0.017. The values varied from –0.050 to 0.035 inches. The team noted that the within-day variation was large.

The team goals were to better center the process and reduce side shift standard deviation by at least 50%. They expected these changes to reduce scrap substantially at the machining operation.

The team used a coordinate measuring machine (CMM) to make the measurements. They checked the system with a gage R&R investigation, carried out over a short time period, contrary to the recommendations in Chapter 7. The observed measurement variation was small relative to the baseline.

The team decided to look for a dominant cause and planned a multivari investigation to rule out a large number of suspects. They considered the four families shown on the diagnostic tree in Figure 11.3.

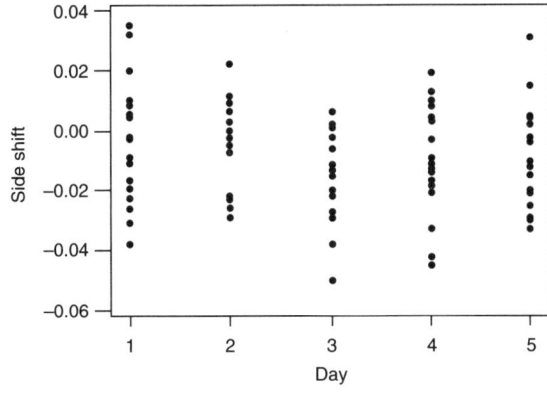

Figure 11.2 Plot of side shift by day.

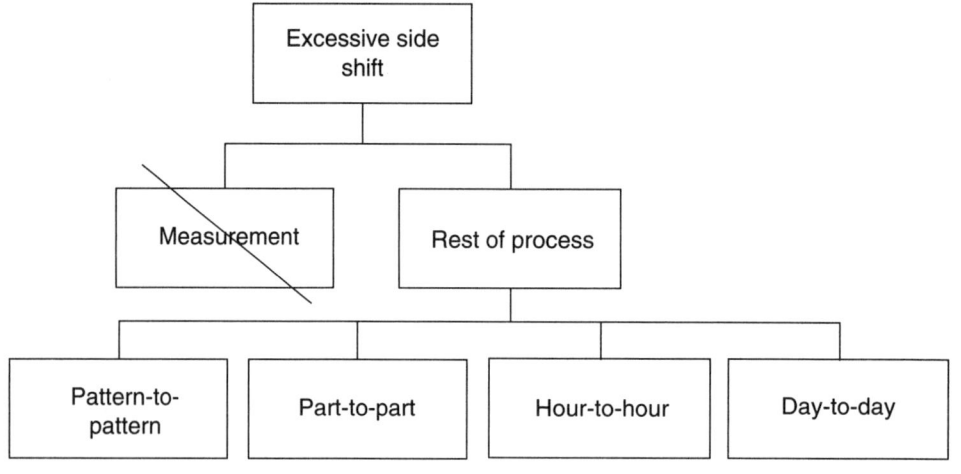

Figure 11.3 Diagnostic tree for side shift.

In order to assess the variation contributed by each of these families, they used the following sampling protocol that required tracking and measuring 96 parts:

1. Sample and measure four parts cast consecutively from each pattern.

2. Repeat step 1 for castings poured at hours one, three, and five of the day shift.

3. Repeat step 2 on two consecutive days.

With this sampling protocol, the team could isolate the contributions to the baseline variation from each family in the diagnostic tree. For example, sampling four consecutive parts showed whether the variation was dominated by a cause that changed quickly from one part to the next, such as the relative positioning of the two halves of the mold onto the guiding pins and bushings.

From the baseline investigation, the team expected the full extent of variation to occur within the two days. The major difficulty in implementing the plan was tracing castings molded consecutively. The team marked the sand molds so that they could find these castings at the end of the shakeout and cleaning process.

Side and end shift for the 96 measured parts are stored in the file *cylinder head scrap multivari*. For side shift, we see close to the full extent of variation in the histogram shown in Figure 11.4, so we know the dominant cause acted during the investigation.

We look at a variety of plots. To simplify the data display, we start by creating a new input called time that combines hour and day. See the section on multivari charts in Appendix C.

In Figure 11.5, the lines join the average side shift at the six times and the dots give the 16 side shift values at each time. We recommend that you always select the option (unfortunately not the default) to plot the individual values on MINITAB multivari charts. From Figure 11.5, we see:

- The time averages are almost constant; that is, there are no systematic differences in the side shift between days or hour to hour within days.

- The variation of side shift within each time is close to the full extent of variation.

Chapter Eleven: Investigations to Compare Three or More Families 153

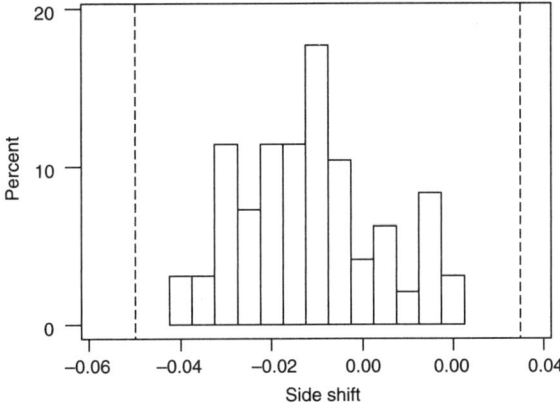

Figure 11.4 Histogram of side shift values from the multivari investigation (full extent of variation in the baseline given by dashed lines).

The observed pattern cannot be explained by a cause that changes slowly from time to time so we are tempted to eliminate the hour-to-hour and day-to-day families. We wait to do so until we look for possible interactions with the other families.

In the left panel of Figure 11.6, we look at a multivari chart with side shift stratified by pattern. From the chart, we see large systematic differences among the patterns. The variation within each pattern is only about 60% of the full extent of variation. The differences in the pattern averages contribute substantially to the baseline variation. We conclude that the pattern-to-pattern family contains a dominant cause.

To deal with the part-to-part family, we need to recognize that variation from consecutive molds is likely to be haphazard. We create a new input called *group* with a different value for each combination of the four patterns and six times. There are four side shift values within each of the 24 groups corresponding to the consecutive castings in that group.

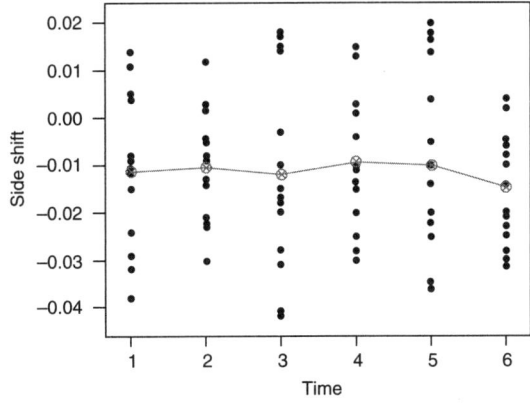

Figure 11.5 Multivari chart of side shift by time.

154 *Part Three: Finding a Dominant Cause of Variation*

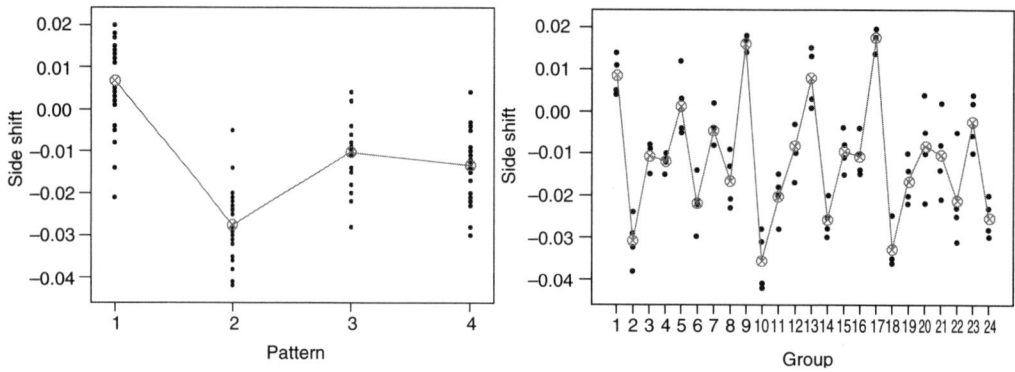

Figure 11.6 Multivari charts for side shift versus pattern on left, side shift versus group on right.

In the multivari chart in the right panel of Figure 11.6, the first six groups correspond to the six time points for the first pattern. The next six correspond to the second pattern, and so on. The chart shows that the variation within each group, as represented by the four dots, is small relative to the baseline variation. Alternatively, we see that the side shift averages vary substantially from group to group. The pattern-to-pattern family is included in the group-to-group family, which explains the group-to-group differences in side shift average seen in the right panel of Figure 11.6. We also see that the within-group variation is roughly constant over all sampling periods, which means that there is no evidence of interaction between the part-to-part family and the other families of causes. We eliminate the part-to-part family from further consideration.[3]

We now look for possible interactions with the pattern-to-pattern family by constructing multivari charts with pattern and time. In Figure 11.7, we show the side shift averages and individual values for each time and pattern. The effects of the time family are small for each

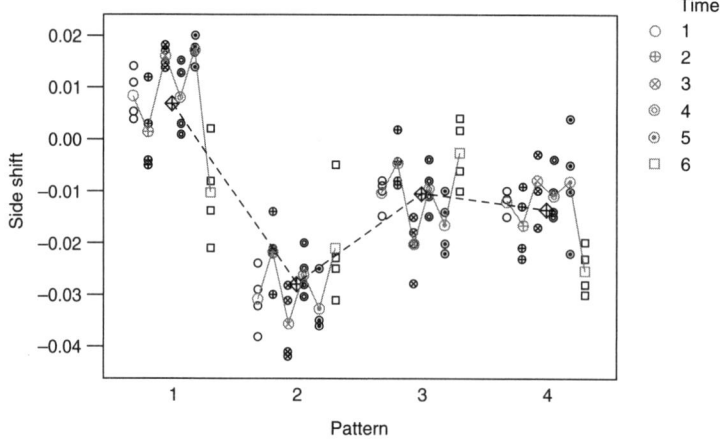

Figure 11.7 Multivari chart of side shift versus pattern and time.

pattern. There is no evidence of an interaction between causes in the pattern-to-pattern and time-to-time families.

We conclude that the dominant cause lives in the pattern-to-pattern family. To reduce the variation, we need to focus on causes within this family.

The team knew how to change the pattern averages by making a one-time adjustment to the dies. From the left panel of Figure 11.6, they saw that if they aligned all the patterns they would almost meet the project goal in terms of reducing side shift standard deviation. Based on the side shift averages in the multivari investigation, they recommended that the dies for the top half of the mold be moved to better center the process and remove the pattern-to-pattern differences. They also planned a weekly capability study to examine side shift by pattern to ensure that the averages did not drift apart again.

Camshaft Journal Diameter

In Chapter 7, we introduced a project to reduce scrap and rework due to excess variation in the diameter of camshaft journals. At final 100% inspection, the four journal diameters (the maximum diameter as the part was rotated) were measured at two locations, front and rear. The diameter specifications were ± 12.5 microns, measured from a target value. There were also specifications on out-of-round and taper for each journal. Camshafts that did not meet the specifications at the final gage were sent to a rework station. There the parts were remeasured and either scrapped or reworked. At the beginning of the project, the average monthly reject rate at the final gage was 4.7% of which 1.5% was scrap. Management set a goal to cut both these rates in half.

The team focused on journal diameter variation. There was no data storage at the final inspection gage. To establish a baseline, the team had two or three camshafts per hour set aside before being measured at the final gage to get a sample of 20 parts per day. At the end of the day, an operator measured and recorded the diameters for the four journals and two locations on each camshaft. This process was repeated for five days. The 800 measured values (from 100 camshafts) are stored in the file *camshaft journal diameter baseline*. The average diameter was 2.41 microns, the standard deviation was 5.00 microns, and the corresponding capability (P_{pk}) was 0.67. The operators had centered the process above zero to avoid scrap at the expense of rework. There were six parts needing rework and one scrapped in the baseline sample. The run chart of diameter over time did not show any clear patterns. The full extent of variation was –12.5 to 17.5 microns. The team set a goal to improve the process capability (P_{pk}) to more than 1.20.

The investigation of the final gage, described in Chapter 7, revealed that there was a significant bias in the head that measured the front diameter on the first journal. The team removed this bias with the expectation that the process variation would be reduced. They decided to repeat the baseline investigation before proceeding. The data are stored in the file *camshaft journal diameter baseline2* and the performance is summarized in Figure 11.8. The average, standard deviation, and P_{pk} for the second baseline investigation were 2.55, 4.53, and 0.73, respectively. Five out of the 100 parts required rework because of oversize diameters. The full extent of variation in the new baseline was –11.8 to 15.9 microns.

Before choosing a working approach, the team decided to look for a dominant cause of journal diameter variation. We give a high-level process map in Figure 11.9.

156 Part Three: Finding a Dominant Cause of Variation

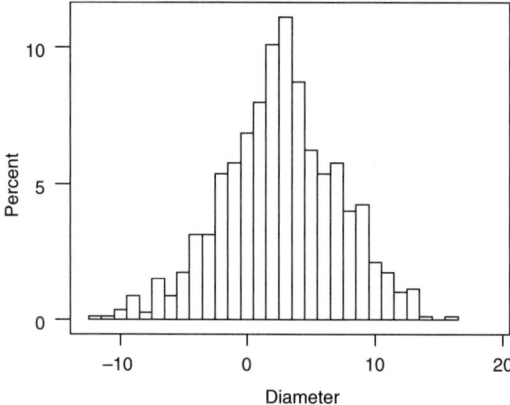

Figure 11.8 Histogram of journal diameter for second baseline investigation.

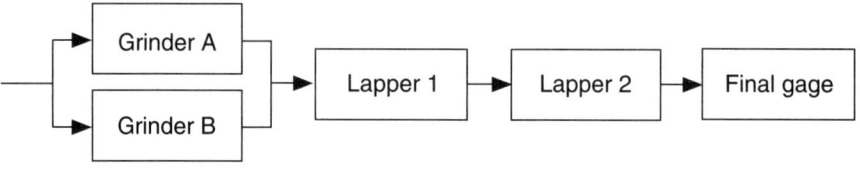

Figure 11.9 Camshaft production process map.

Blank parts arrived in batches from a supplier and were haphazardly assigned to one of the two grinders. The grinders used a feedback control system in which 1 part in 10 was measured (journal 1 only) and the grinders were adjusted if the diameter was out of specification. The within-process specification limits at the grinder were set based on the assumption that the lappers would reduce the diameter by 24 microns. Each grinder had its own gage.

The team planned a multivari investigation to examine the following families:

Family	Description
Position-to-position (within-part)	Causes that explain differences in diameter from one position to the other within the same camshaft
Grinder-to-grinder	Causes that explain differences in diameter between grinders
Part-to-part	Causes that explain differences in diameter among consecutive camshafts
Hour-to-hour	Causes that explain differences in diameter from one hour to the next
Batch-to-batch	Causes that explain differences in diameter from one batch of incoming blanks to the next

They decided to use three batches (about one day's production) of blank parts and to sample parts once every two hours. They selected five camshafts from each grinder within

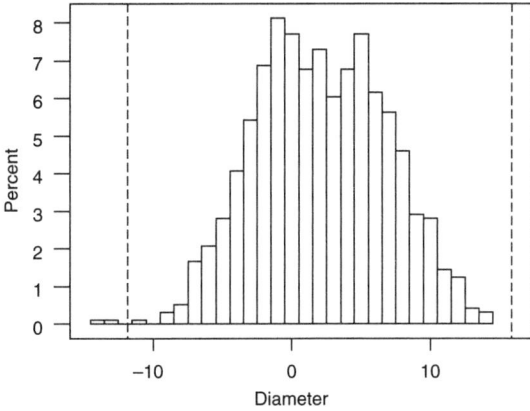

Figure 11.10 Diameter variation in the multivari (dashed lines give the full extent of variation).

an adjustment cycle (part numbers 1, 3, 5, 7, and 9). These parts were specially marked so that they could be found at the final gage. They planned to collect 120 camshafts and make 960 diameter measurements in total. The team carried out the plan without any difficulties.

The data are recorded in the file *camshaft journal diameter multivari*. We see the full extent of variation in the histogram shown in Figure 11.10. We know the dominant cause acted during the investigation.

We give the multivari charts for position, grinder, and the combined hour and batch families in Figure 11.11. We expect these families to show systematic differences in the averages if they contain a dominant cause. There is a large difference between the two grinder averages indicating a possible dominant cause. There is no evidence of a dominant cause acting in the other families.

Since a dominant cause appears to act in the grinder-to-grinder family, we look for an interaction with the position or batch/hour families using the multivari charts in Figure 11.12. Since the effect of grinder does not appear to depend on position or batch/hour, we see no evidence of any interactions.

We expect the part-to-part variation to be haphazard.[4] To display the magnitude of this variation, we create a new input called group that indexes the 192 sampling points (8 positions by 2 grinders by 3 batches by 4 hours). Since there appears to be a large cause in the grinder-to-grinder family, we define the group input so that the first 96 values correspond to grinder *A* and the second 96 values correspond to grinder *B*.

We give the multivari chart of diameter by group in Figure 11.13.

This chart is difficult to interpret because there are so many groups. However, we can see that the variation around the averages is large, which indicates substantial part-to-part variation. The magnitude of this variation is roughly the same for each group of consecutive diameter measurements at a particular position and from grinder to grinder, which indicates a lack of interaction. We find ANOVA useful here to quantify the effects and augment the multivari charts.[5] We conclude there are dominant causes within the grinder-to-grinder and part-to-part families.

The team concentrated their efforts on these two families in the search for the dominant causes. The team was surprised by the grinder-to-grinder differences. They had not seen this

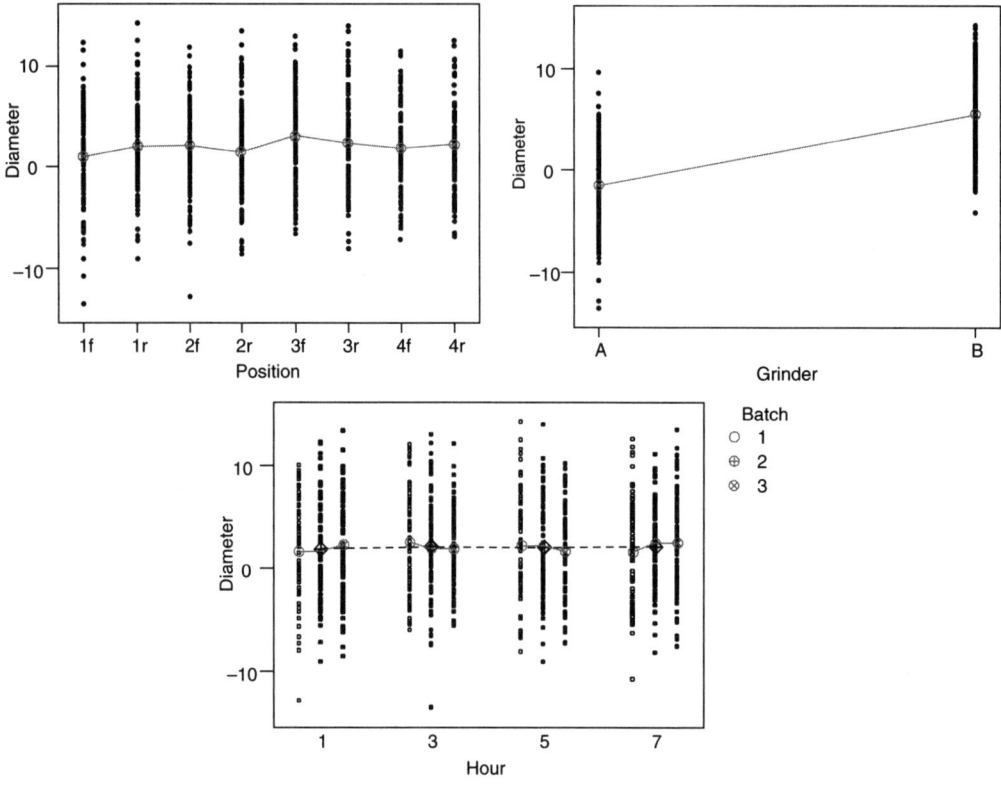

Figure 11.11 Multivari charts of diameter by position, grinder, and batch/hour.

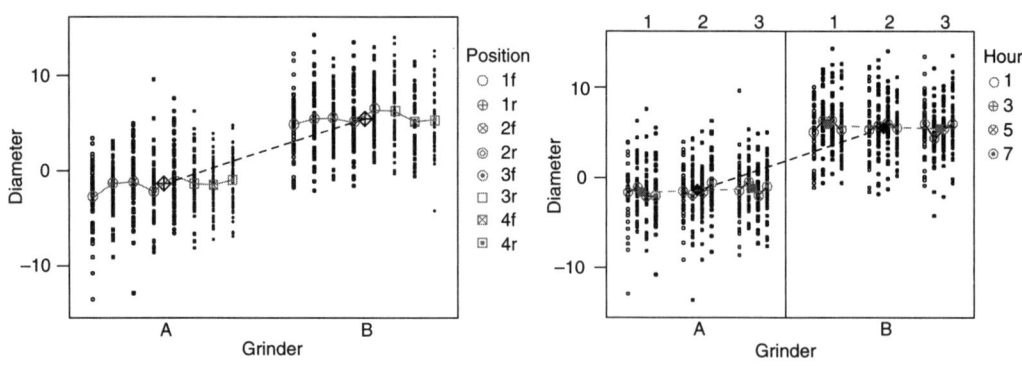

Figure 11.12 Multivari charts of diameter by grinder versus position and batch/hour.

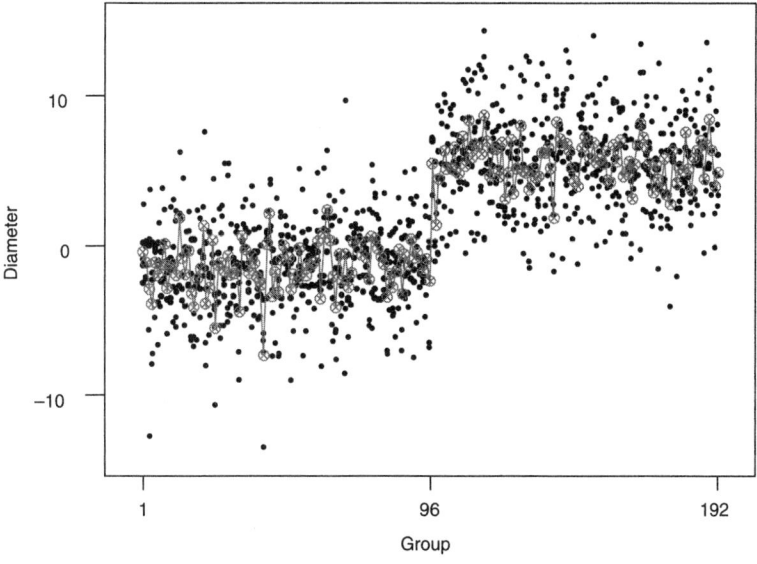

Figure 11.13 Multivari charts of diameter by group.

difference in regular production data because normally there was no traceability to the grinder.

Fascia Cratering

Management assigned a team to reduce scrap and rework on front and rear fascias because of cratering. This was a new product, and during the startup period the reject rate was 8.9%. The goal was to reduce this rate to less than 1%. The crater defects were not visible directly after molding but they could be clearly observed after the fascia was primed. The crater rejects occurred over all shifts and were haphazardly distributed over time. The team developed and checked a measurement system to count the number of craters after priming.

In a baseline investigation, the team found that about 80% of the fascias had no craters, 10% had fewer than 25, and the remaining 10% had more than 25 craters. They decided to search for a dominant cause of the craters. From the baseline sample, the team constructed a defect concentration diagram that revealed that 75% of the craters were located in the front half of the fascia. However, the team could not see how to use this knowledge to rule out any family of causes.

The fascias were produced on two different molding machines. The team planned a multivari investigation to examine the fascia-to-fascia, machine-to-machine, and time-to-time families. They sampled five consecutive fascias from each of the two machines every four hours. They continued the sampling for three shifts (24 hours) resulting in a sample of 60 fascias. To keep track of the consecutive parts and machine numbers, they marked each fascia on the inside immediately after molding.

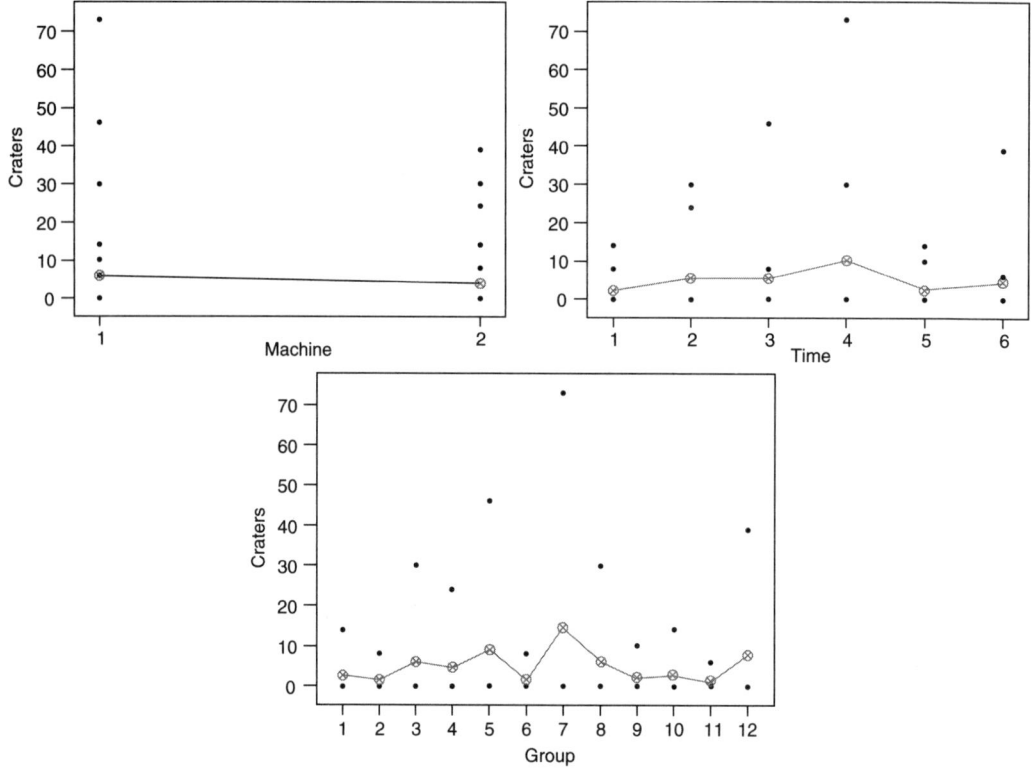

Figure 11.14 Multivari chart from first fascia cratering multivari investigation.

The data are given in *fascia cratering multivari*. The distribution of the number of craters matches the baseline closely. To look at the fascia-to-fascia (part-to-part) family, we create a new input group, with a different value for each combination of machine and time. We give multivari charts for the time-to-time, machine-to-machine, and fascia-to-fascia families in Figure 11.14.

There is no evidence that the dominant cause acts in the machine-to-machine or time-to-time families. There is large fascia-to-fascia variation in some of the groups and little in others.

An observant team member noticed that there were only two values plotted for each group of five fascias. When the team examined the data more closely, they saw that there was only one fascia in each group with a value different from 0. This was a very strong clue about the dominant cause because the team could only think of one process step where something happened only once in every five parts. The operator sprayed mold release every 10 pieces. After five pieces he applied a minor spray. Mold release was used to prevent tearing when the fascia was removed from the mold.

In a follow-up investigation, the team sampled 10 consecutive fascias in the molding sequence at three different times in the morning shift. The data are given in the file *fascia cratering multivari2*. The team confirmed mold release as the dominant cause of cratering as shown in the Figure 11.15. With knowledge of the dominant cause, the team needed to choose a working approach. We continue the story in Chapter 14.

Chapter Eleven: Investigations to Compare Three or More Families 161

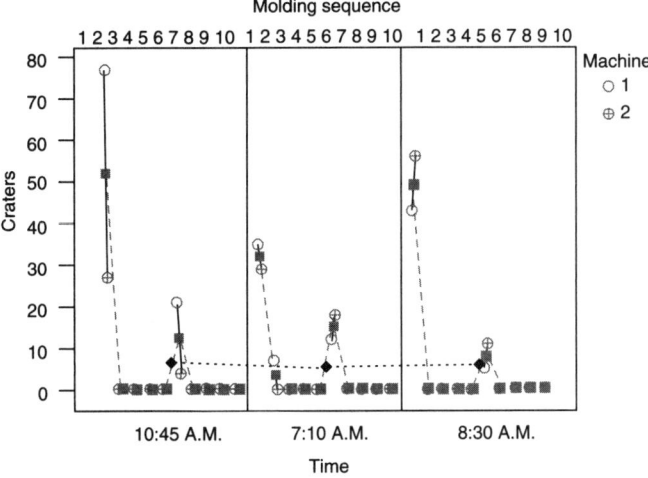

Figure 11.15 Multivari chart from second fascia cratering multivari investigation.

Note that in this example the fascia-to-fascia family of causes results in systematic variation and not haphazard variation as in most applications where we have a part-to-part family.

Multivari Investigation Summary
Select families to investigate by grouping the remaining suspect dominant cause. **Question** For the current process, which of the selected families is home to a dominant cause? **Plan** • Specify a study population long enough to capture the full extent of variation. • Select a sample using a systematic sampling protocol designed to capture the variation due to the chosen families. • Spread out over the study population. **Data** Record the output value and the corresponding families, one output per row. **Analysis** • Plot a histogram of the data. Check that the range of output values covers most of the full extent of variation. • For families likely to have systematic effect, construct the single-family multivari charts, plotting individual output values.

- For the part-to-part family (or other families expected to have haphazard effect), create a new input group so that the effect of the family is seen in the variation within each value of group. Construct the multivari chart with *group*.
- For each family identified as having a large effect, construct two family multivari charts to look for evidence of interaction.

Conclusion

- Identify the dominant family (or families if the dominant cause acts in two or more families).

Comments

A multivari investigation is a powerful tool for assessing the contribution of families defined based on time or location.

In planning a multivari investigation, keep the number of families to five or fewer so that the sampling plan and analysis are not too complex. The camshaft journal diameter multivari illustrates the difficulties.

In a multivari investigation, we do not use random sampling; instead we select parts deliberately to estimate the effects of the specified families. We may need to sample additional parts beyond those collected for the normal control or monitoring of the process.

For any family in a multivari investigation, we need traceability. In the cylinder head example, we were interested in the mold-to-mold family, that is, variation in consecutive pieces from the same pattern. To assess this family, we must be able to identify the heads produced consecutively from each pattern when we measure the output. If we cannot trace parts through the process easily, then for time-based families, we need first-in, first-out discipline at each process operation so that the time sequence is not lost when we select parts at the end of the process.

Multivari investigations are not effective for binary output characteristics unless the rate of defectives is high. In the block leaker project, introduced in Chapter 1, the baseline rate of leaking blocks was 2.2%. The team initially conducted a multivari investigation in which five consecutively poured blocks were tracked through the finishing process every two hours for several days. After 160 blocks, they found only one leaker. There was no useful information in the multivari investigation for this rare defect.

Sometimes the recommended analysis process fails because we do not find a dominant family when we look at one family at a time. In this case, we recommend constructing all two-family multivari charts and looking for interactions.

If the multivari charts fail to reveal the dominant family, we suggest a formal analysis of variance (ANOVA) to quantify the relative contributions of each family.[6] We may then decide to search for causes within one or more of the families with the largest contributions, or we may abandon the partition of causes in these families and start over with a new set of families.

11.2 COMPARING FAMILIES DEFINED BY PROCESSING STEPS

Most manufacturing processes consist of several steps. We can form families of causes based on inputs that change within each of these steps. In Chapter 10 we looked at some examples where the process was split into two families, causes acting upstream and downstream of a particular point in the process. Here we consider applying this idea repeatedly.

We consider two types of plans:

- Measure the output characteristic on the same parts after each processing step.

- Allow some parts to skip processing steps and measure the final output characteristic.

The first type of plan can be applied more commonly than the second. Consider the following examples.

V6 Piston Diameter

In Chapter 5, we described a problem to reduce variation in the diameters of machined aluminum pistons. In the baseline investigation, the final diameter after Operation 310 varied between 581 and 601 microns, measured from a fixed value. In searching for a dominant cause, the project team decided to define families based on a number of processing steps as shown in Figure 11.16. The processing steps in Figure 11.16 include all those that affect diameter.

To eliminate as many processing steps as possible, the team tracked 96 pistons through the process, measuring the diameter after Operation 200 and after each of the steps shown in the process map. In total they measured each piston diameter six times. The team carried out the investigation over three days, 32 pistons at a time. During this period, the process operators collected the marked pistons for measurement after each operation.

From the baseline investigation, the team expected that they would see close to the full extent of variation over the three days. They decided to use only 32 pistons each day to limit the disruption to the normal operations. The team used the in-process gages for the measurements after Operation 200 to Operation 260 and the Operation 310 gage for the final three measurements.

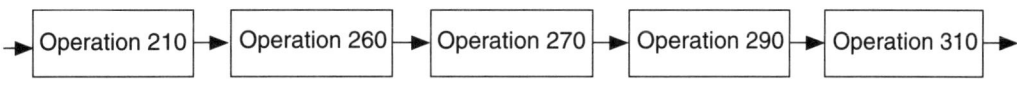

Figure 11.16 Piston machining process map.

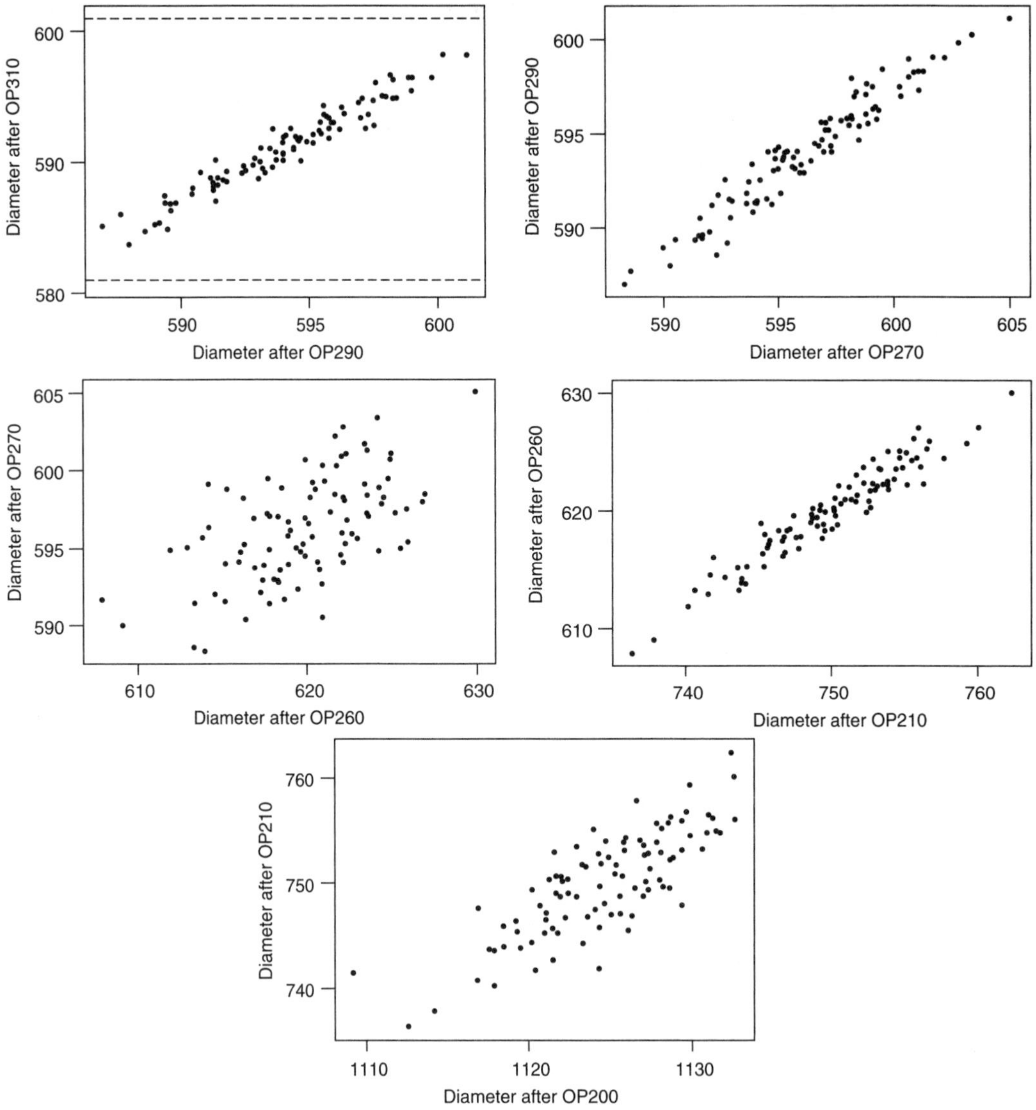

164 Part Three: Finding a Dominant Cause of Variation

Figure 11.17 Scatter plots of outgoing versus ingoing diameter (by operation; dashed horizontal lines give the full extent of variation).

The team converted all measurements, given in the file *V6 piston diameter variation transmission*, to the same units as at the final gage. The team produced the scatter plots of the incoming and outgoing diameter for each operation, as shown in Figure 11.17. They added horizontal lines to show the full extent of variation to the plot that shows the final diameter.

We interpret the scatter plots in Figure 11.17 starting from the end of the process. From the top left diagram, we see the full extent of variation in the final output. We also see that most of the variation in the final diameter is transmitted through Operation 310. If we could eliminate the variation in the diameter after Operation 290 so that the output is, say, 595, then from the graph we see there would be very little variation in the diameter after Operation 310. That is, little diameter variation is added in Operation 310; most of the variation is transmitted from upstream. We can eliminate Operation 310 as the home of a dominant cause. From the top right plot, we see similarly that most of the variation in the diameter after Operation 290 is due to variation at Operation 270 and, hence, we can also eliminate Operation 290.

The left plot in the second row of Figure 11.17 has a different pattern. At Operation 270 there is substantial variation added to the process. If the diameters after Operation 260 were fixed at, say, 620, there would still be substantial variation after Operation 270. Hence we eliminate all operations upstream of Operation 270 and conclude the dominant cause of the variation in the final diameter is in the Operation 270 family. Next, the team focused their efforts on Operation 270 to isolate the dominant cause of variation.

In this example, it was possible to measure the diameter of each piston at various points throughout the process. The pistons were traceable through the process by implementing special measures during the investigation. Also, in this example, the team measured the diameter after each operation in a single investigation. They could have adopted another strategy and, for example, measured the diameter after Operation 270 and after Operation 310 only. Then, they would repeat the investigation, splitting the remaining family. This way they would compare only two families at a time, as in Chapter 10. In the example described here, it was easier to track pistons through in a single investigation process.

We can use regression analysis to quantify the amount of variation added and transmitted at each stage if the pictures do not tell a clear story.[7]

Roof Paint Defect

In a painting process, there was a defect, called a *line*, on the roof of a finished car. The defect rate was about 10% and the repair costs were high. In an earlier application of the algorithm the team discovered that by reducing the film build (thickness) of the clear coat, the last step in the multistep painting process, the defect rate was reduced to less than 1%. However, this process change resulted in an inferior overall appearance of the painted surface. The team decided to reapply the algorithm to find a solution to the line defect problem that would allow a higher film build.

The team used the baseline defect rate of 10%. The goal was to achieve a defect rate of less than 1% with the original specifications for the clear coat film build.

The team decided to look for a dominant cause of the defect. They could not restore the clear coat film build specification in the production process because of the high cost of rework. As a consequence, the team decided to use test panels instead of car roofs as the study units because they could apply a high film build to the panels as in the original process. They mounted the panels on the roof of a dummy vehicle where the line defect normally occurred. To check this procedure, the team ran a set of panels through the process with the clear coat film build above its original level. Under these conditions, the defect occurred on almost every panel. The team was confident that knowledge acquired using the panels could later be applied to vehicles. In our language, they felt that there would be little study error.

The painting process has five major steps as seen in Figure 11.18.

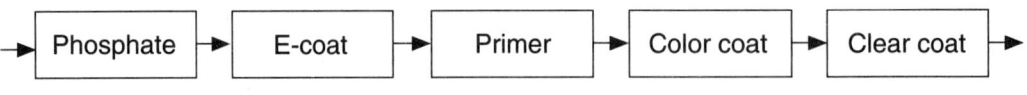

Figure 11.18 Painting process map.

The team decided to search for a dominant cause by first isolating where in the process the defect occurred. They divided the causes into five families corresponding to the major processing steps.

In the first investigation, the team masked three panels until after the e-coat step. Three other panels were processed normally. The defect occurred on all six panels. The team eliminated the incoming metal, the phosphate step, and the e-coat step from consideration. In a second investigation, they processed another six panels normally up to the end of the e-coat step. Three of the panels were processed through the primer stage but not the color coat. The other three were not primed but color coated as normal. Clear coat was applied to all the panels. Again the defect was present in all six panels. The team concluded that the primer and color coat steps could be eliminated and hence the clear coat process was the home of the dominant cause. This was not surprising given the first solution to the problem.

The two major substeps within the clear coat process were paint application and oven cure. To split these two steps, three panels were removed from the test car after painting and were cured in a laboratory oven. None showed the defect. The remaining three panels were processed normally and as expected, all showed the defect. The dominant cause lived in the clear coat cure oven. The team continued to split the process using similar trials. Eventually, they isolated the airflow from exhaust ducts within a particular zone of the oven as the dominant cause of the line defect.

The team then carefully tested a minor modification to the ducts that would change the airflow pattern over the roof of painted vehicles. Twenty vehicles were painted in the modified process with high clear coat film build. There were no line defects and no other noticeable negative side effects. The change was made permanent. The clear coat film build was increased with a marked improvement in appearance. There were no occurrences of the line defect in the new process.

In this problem, the team repeatedly split the process and homed in on the dominant cause. Because of the nature of the process and the defect, they could skip or alter steps of the process and still produce painted panels with or without the defect. This would not be possible in many processes.

Variation Transmission Investigation Summary

Question

For the current process, which process step is the home of a dominant cause?

Plan

- Specify a time frame over which we expect to see the full extent of variation in the output.
- After the first process step, select a sample of 30 or more parts spread over the time frame.

- Measure the characteristic corresponding to the output on each part in the sample after each process step.

Data

Record the measured output values, one row for each part.

Analysis

- For each process step (after the first), plot the output after the step versus the output before that step.
- For the plot showing the output from the final process step, add horizontal lines showing the full extent of variation.

Conclusion

- If the full extent of variation is not observed, the process step containing the dominant cause cannot be identified.
- Starting at the last process step, eliminate any step where most of the variation is transmitted. The first process step where the variation is not transmitted is the home of the dominant cause.

11.3 COMPARING COMPONENT FAMILIES

In Chapter 10, we looked at problems defined in terms of the function of an assembled product. We separated the contribution of the assembly and the component families by repeatedly disassembling and reassembling two assemblies. If the dominant cause lies with the components, we can conduct further investigations to pinpoint the particular component family that is home to the dominant cause. In this section we look at such investigations.

We call the individual pieces that make up the assembly *components*, a specified set of components a *group*, and all components known not to contain the dominant cause the *housing*. To start, there may be no housing if all components are suspect. The number of components in the housing grows as we eliminate components.

To compare component families, we find two assemblies with opposite and extreme performance in terms of the output of interest; that is, we use leverage as discussed in Section 9.3. With this plan the output of the two assemblies covers the full extent of variation. Then, we repeatedly swap groups of components between the two assemblies and remeasure the output.

We demonstrate the conduct, analysis, and interpretation of a component swap investigation with two examples.

Headrest Failure

Management initiated a project to reduce the frequency of customer complaints about seat headrests that would not stay in a set position. In focusing the problem, the team demonstrated that the problem could be solved by eliminating headrests that required a force of less than 35 Newtons (N) to move through their full range of motion.

The team carried out a baseline investigation. To meet their goal, they wanted to reduce variation in force rather than increase the average force to avoid headrests that would be very hard to move; that is, they wanted to change only those headrests with low output. They assessed the force measurement system and found that there was little contribution from this system to the baseline variation. They decided to search for a dominant cause of force variation.

The team selected two seat assemblies, one with high force (70.3 N) labeled H, and one with low force (25.8 N) labeled L, relative to the baseline. The seat assembly consists of a headrest and a seat. To assess the assembly process (as in Chapter 10), the team removed and reinstalled each headrest three times in its original seat. The force was measured after each reinstallation. The data are:

Original force	After first reassembly	After second reassembly	After third reassembly
H (70.3)	72.5	74.2	71.9
L (25.8)	25.9	24.3	25.4

Because there were only small changes after reinstallation, the team ruled out the assembly process and concentrated on the components. Next, they swapped the two headrests between the seat assemblies and measured the force. For this and subsequent component swaps we display the data in a table where:

- Diagonal cells contain the force measurements from the assembly-reassembly phase.

- Off-diagonal cells contain the force measurements after the components are swapped.

For the example the data are:

Seat	Headrest	
	H	L
H	70.3, 72.5, 74.2, 71.9	74.6
L	26.7	25.8, 25.9, 24.3, 25.4

Changing the headrest does not change the force. We say the *performance follows* the seat, since the seat from the original high (low) force assembly results in high (low) force even with a different headrest. The team eliminated the headrest family and looked at the seat in more detail.

The seat was assembled from three components: guides, springs, and the actual seat. The team disassembled and reassembled the two seats into the three components three times and measured the force. The data are shown in the diagonal cells of the following table. The team eliminated the assembly process for the three components as the home of the dominant cause. During the seat disassembly-reassembly investigation, the team noticed differences in the shape of the springs in the two assemblies. They next swapped the springs in the two seat assemblies and measured the force. The data are given in the table.

Other components	Spring	
	H	L
H	71.4, 72.5, 72.3	25.5
L	75.2	25.9, 25.5, 24.3

The dominant cause acts in the spring family of causes. The diagnostic tree in Figure 11.19 summarizes the search for a dominant cause.

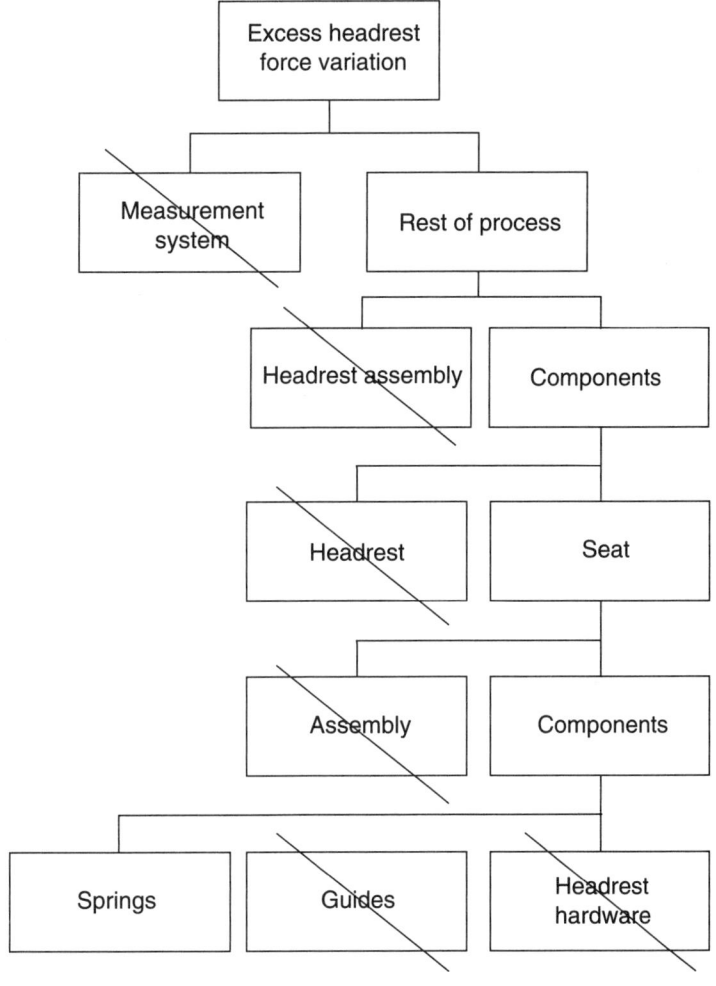

Figure 11.19 Diagnostic tree for headrest failure example.

The team worked with the supplier to understand the differences in the springs. They discovered that the supplier had reworked the springs in the low-force seats. Together, the supplier and customer developed a change of the rework process so that the force associated with the reworked springs exceeded 35 N. Customer complaints about the headrest moving were virtually eliminated.

Power Window Buzz Noise

There were frequent complaints about a noise fault called *buzz* in power window regulators (the motor and linkage that opens and closes the window). The customer, an assembly plant, demanded that the supplier implement 100% inspection to prevent noisy regulators from reaching the assembly plant. Management assigned a team the goal of reducing or eliminating the defect so that the costly inspection could be removed.

Trained listeners measured buzz subjectively on a scale of 1 to 7. They convinced their customer that if the buzz score was less than 4 on all regulators, then they could eliminate the 100% inspection.

The team did not carry out formal baseline or measurement system investigations. They had informal baseline information from past experience since they knew the buzz noise level varied between 1 and 7. They could ensure the dominant cause was acting in any investigation by selecting regulators with noise levels at the extremes. The decision not to check the measurement system was more risky. Large differences in measured noise for equally noisy regulators would make finding a dominant cause difficult. To address this concern, the team used a single listener to assign the noise levels for all regulators in the subsequent investigations.

They decided to search for the cause of the buzz. They divided the causes into two families, assembly and components. They selected a noisy regulator (score 7) and a quiet regulator (score 1). They could take the regulator apart without damaging or changing any of the major components. There was no change in buzz score when they disassembled and reassembled the two regulators three times each using the normal assembly sequence. The team eliminated the assembly process and concentrated on the components family.

The regulator has six components: arm one, arm two, spring, motor, back-plate, and sector. The team felt the motor was the likely home of the dominant cause of noise. They swapped the motors in the two regulators and measured the noise. The motor with the components from the originally quiet regulator had a score 7 and the motor with the other noisy components had a score 1. Indeed, the dominant cause of buzz lived in the motor.

The motor was a purchased part and comprised 18 components that could be disassembled without damage. The team formed two groups of nine components each, labeled G1 and G2, based on their limited knowledge of how the motor worked and the ease of disassembly. They did not know the details of the assembly process used by the motor supplier, so they ignored the motor assembly family for the moment. They next swapped all of the G1 components simultaneously and remeasured the buzz. The results were:

	G1	
G2	Noisy	Quiet
Noisy	7	7
Quiet	2	1

The entries 7 and 1 on the diagonal are the initial buzz measurements from the two original regulators. The off-diagonal entries are the buzz measurements for the regulators with all components in G1 swapped. Note that to measure buzz, the motors were assembled together with the other five components from the original quiet regulator. Since all components other than the motor had been eliminated as possible homes of the dominant cause, the choice of housing for the motor was not important.

The buzz followed G2 so all components in G1 were eliminated from consideration. Now G2 was split into two groups, G21 and G22, and the team found that the dominant cause lived in G22, a group of four components, three gears in the drive train, and the armature shaft.

When the first and second gears were swapped together, the buzz only occurred for the combination of all the G22 components from the originally noisy motor. The data were:

	First and second gears	
Gear 3 and armature	Noisy	Quiet
Noisy	7	3
Quiet	2	1

Now the picture is less clear. This pattern suggests a dominant cause involving one (or both) of the first and second gears, and one (or both) of the third gear and the armature. In other words, the dominant cause involves one or more components in each of the two remaining families.

The team returned the components to their original motor. They next simultaneously swapped the second and third gears with the results:

	Second and third gears	
First gear and armature	Noisy	Quiet
Noisy	7	1
Quiet	7	1

The buzz followed the second and third gears together. The dominant cause of buzz is an interaction between these two gears. Swapping only one of these two gears does not produce a complete switch in the buzz. We show the diagnostic tree for the search to this point in Figure 11.20.

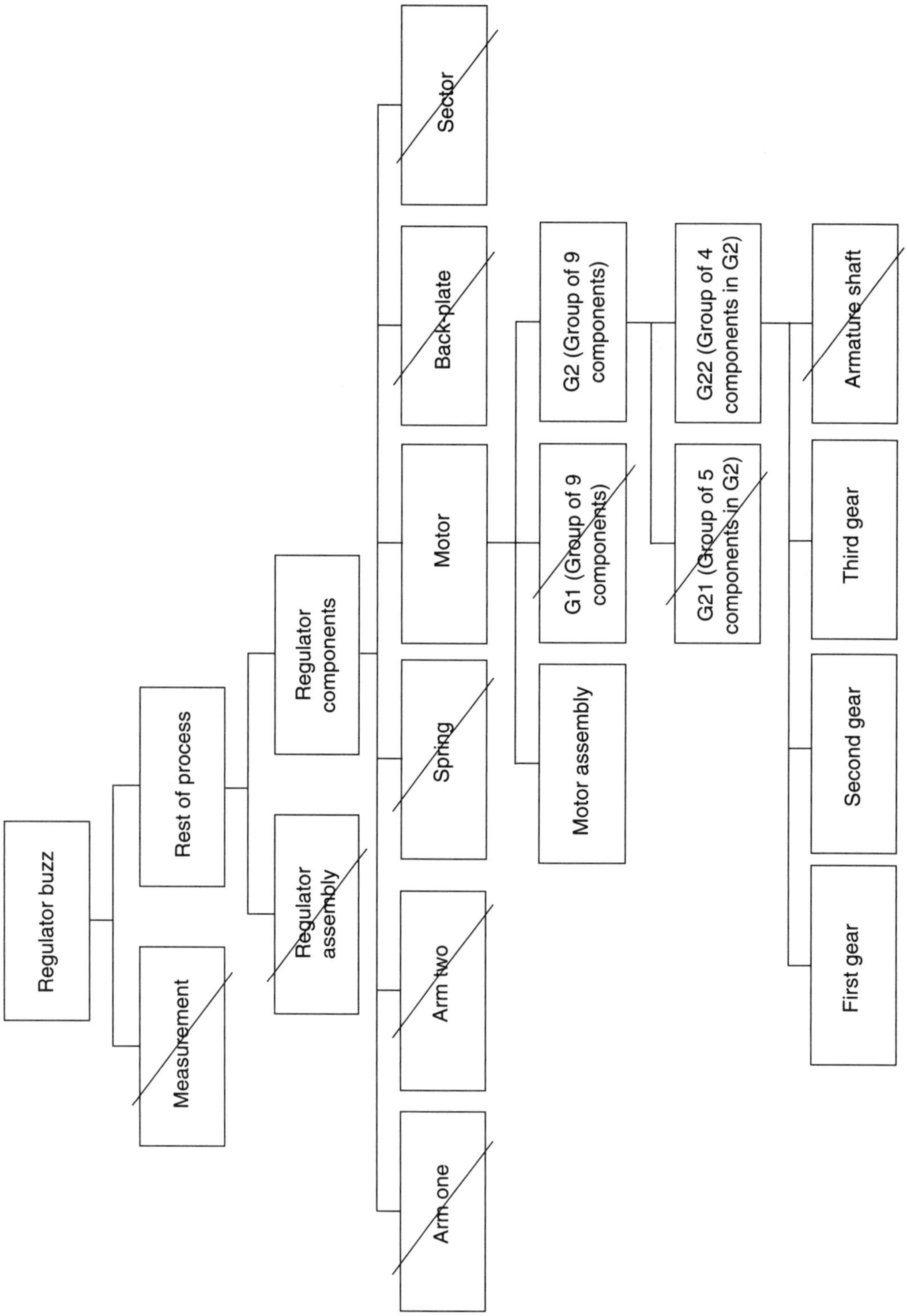

172 Part Three: Finding a Dominant Cause of Variation

Figure 11.20 Diagnostic tree for power window buzz noise.

The team focused their search for the dominant cause on the second and third gears. They approached the motor supplier with their findings. The supplier provided a new quieter type of motor to solve the problem. Given that this solution does not require an understanding of the cause of motor noise, the team would have been better off talking with the supplier much earlier.

Interpreting the Results

In any component swap investigation where we divide an assembly into two groups of components denoted by G1 and G2, there are a number of possible results when we swap G1 and G2. Consider the following hypothetical example based on the window buzz problem. We always start with the original measurements and the results from the repeated disassembly and reassembly, if they are available. Suppose we have:

	G1	
G2	Noisy	Quiet
Noisy	7, 6, 7, 6	
Quiet		1, 2, 2, 1

The first numbers in the diagonal cells correspond to the original values of the extreme assemblies that we selected for the investigation. When we repeatedly disassemble and reassemble, we see the variation due to the assembly process. In this case, we eliminate the assembly process as the home of the dominant cause.

We obtain the off-diagonal elements by swapping the components in G1 between the two assemblies. If the performance follows one of the two groups of components, the interpretation of the results is straightforward. For example, we might have the data:

	G1	
G2	Noisy	Quiet
Noisy	7, 6, 7, 6	1
Quiet	7	1, 2, 2, 1

or

	G1	
G2	Noisy	Quiet
Noisy	7, 6, 7, 6	7
Quiet	2	1, 2, 2, 1

In the left table, G1 is the home of the dominant cause; in the right table, the dominant cause acts in G2. In other cases the performance does not follow either group. For instance, consider the following four possible results:

	G1	
G2	Noisy	Quiet
Noisy	7, 6, 7, 6	1
Quiet	1	1, 2, 2, 1

	G1	
G2	Noisy	Quiet
Noisy	7, 6, 7, 6	4
Quiet	4	1, 2, 2, 1

	G1	
G2	Noisy	Quiet
Noisy	7, 6, 7, 6	7
Quiet	4	1, 2, 2, 1

	G1	
G2	Noisy	Quiet
Noisy	7, 6, 7, 6	4
Quiet	1	1, 2, 2, 1

In all of these cases the dominant cause involves (at least) one component from both G1 and G2. We need to re-form the groups to determine where the dominant cause lies. We discuss the situation where a dominant cause involves two or more components in more detail later in this section. We also explore the general issue of dealing with a dominant cause involving two (or more) inputs in Chapter 14.

Summary

Suppose we have:

- Two assemblies with opposite and extreme output relative to the full extent of variation
- Assemblies that can be disassembled and reassembled without damage
- Elimination of the assembly family as the possible home of a dominant cause (see Chapter 10)

We search for the component that is the home of the dominant cause with a series of small experiments where we divide the components into two groups and swap one group. We analyze the data by looking at two-way tables. We assume that a dominant cause lives with one or at most a pair of components. If we swap this component or pair, the output will move across most of its full extent of variation.

The component swap procedure has two parts. In the first we assume the dominant cause is in a single component family.

Component Swap Procedure

1. Divide the (remaining) components into two groups.
2. Swap one of the two groups of components, as shown in Figure 11.21, and measure the output for the two new assemblies.
3. Interpret the results. If the performance:
   - Follows one of the two groups, eliminate the other group. Stop when a single component remains; otherwise go back to step 1.
   - Does not follow one of the two groups, the dominant cause lives with a pair of components, one from each of the two groups. Go to the component swap add-on procedure.

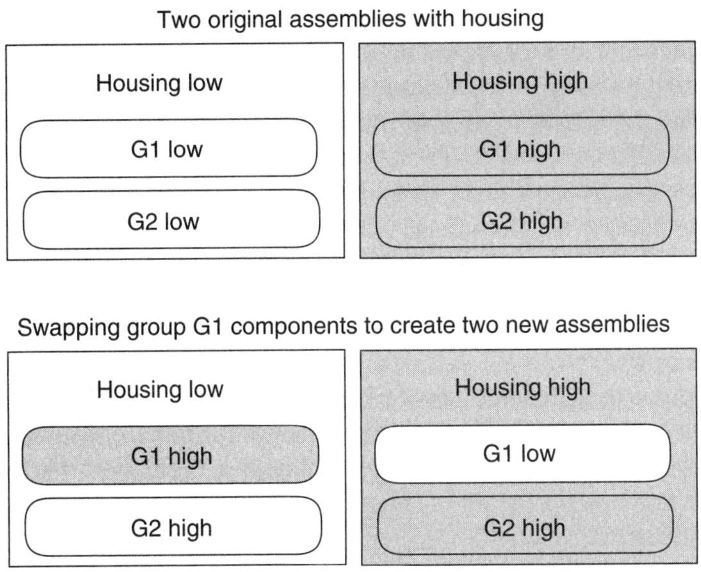

Figure 11.21 Illustration of swapping the group of components labeled G1.

The procedure is more complicated if the dominant cause involves two components.

> **Component Swap Procedure Add-on**
>
> A1. Divide the remaining components into two groups in a new way.
> A2. Swap one of the two groups of components, as shown in Figure 11.21, and measure the output for the two new assemblies.
> A3. Interpret the results. If the performance:
> - Follows one of the two groups, eliminate the other group. Stop when a single pair of components remains; otherwise divide the remaining components into two groups and go back to step A2.
> - Does not follow one of the two groups, go back to step A1.

This procedure will fail if there is no dominant cause or if the dominant cause involves three or more components.

Comments

A key requirement for component swap plans is that the product can be disassembled and reassembled without damaging the parts. To avoid study error, the reassembly process should match the assembly process in normal production as closely as possible. Once we have ruled out the assembly process, we carry out the component swap investigation offline to avoid interference with regular production.

By exploiting leverage, we use only two assemblies chosen to reflect the full extent of variation. There is a risk we may select assemblies that are extreme due to a different failure mode, and hence with a different dominant cause, than that of the problem we are trying to address. To alleviate this risk we recommend confirming the conclusion with an experiment that uses several extreme assemblies. For instance, to confirm the conclusion in the headrest failure example, we can put good springs in five seat assemblies that originally required less than 35 Newtons force to move through their full range of motion. If the new springs increase the force above 35 Newtons, we confirm the spring as the dominant cause.

Since all components that make up the housing have been eliminated as a possible home of the dominant cause, we can use a single housing for further disassembly-reassembly and component swap investigations. We employed this idea in the power window buzz noise example. Note that at the start of an investigation there is often no housing because all components are suspect.

In the recommended component swap procedure, we assume the assembly family has been eliminated using an investigation where we disassembled and reassembled down to the individual components. One advantage of this plan is that we can use the production assembly process. Alternatively, suppose we divide the assemblies into a number of subassemblies (groups of components). In that case, we start by disassembling and reassembling down to the subassembly level. Then, if this assembly process is eliminated, we apply the component swap procedure to the subassemblies. A complication arises if the proposed component swap procedure indicates the dominant cause is an interaction between two groups of subassemblies. This interaction could be due to either components or the assembly process for the subassemblies.

To illustrate, consider the window buzz noise example. The team found the dominant cause acted in the motor, which had 18 individual components that could be grouped into two subassemblies. Suppose the team started by disassembling and reassembling each extreme motor into the two subassemblies and that this part of the assembly family was eliminated. Next, the team would swap one of subassemblies between the two motors. Suppose the dominant cause was found to act within one of the subassemblies. Then the next step would be to check the assembly family for that subassembly before proceeding to swap components from the identified subassembly.

There are many other component swapping plans.[8] For example, we can use more than two groups of component groups at each stage.[9] We strongly recommend dividing the remaining components into only two groups at each stage to keep the procedure simple and to maximize the number of components eliminated with each swap.

Chapter Eleven: Investigations to Compare Three or More Families

Key Points

- We use a multivari investigation to compare time and location families. We design a systematic sampling protocol to isolate the contribution of each family to the baseline variation.
- We use a variation transmission investigation to eliminate process step families. We trace parts through the process and measure the output characteristic after each process step.
- We use component swap to eliminate families related to an assembly process and the corresponding components. We repeatedly disassemble, reassemble, and swap components in an organized way.

Endnotes (see the Chapter 11 Supplement on the CD/ROM)

1. If we cannot isolate a dominant family or families from the multivari charts, we can use analysis of variance (ANOVA) to quantify the contribution of each family to the overall variation.
2. In multivari investigations, we expect families such as part-to-part to show haphazard rather than systematic variation. We need to plot the data carefully to detect the effect of such families. We provide more detail on assessing this type of variation in the supplement.
3. See note 2.
4. See note 2.
5. See note 1.
6. See note 1.
7. We can use regression analysis to separate the variation transmitted and added at each operation if we can measure the output characteristic before and after the operation. We elaborate in the supplement.
8. We can conduct investigations to compare component families in different ways. We discuss a specialized tool first introduced by Dorian Shainin and some other related methods in the supplement.
9. We propose component swap investigations that involve dividing all remaining components into only two groups. In the supplement, we consider plans that divide components into three groups.

Exercises are included on the accompanying CD-ROM

12

Investigations Based on Single Causes

I walk slowly, but I never walk backward.
—Abraham Lincoln, 1809–1865

In this chapter, we look at investigations to search for a dominant cause based on single causes, that is, individual varying inputs. These plans are not particularly useful early in the search because they do not result in the elimination of families with a large number of causes. However, we can use these investigations for available data or when the family of remaining suspects is small. With these plans, we measure the output and selected suspects for a number of parts.

Large variation in a measurement system for an input can mask a dominant cause. Ideally, we would assess the measurement systems used to determine these inputs—see Chapter 7 for details. This is often not done, either because of prior experience with the measurement system or due to lack of resources.

In this chapter, we distinguish between plans for continuous and binary outputs because they lead to different analysis tools.

12.1 GROUP COMPARISON: COMPARING PARTS WITH BINARY OUTPUT

We use a *group comparison* to examine the effects of individual causes when the output is binary. To start, we select a number of parts with each of the two output values. Then we measure as many inputs as possible as suspects. Recall that a suspect is a cause that has not been ruled out in the search for a dominant cause. The values for a dominant cause will be substantially different for the two groups of parts. We look at two examples.

Engine Block Leaks

In the engine block leak example discussed in Chapter 1, the project was divided into three separate problems to address different failure modes. The blocks that leaked from the left rear intake wall had a visible defect. Close inspection with a microscope revealed that the defect was a sand inclusion. The team also observed that wall thickness varied from block to block.

The team planned a new investigation. Whenever they found an intake wall leaker, they set aside a nonleaking block cast in the same hour. They collected 100 blocks for each group. Then, for each of the sampled blocks, they measured thickness in inches at six locations in the left rear intake wall. The data are given in the file *engine block leaks comparison*.

To analyze the data, we construct box plots of wall thickness at each location for leakers and nonleakers. We show the results for two locations in Figure 12.1.

The right panel of Figure 12.1 shows a difference in average wall thickness between leakers and nonleakers at location four. There was little difference for the other locations as illustrated in the left panel of Figure 12.1. The team concluded that wall thickness at location 4 was a dominant cause of rear intake wall leaks.

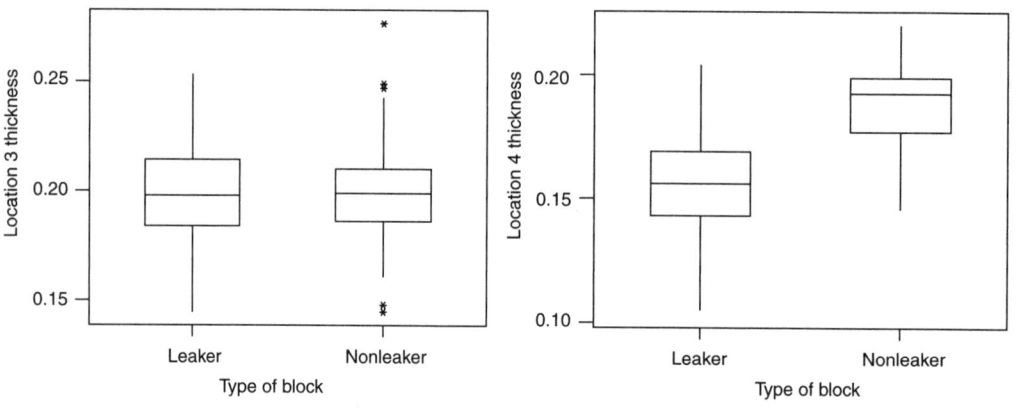

Figure 12.1 Box plots of wall thickness by block type at locations 3 and 4.

Window Leaks

Based on the results of customer surveys, the management of a truck assembly plant identified rear window water leaks as a substantial customer concern and assigned a team to address the problem. In the plant, 10 trucks per day were leak-tested in a special chamber that simulated extreme conditions. Using one month's data to establish a baseline, the team found that the rear window leak rate was 8% in the aggravated test. The team assumed that if they reduced this rate there would be substantial reduction in customer complaints. Using Pareto analysis by location, the team identified the upper ditch as the source of about 50% of the leaks. They set the problem goal to eliminate upper ditch rearindow leaks.

The team showed that the measurement system was acceptable for classifying trucks as either leaking or not and for determining the location of the leak. However, the system could not consistently measure the severity of the leak. For this reason, the team used the

Chapter Twelve: Investigations Based on Single Causes

measurement system only to classify a truck as a leaker or a nonleaker of the upper ditch rear window.

The team decided on a group comparison with eight leakers and eight nonleakers. They selected eight trucks that had failed the leak test with an upper ditch leak. Obtaining trucks that did not leak was more difficult since there was great pressure to immediately ship any good truck. The team found eight nonleakers from trucks that had been set aside for other problems.

They measured nine input characteristics thought to be related to water leaks. The data are given in the file *window leaks comparison*. The team plotted the data by group for each of the nine suspects. We show the result for primary seal fit in the left panel of Figure 12.2. There was no clear separation between leakers and nonleakers for any input. They also created scatter plots for all pairs of suspects with a different plotting symbol for leakers and nonleakers.[1] The plot for primary seal gap and quality of plastisol application, in the right panel of Figure 12.2, shows that all leaking trucks had both a large primary seal gap and poor plastisol application, while that combination never occurred for nonleaking trucks.

The team concluded that the dominant cause of upper ditch leaks was the combination of a large primary seal gap and poor plastisol application. In addition, they eliminated the other seven inputs from consideration. This conclusion was based on a small number of trucks and required verification.

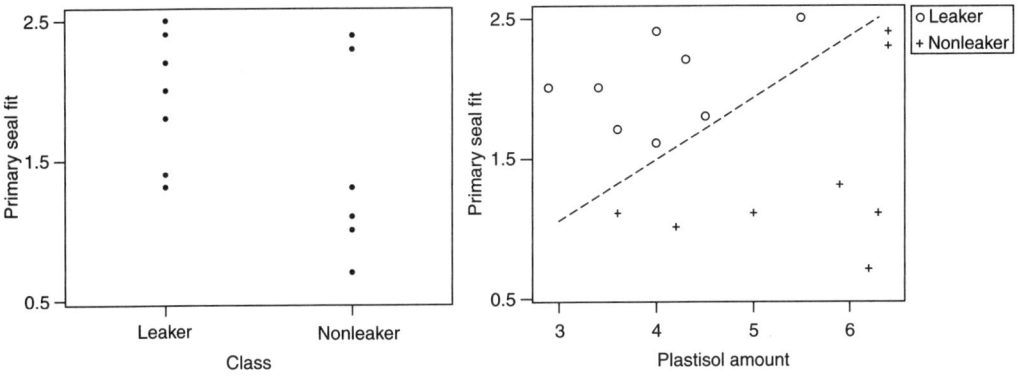

Figure 12.2 Plot of primary seal fit by group (left) and by plastisol amount (right).

Group Comparison Summary

Select as many suspects as possible that can be determined after measuring the output.

Question
In the current process, which, if any, of the suspects is a dominant cause?

Plan
- Select a group of at least six parts for each of the two output values.
- Measure all suspects for each part.

> **Data**
>
> Record the output (group) and input values, one row for each part.
>
> **Analysis**
>
> - Plot the values of each suspect stratified by group.
> - If no clear dominant cause is evident, create scatter plots of all possible pairs of inputs using different plotting symbols for each group.
>
> **Conclusion**
>
> - If there is clear separation of the suspect values between the groups, the suspect is the dominant cause.
> - If there is a separating line on a scatter plot that divides the groups, a combination of the two suspects is the dominant cause.

Comments

Group comparison is especially useful if the problem is defined in terms of a rare defect. Low-frequency problems are difficult to solve because it is hard to get information about the dominant cause of the defect. We can collect the two groups of parts over time until we have sufficient numbers to make the comparison.

We sometimes try a group comparison early in the search for a dominant cause where we measure many continuous characteristics on each part in the two groups. If we can find one characteristic that separates the defectives and the good parts, we can reformulate the problem in terms of the continuous characteristic. The reformulated problem may be much easier to solve because we can now see variation in the output on each part.

We can use group comparison if the output is binary or if the output is continuous and we select parts from the extremes to get the full extent of variation. In this instance, we create two groups and ignore the measured values of the output. We need to be careful if we use leverage in this way. One danger is that the extreme parts used in the investigation may be due to different failure modes and thus have different causes. This highlights the importance of focusing the problem so that there is only a single dominant cause. Suppose, in the engine block leaks example, the team had not focused the problems based on the location of the leak. If they had then carried out the group comparison described previously, they would not have seen large differences in wall thickness between the two groups, because the leakers group would likely contain blocks that leak at other locations.

In a group comparison, the groups are formed using the output. As a result, we can only compare inputs that can be measured or determined after the part is produced. For instance, in the window leaks example, it was not possible to compare process inputs such as window installer or machine settings because these cannot be determined after a truck is leak tested.

We require at least six to eight parts per group to avoid falsely identifying a dominant cause. We examine some alternate plans and analyses in the supplement to this chapter.[2]

12.2 INVESTIGATING THE RELATIONSHIP BETWEEN INPUTS AND A CONTINUOUS OUTPUT

For an *input/output relationship investigation*, we measure the output and several inputs on each part. We present three examples.

Crossbar Dimension

There was excess variation in a key crossbar dimension of an injection-molded part. In the baseline investigation, the team estimated the standard deviation and the full extent of variation of the dimension to be 0.46 and –0.3 to 2.0 thousandths of an inch. The problem goal was to reduce the standard deviation to less than 0.25.

The team showed that the measurement system was highly capable. They decided to search for a dominant cause. They conducted a multivari investigation where five consecutive parts from the single mold were sampled every 30 minutes for four hours. See the exercises for Chapter 10. The team found that the time-to-time family contained the dominant cause.

Next they planned an input/output relationship investigation. Forty shots of the process were selected haphazardly over a two-day period. For each shot, the team measured the crossbar dimension of the part and recorded five inputs: die temperature, nozzle temperature, barrel temperature, hydraulic pressure, and cavity pressure. All these suspects were thought to exhibit time-to-time variation.

The data are given in the file *crossbar dimension input-output*. The first step in the analysis is to check that the output varied over the full extent of variation as in the baseline. We see from the summary that this is the case.

Variable	N	Mean	Median	TrMean	StDev	SE Mean
dimension	40	0.8340	0.7618	0.8353	0.5467	0.0864

Variable	Minimum	Maximum	Q1	Q3
dimension	-0.1611	1.9154	0.4282	1.3649

The second step is to fit a regression model[3] (Montgomery et al., 2001) that includes all of the inputs simultaneously. See Appendix E for the MINITAB directions. The residual standard deviation is 0.2515, which is substantially less than the baseline standard deviation. The residual standard deviation is an estimate of the variation in crossbar dimension if we could hold all of the inputs in the regression model fixed. The small value of residual standard deviation indicates that one or more of the inputs is a dominant cause of the variation.

In the third step, we look at the scatter plots of crossbar dimension versus each of the inputs.[4] Two of the plots are shown in Figure 12.3.

There is a strong relationship between barrel temperature and crossbar dimension. If we could hold barrel temperature fixed at 77°, for example, we see from the plot that crossbar dimension would vary only by about 0.5. There is strong evidence that barrel temperature

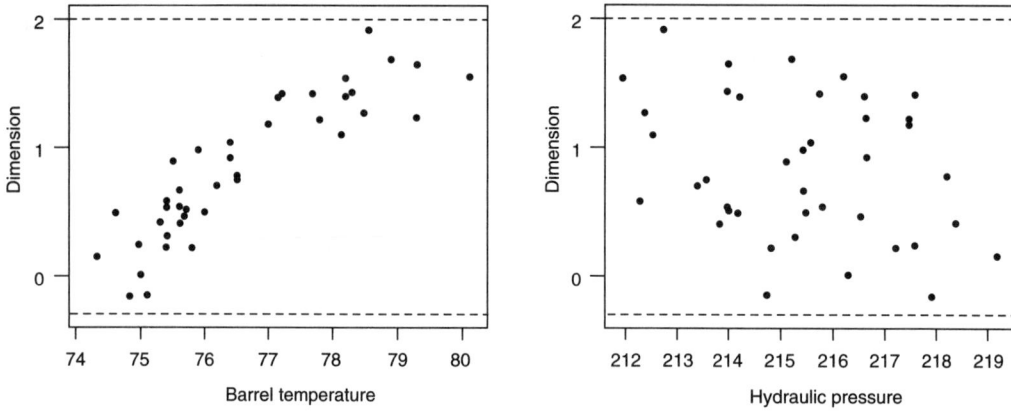

Figure 12.3 Plot of crossbar dimension versus barrel temperature and hydraulic pressure.

is a dominant cause of crossbar dimension variation. In Chapter 13, we describe an experiment to verify this conclusion. There is no strong association of crossbar dimension with the other inputs (as illustrated for hydraulic pressure in the right panel of Figure 12.3) and we eliminate them from further consideration.

If the relationship between the input and output is roughly linear, as is the case here, we can quantify the contribution of the input to the output variation using a regression model. The residual standard deviation represents the remaining (unexplained) variation in the output if we could hold the input fixed. We compare the residual standard deviation to the baseline to assess the contribution of the input.

Using MINITAB, we fit a regression model for crossbar dimension as a function of barrel temperature. The results of the analysis are:

```
The regression equation is
dimension = -23.9 + 0.323 barrel temp

Predictor         Coef      SE Coef         T         P
Constant       -23.898        2.075    -11.52     0.000
barrel temp    0.32293      0.02709     11.92     0.000

S = 0.2544     R-Sq = 78.9%    R-Sq(adj) = 78.3%

Analysis of Variance

Source           DF         SS         MS         F         P
Regression        1     9.1979     9.1979    142.13     0.000
Residual Error   38     2.4591     0.0647
Total            39    11.6570
```

The residual standard deviation is 0.25, much smaller than the baseline value 0.46. If the team could eliminate the effect of barrel temperature they would meet their goal.

Truck Pull

Consider again the truck pull problem introduced in Chapter 1. The team decided to focus on right caster since caster variation had a much larger effect on pull than camber variation. The baseline standard deviation for right caster was 0.24°. Previous investigations ruled out the measurement system. The assembly process was difficult to investigate so the team set that family aside for the moment and concentrated on the component families. A feedforward controller compensated for variation in the frame geometry. This left the other components of the alignment system, namely the knuckle and upper and lower control arms, as possible homes of the dominant cause. The search for the cause is illustrated in the diagnostic tree in Figure 12.4.

At this point the team had several choices. They could have tried assessing the assembly and component families, as discussed in Chapter 11, but they could not use the production assembly process. Moreover, at the time of the investigation, there was a proposal to bar-code the control arms with dimensional data that could be fed into the feedforward controller already in use to compensate for the frame geometry (see Chapter 16). With this proposed process change, all control arms would be measured and bar-coded at the supplier's plant. Before implementing such an expensive proposal, the team wanted assurance that it would be worthwhile. If the control arm inputs were not dominant causes, there would be little reduction in pull variation from this costly change.

The team decided to explore specific characteristics of the control arms and knuckle. Based on the proposal for the feedforward control scheme, they selected dimensional characteristics of the components, one for each control arm and two for the knuckle, thought to affect caster.

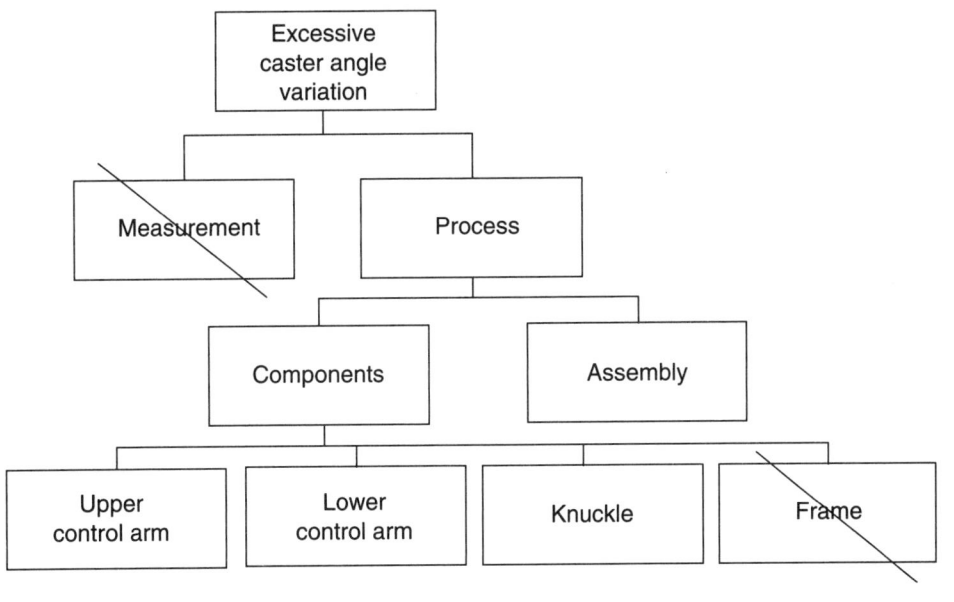

Figure 12.4 Diagnostic tree for search for the cause of excessive caster variation.

186   Part Three: Finding a Dominant Cause of Variation

For the investigation, the team selected 30 sets of components (two control arms and a knuckle) from regular production over six days. From the baseline investigation, the team expected to see the full extent of variation in caster angle with this sampling scheme. They measured the four inputs on each set of components and the right caster angle on the assembled truck. The data are given in the file *truck pull input-output*. The standard deviation of the caster angle in the investigation is 0.22, somewhat smaller than the baseline standard deviation.

We fit a regression model to describe the relationship between caster angle and the four inputs. The results from MINITAB are:

```
The regression equation is
caster = 68.5 + 1.57 lower ball + 1.48 upper ball + 0.242 U reading
          -0.184 L reading

Predictor         Coef      SE Coef        T        P
Constant         68.546       9.922      6.91    0.000
lower ball        1.5710      0.6905     2.28    0.033
upper ball        1.4779      0.6839     2.16    0.042
U reading         0.24187     0.02282   10.60    0.000
L reading        -0.18446     0.02844   -6.49    0.000

S = 0.1620     R-Sq = 87.8%    R-Sq(adj) = 85.6%
```

The residual sum of squares is $s = 0.162$, a moderate reduction from the baseline variation 0.24. Next we look at the scatter plots of the inputs by caster angle. We cannot see any strong relationships between caster angle and the individual component dimensions in Figure 12.5.

To quantify the contribution of each component dimension, we fit regression models with one input at a time. We rank the inputs based on the residual standard deviation in Table 12.1.

Table 12.1 Component dimensions ranked by residual standard deviation.

Input	Component	Residual standard deviation
U reading	Upper control arm	0.19
L reading	Lower control arm	0.21
Upper ball	Knuckle	0.21
Lower ball	Knuckle	0.22

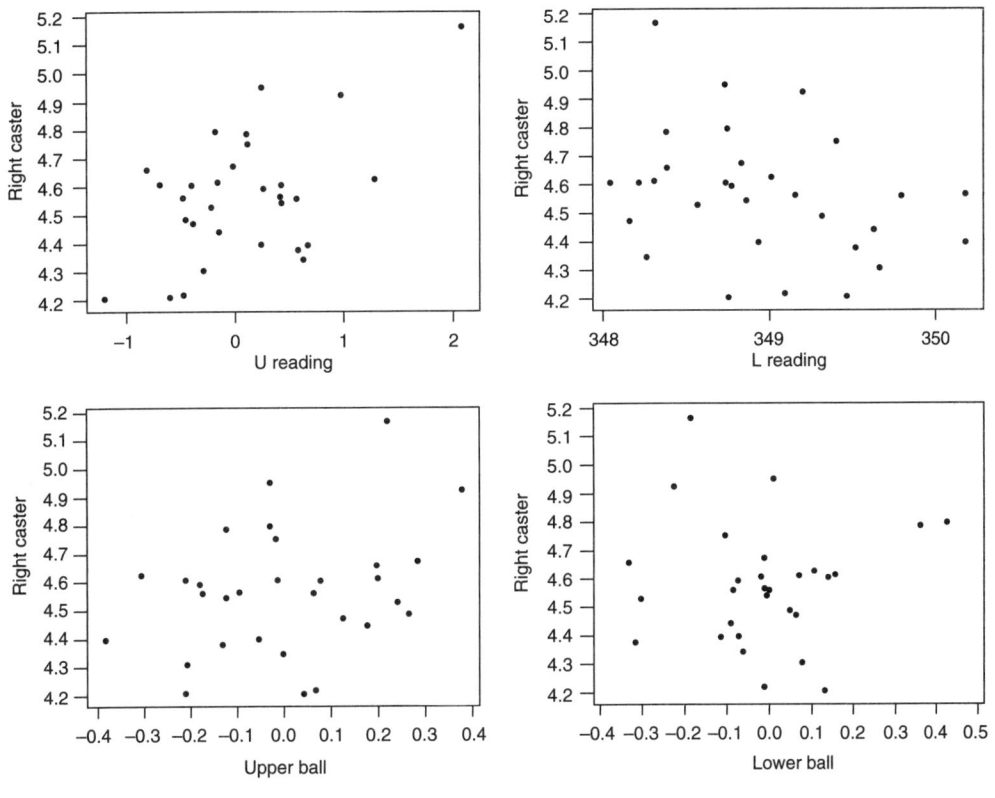

Figure 12.5 Scatter plots of right caster versus the component characteristics.

None of the component dimensions is a dominant cause when considered singly. When we fit regression models with pairs of inputs, we see in Table 12.2 that the two control arm dimensions together produce the smallest residual standard deviation.

We conclude that if we could completely remove the effects of the two control arms by using the feedforward controller, we would reduce the caster variation by about 20%. This reduction could be achieved only if the feedforward controller worked perfectly. Also, the sample size in this investigation is small and there is considerable uncertainty in the estimates of the standard deviations.

The team used these results to argue that the proposed bar-coding of the control arms would not be cost-effective. They decided instead to investigate the feasibility of feedback control on caster angle (see Chapter 17).

Table 12.2 Pair wise component dimensions ranked by residual standard deviation.

Input pairs	Residual standard deviation
U reading, L reading	0.174
U reading, Upper ball	0.177
U reading, Lower ball	0.195
L reading, Upper ball	0.212

Manifold Sand Scrap

In the production of cast-iron exhaust manifolds, a foundry attributed many defects to the sand system used to create the molds. Historically, the sand-related defect rate was 2% with substantial variation shift to shift. Management assigned a project team to reduce sand-related scrap.

The team used a run chart of the scrap rate by shift as a baseline. This chart was produced daily as part of the management information system. They did not formally investigate the measurement system used to determine whether a manifold had a sand-related defect. They believed from past experience that this system was reliable. They decided to search for a dominant cause of the defects.

Many characteristics of the molding sand such as temperature, compactness, permeability, moisture level, green strength, and percent friability were routinely measured during production. The team decided to use the available data to determine if any of these individual sand characteristics was a dominant cause. They chose not to use broad families of causes as is usually done early in the search because the data were already available and the cost was low.

It was not easy to get the data into a usable form. The sand characteristics were not measured for each casting. There were substantial and varying time lags between the measurement of a sand characteristic and the use of the sand to make a mold. For each casting, the time of casting was known only up to the nearest hour. Because of these traceability difficulties, the team used the hourly scrap rate as the output and the average of the sand characteristics over the hour as the inputs.

The team selected 91 hours of production and made the linkages between hourly scrap rate and average sand characteristics. In each hour, the plant produced between 30 and 180 castings. The data are given in the file *manifold sand scrap input-output*. The average and standard deviation of the hourly scrap rate were 0.017 and 0.023. The average rate 1.7% matches the baseline well.

To analyze the data, the team ignored the changes in volume and focused solely on the proportion of sand scrap in each hour. Fitting a regression model to the hourly scrap rates with all of the sand characteristics, the residual standard deviation was 0.022, a very small reduction from the baseline value 0.023. None of the sand characteristics was a dominant cause. This is confirmed by looking at scatter plots of the sand scrap proportion versus the inputs. Some of the plots are given in Figure 12.6.

Chapter Twelve: Investigations Based on Single Causes

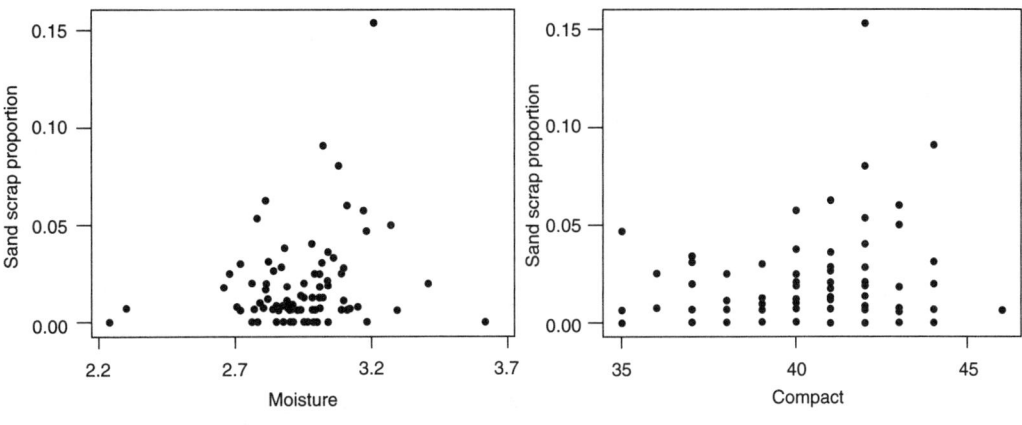

Figure 12.6 Scatter plot of sand scrap proportion versus moisture and compactness.

The team noticed in the plot of hourly scrap rate versus temperature, shown in Figure 12.7, that there was a nonlinear relationship. Using MINITAB, they fit a quadratic regression model that included the square of the temperature. Part of the results are shown as follows:

```
The regression equation is
sand scrap proportion = 6.16 - 0.121 temperature +0.000596 temp. squared

Predictor            Coef       SE Coef          T         P
Constant           6.1647        0.9629       6.40     0.000
temperature       -0.12116       0.01911     -6.34     0.000
temp. squared   0.00059610    0.00009476      6.29     0.000

S = 0.01922      R-Sq = 33.0%      R-Sq(adj) = 31.4%
```

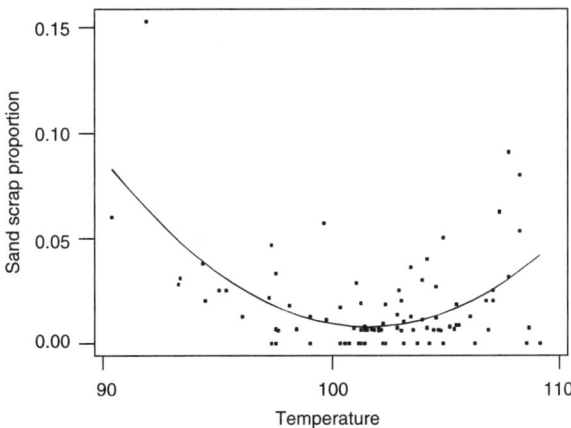

Figure 12.7 Sand scrap proportion versus sand temperature with quadratic fit.

While the observed pattern makes physical sense, temperature is not a dominant cause. Holding sand temperature fixed (which would be very expensive) would not reduce the sand scrap proportion substantially.

In the end the team was forced to conclude the investigation was a failure. They were not able to find a dominant cause, and they could not eliminate any of the suspects because the linkage between sand characteristics and the defect rate was poorly established.

This example demonstrates the risk of using an investigation based on individual causes early in the search for a dominant cause. Even if the team could have eliminated the measured sand characteristics as dominant causes, they would have made little progress, as there were a large number of remaining suspects.

Input/Output Relationship Investigation Summary

Select as many suspects as possible that can be traced to the output.

Question

For the current process, which, if any, of the suspects is a dominant cause?

Plan

- Select a time frame in which we expect to see the full extent of variation in the output.
- Select 30 or more parts spread across the time frame.
- For each part, measure the output and all suspects.

Data

Record the output and suspect values, one row for each part.

Analysis

- Use regression to model the output as a function of all suspects simultaneously. For categorical suspects, use indicator variables.[5]
- Plot the output versus each one of the suspects. For continuous suspects, fit the corresponding simple regression model to quantify any strong linear relationships. For categorical suspects, use one-way ANOVA.

Conclusion

- If the residual standard deviation in the first regression model is much less than the baseline standard deviation, one of more of the suspects is a dominant cause.
- If there is a dominant cause, identify it from the scatter plots.

Comments

In an input/output relationship investigation, we can select parts in one of three ways:

- Using some time- and/or location-based sampling scheme
- Based on the values of the inputs
- Based on the output values

In all cases, the goal is to get the full extent of variation so that we know the dominant cause has acted.

With the first option, we need to have a large enough sample spread across appropriate times and locations to meet the goal. We can use the results of the baseline investigation to help to define the sampling scheme. This option is the most common and was used in the examples discussed in this section.

The second option requires more effort. For example, in the truck pull example, the team had information about the historical variation in the four inputs. They could have selected components with values for these dimensions at both ends of their historical range. This would have required extra measurement to find the appropriate components. Using extreme values for the inputs, we employ leverage and can use a smaller sample size. If we do not see the full extent of output variation with this plan, none of the selected inputs is a dominant cause.

The final option corresponds closely to the group comparison plan. For example, in the window leaks problem, had the team been able to measure the severity of the leak, they could have fit a regression model to the data. By choosing parts with extreme output values, we are sure to see the full extent of variation. However, we may not be able to determine the values of many suspects due to lack of traceability.

We need to be careful interpreting the MINITAB regression results. In the truck pull example, each of the four inputs is *statistically significant* because the *p*-value is small (less than 5%). This does not mean that any of the input is a dominant, or even large, cause, as we saw in the example. The same difficulty occurs in group comparisons when the input averages for the two groups are statistically *significantly different*, and yet there is no evidence that the input is a dominant cause. Hypothesis tests are not a useful tool for identifying a dominant cause. See further discussion of this issue in the Chapter 10 supplement.

Regression models can accommodate binary outputs, discrete inputs, quadratic terms, interactions between inputs, and so on.[6] We can transform or combine the input values. We need to be careful using these models because we can be misled by nonlinear relationships, outliers, and influential observations.

 **Key Points**

- Group comparison and input/output investigations assess single causes and are most useful in the latter stages of the search for a dominant cause.
- With a binary output, we can compare inputs measured on parts using a group comparison. We use plots of the individual values or box plots to look for large differences in the inputs for the two groups of parts defined by the output values.
- With continuous output, we use regression models and scatter plots to show the input/output relationships and isolate dominant causes. We calculate the residual standard deviation from a simple regression fit to assess whether a particular suspect is a dominant cause.

Endnotes (see the Chapter 12 Supplement on the CD-ROM)

1. We show in the supplement that for a group comparison or an input/output investigation, there is a quick way to make all the desired scatter plots in MINITAB.
2. In the chapter supplement we compare group comparison with paired comparison as suggested in Bhote and Bhote (2000).
3. Regression analysis is a flexible analysis tool that has been extensively studied in the statistical literature. In the supplement, we explore some of the useful extensions in more detail and give references to further work.
4. See note 1.
5. See note 3.
6. See note 3.

 Exercises are included on the accompanying CD-ROM

Chapter 13
Verifying a Dominant Cause

Approach each new problem not with a view to finding what you hope will be there, but to get the truth, the realities that must be grappled with. You may not like what you find. In that case you are entitled to try to change it. But do not deceive yourself as to what you do find to be the facts of the situation.

—Bernard M. Baruch, 1870–1965

We recommended the method of elimination and a series of simple observational investigations to isolate a dominant cause. Before proceeding with the next stages of the Statistical Engineering algorithm, we want to be sure that the suspected cause, here called a *suspect,* is dominant. We call this *verification.*

In many applications of the algorithm, we have sufficient evidence from the search to be sure we have found a dominant cause and we require no further verification. For example, in the fascia cratering problem discussed in Chapter 11, the team found that craters occurred only on every fifth and tenth fascia taken from the mold. They concluded that the dominant cause was the application of mold release. They did not verify this conclusion because no other cause matched the observed pattern of craters. Similarly, in the V6 piston diameter example (Chapter 11), the team concluded that a dominant cause of diameter variation at the final gage was the diameter after Operation 270. The variation transmission investigation showed that pistons with large (small) diameters after Operation 270 were large (small) at the final gage. The team could explain the observed pattern in only one way and decided not to verify their conclusion.

Why do we need to verify? In the search, we might have inadvertently ruled out a family that contains the dominant cause. More commonly, we may have selected the suspect from the remaining family of causes using our best judgment. If we are wrong, there may be other causes in the family that are dominant. Consider the problem of excess crossbar dimension variation discussed in Chapter 12. There, the team concluded that barrel temperature was the dominant cause based on the results of an input/output relationship investigation. They decided that verification was necessary because it was possible that the actual dominant cause was another (unidentified) cause in the same family (time-to-time) as barrel temperature.

194 *Part Three: Finding a Dominant Cause of Variation*

To verify that a suspect is a dominant cause, we use an experimental plan, often called a *designed experiment,* where the value of the suspect is deliberately manipulated. Good references on experimental plans include Box et al. (1978), Ryan (1989), Wheeler (1990), and Montgomery (2001).

Experimental plans are also important tools, as we will see in later chapters, for helping to assess the feasibility and determine how to implement several of the variation reduction approaches. In this chapter, we introduce the language and principles of experiments. We start with plans to verify a single suspect dominant cause. We then introduce more complex plans used to isolate a dominant cause from a short list of suspects.

13.1 VERIFYING A SINGLE SUSPECT DOMINANT CAUSE

To verify a single suspect, we plan a simple experiment where only the suspect is deliberately varied. For illustration, we use the engine oil consumption problem discussed in Chapter 10. Using stratification, the team had determined that a dominant cause acted in the plant-to-plant family. Based on knowledge of the engine, one of the few plant-to-plant differences that could explain the variation in oil consumption was a different supplier of valve lifters. The valve lifters were suspects because changing the valve lifters on a returned engine eliminated the oil consumption. Further application of the method of elimination pointed to a clearance dimension in the lifter as a primary suspect. The team decided to use an experimental plan to verify that the valve lifter clearance dimension was the dominant cause of oil consumption.

We introduce some terminology used in designed experiments. We plan to deliberately change one or more inputs. We call the different values the *levels* of the inputs. In the oil consumption example, clearance was the single suspect. The team decided to use two levels for clearance, corresponding to the high and low ends of its known historical range. We call these the *low* and *high level* of the suspect input.

The suspect is a dominant cause if the output moves across its full extent of variation when we change the level of the suspect from low to high.

An *experimental run,* usually shortened to a *run,* corresponds to setting the level of the suspect, running the process and measuring the output characteristic. In the example, a run corresponds to installing a set of lifters with low (or high) clearance into an engine, putting the engine through the accelerated dynamometer test, and measuring the oil consumption. We carry out one or more runs for each level of the suspect. If there is more than one run for a given level, we call this *replication.* To achieve *balance,* we use an equal number of runs for each level.

We normally require a minimum of three or four *replicates* at each level of the suspect. More replicates will give more reliable conclusions but increase the cost of the experiment. Within a run, the process may produce a number of parts, some of which are measured. We call these parts *repeats*. We recommend the same number of repeats per run to preserve the balance. The connection between replicates, runs, and repeats is illustrated in Figure 13.1.[1]

In the example, the team found three sets of lifters (each engine requires a set of eight lifters) with low clearance and three sets with high clearance by measuring clearance on incoming lifters from the poor supplier. In turn, they installed each set in an engine and measured the oil consumption. The experiment had six runs with no repeats. There were three replicates for each of the two levels of clearance.

Chapter Thirteen: Verifying a Dominant Cause

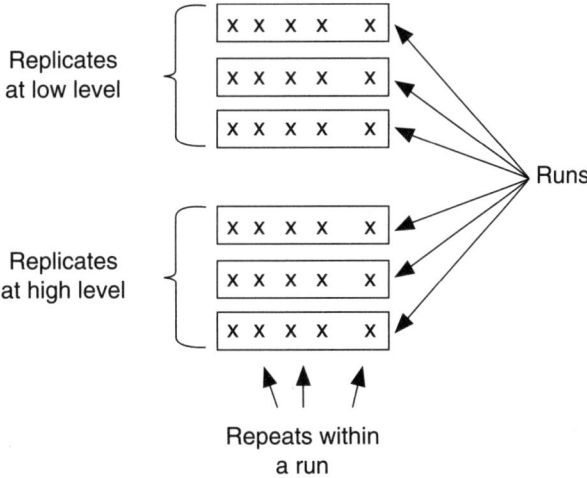

Figure 13.1 Runs, replicates, and repeats for an experiment with a single suspect at two levels.

To protect against some other input changing systematically as we change the suspect, we randomize the order of the runs, if feasible.[2] This use of randomization is one of the main differences between observational and experimental investigations. We can randomize the order of replicates but not repeats. For example, we give the random order and the oil consumption for each run in Table 13.1.

Table 13.1 Valve lifter clearance experiment and results.

Level	Average lifter clearance	Run order	Oil consumption (grams per hour)
Low	5.0	4	23
Low	7.0	3	24
Low	8.5	6	29
High	21.0	1	76
High	24.0	2	113
High	25.0	5	120

To analyze the results, we rely on tables of averages and graphical summaries such as box plots or scatter plots. If the suspect is a dominant cause of variation, the output characteristic should vary over most of its full extent of variation when the value of the suspect changes from its high to low level. In Figure 13.2, we see that changing the valve lifter clearance has a large consistent effect on the oil consumption. The team concluded that valve lifter clearance was a dominant cause of the variation in oil consumption and that low clearance values led to less oil consumption.

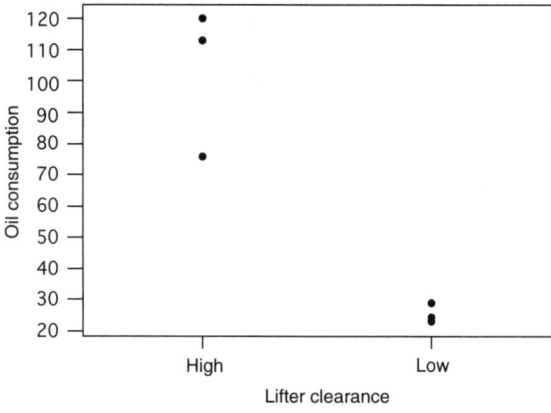

Figure 13.2 Oil consumption by lifter clearance level.

There were several limitations to this conclusion. First, the team did not have a good idea of the baseline in terms of the accelerated dynamometer test. They had measured only one engine from the field failures. Repeated testing of this engine in the dynamometer gave an average oil consumption of 117 grams per hour. The team was reassured because the oil consumption at the high clearance level was close to this average. Second, the verification experiment used only a single engine. The clear results for this engine may not carry over to others. Finally, the team did not have a direct connection between oil consumption in the field and in the result of the accelerated test on the dynamometer. In spite of these limitations, they proceeded, assuming that lifter clearance was the dominant cause.

Crossbar Dimension

In the crossbar dimension example, discussed in Chapter 12, the team identified barrel temperature as the dominant cause. For the verification experiment, the team chose the low level for barrel temperature as 75° and the high level as 80°. This covered close to the full range of barrel temperatures seen in the earlier investigation.

Barrel temperature could be controlled for the experiment and changed in a few minutes. As a result, the verification experiment used only two runs. The barrel temperature was set, 25 parts were made to ensure the temperature had stabilized, and the next 10 parts were selected and measured. There are two runs with 10 repeats per run and no replication.

The data from the experiment are given in the file *crossbar dimension verification* and are presented in Figure 13.3. Barrel temperature has a large effect on crossbar dimension relative to the full extent of variation (given by the dashed lines). The team concluded that barrel temperature was the dominant cause of crossbar dimension variation.

The lack of randomization was not an important limitation here since previous investigations had shown that the dominant cause acted in the time-to-time family. Over the short time frame of the verification experiment, it is unlikely that we will see the full range of variation in the output unless barrel temperature is a dominant cause. There is insufficient time for other causes in the time-to-time family to change substantially.

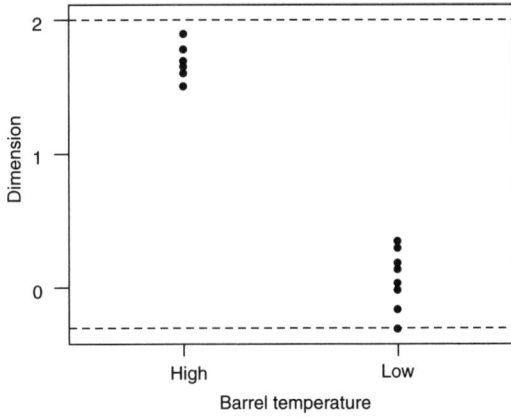

Figure 13.3 Crossbar dimension verification experiment results (dashed lines show the full extent of diameter variation).

The lack of replication may have serious consequences. If the dominant cause is an interaction between barrel temperature and some other cause in the time-to-time family, we may see little effect on crossbar dimension when we change the barrel temperature, depending on the unknown level of the second cause. To remove this uncertainty, we need to take the expensive step of replicating the whole experiment over several time periods, allowing the second cause time to change.

13.2 ISSUES WITH SINGLE SUSPECT VERIFICATION EXPERIMENTS

There are many issues in the planning and analysis of an experiment to verify a single suspect as a dominant cause.

Do We Need to Verify the Suspect?

In deciding if verification is necessary, the team should consider:

- The risk that the suspect is not a dominant cause
- The cost and difficulty of conducting the verification experiment

To assess the risk, the team should think about how they identified the suspect and the size of the remaining family. The risk is high if the team has selected the suspect from a large family of remaining possibilities based on uncertain process knowledge.

The cost of verification may be high, especially if the levels of the suspect are difficult to control. In the engine block porosity problem discussed in Chapter 10, the team noticed that severe porosity occurred when there was downtime in the operation. The team compared the porosity levels before and after breaks and identified two suspects, iron-pouring temperature and the addition of ladle wash. During work stoppages, iron that remained in the six pouring ladles cooled off since there was no external heat source. At the start of the break, ladle wash was added to the ladles to protect the refractory (surface). Since the wash was water-based, the team suspected it was a source of porosity when the process was restarted. The team could not cheaply manipulate the pouring temperature, but they could change and control the level of ladle wash. They added the full amount of wash to three ladles selected at random and half the amount to the other three for two lunch breaks. In each case, they measured the porosity of the first 30 blocks poured (five from each ladle). The data are given in the file *engine block porosity verification*, and the results are discussed in the Chapter 13 exercises. Based on the results and process knowledge, the team eliminated ladle wash as a suspect. They declared pouring temperature a dominant cause, even though they did not verify it.

Choosing the Study Population

Most verification experiments are conducted over a short time using only a few parts. In the oil consumption example, the team used the same engine for all experimental runs. They assumed that the conclusions about valve lifter clearance for this engine would apply more broadly. The team could have strengthened the conclusion by doing the same six-run experiment with three different engines. This change to the plan would have increased the cost and complexity of the experiment.

The more that we know about the nature of the output variation, the easier it is to select a study population. For example, if we know the time-to-time family contains the dominant cause, we can conduct the experiment over a short time period in which the process does not vary materially. There is a small chance of another cause in the time-to-time family changing substantially during the experiment. However, since the suspect is also in the time-to-time family, we may have difficulty manipulating its levels within this short time period.

Choosing the Levels of the Suspect

We select the two levels of the suspect for the verification experiment at the low and high end of its range of values in the regular process. If the suspect is a dominant cause, then changing from the low to high level will produce the full extent of output variation. To determine the levels, we have to know the range of variation of the suspect cause. To acquire this knowledge, we may have to carry out a small investigation on the suspect.

There is some risk in making the levels of the suspect too extreme. First we may induce a different failure mode into the process. Second, we may fool ourselves in concluding that the suspect is a dominant cause because the very extreme levels, which rarely occur in the regular process, may induce the full extent of variation in the output.

For a verification experiment, we strongly recommend using only two levels per suspect. Extra levels give little additional information about whether or not the suspect is dominant and increase the complexity of the experiment.

Randomization and Replication

In some cases we conduct a verification experiment over a short time frame without the protection provided by randomization and replication.[3] In planning the experiment—that is, defining a run, choosing the number of runs, and so on—we need to assess the risk that a dominant cause, other than the suspect, acts within the time of the experiment. The nature of the variation in the output over time is a key piece of information to help assess this risk. If the dominant cause acts over the long term, as in the crossbar dimension example, we can plan a verification experiment over a relatively short time with two runs, one at each level of the suspect. If the dominant cause acts in the part-to-part family, we can use two runs with a moderate number of repeats. That way, if the suspect is not a dominant cause, the true dominant cause (that acts in the part-to-part family) will have time to generate close to the full extent of variation in the runs at both the high and low level of the suspect.

When we do not know the nature of the variation over time, as in the oil consumption example, we need to be careful. We should use replication—that is, several runs for each level of the suspect—and randomize the order in which the runs are conducted. The key question is, "Is there some other unknown cause that might change from run to run in a way that matches the pattern of change of the suspect?" If the answer is yes, then we randomize the order of the runs with at least three replicates per level to reduce the risk.

What If the Output Is Binary?

If the output is binary, we suggest many repeats within each run of the experiment. We can then see if changing the suspect produces a large change in the proportion of defectives within each run. We can assess the importance of the change by comparing the proportions to the baseline. We need runs with many repeats to estimate the proportion of defectives at each level of the suspect. We may have difficulty holding the normally varying suspect constant for a long run.

Is the Cause Dominant?

In the analysis of a verification experiment, we always check that the observed variation in the output characteristic is a substantial proportion of the full extent of variation seen in the baseline. Otherwise the identified suspect is not a dominant cause. We do not recommend a formal hypothesis test to see if changing the suspect produces a statistically significant change in the output. Such a change may be too small to be helpful.

What If the Suspect Is Not a Dominant Cause?

What should we do if the results of the experiment show the suspect is not dominant? First, review the plan and conduct of the experiment in light of the information gathered during the search for the suspect. Some possible questions are:

- What is the family of remaining suspects? Are there other suspects in this family? Are these suspects eliminated by the experiment?
- Could we have missed an interaction with another cause that did not change during the verification experiment?

If the answers to these questions are not helpful, then we have few options. First, review the diagnostic tree. Did we rule out any families or causes during the search without sufficient evidence? If the answer is yes, we may have to reinitiate the search. Second, review the approaches (Chapter 8) and choose a working approach that is not cause-based. If none seem feasible, then we may decide that there is no dominant cause for this problem as formulated and go back to the beginning and reformulate the problem. Finally, we may abandon the problem and accept that Statistical Engineering has failed.

Care in Running an Experiment

When conducting an experiment on a production process, the team requires excellent communication with all parties involved. In practice, things can go wrong, such as:

- Lost runs or repeats within a run
- The level of the suspect may vary within the run or be held at the wrong level over the run
- Unsuspected consequences because of the intervention in the process

More advice on planning and conducting experiments is given in Hahn (1984), Coleman and Montgomery (1993), and Robinson (2000).

13.3 VERIFYING A DOMINANT CAUSE FROM A SHORT LIST OF SUSPECTS

When verifying a dominant cause from a short list of suspects, the experimental plan is more complicated, but many issues are the same as in Section 13.2. Since we plan to investigate all of the suspects simultaneously, we call a particular combination of the input levels used in a run a *treatment*. When the treatments are combinations of the levels of two or more inputs, we call the plan a *factorial experiment*. In other contexts, the inputs are called *factors*, which explains this name. In the plan for the experiment, we must:

- Choose the high and low levels for each suspect.
- Determine the treatments.
- Define a run and determine the number of repeats.
- Decide how much replication is necessary.
- Decide whether randomization of the run order is worth the cost.

We use two levels for each suspect, selected near the extremes of the normal range of variation of the input. If we have three suspects each with two levels, there are eight possible treatments, as shown in Table 13.2. If we have two suspects, there are four possible treatments; with four suspects, there are 16 possible treatments.

We recommend using all the possible treatments in the experimental plan. This is called a *full factorial experiment*. With three suspects, this means we need at least eight runs. Note that even if we do not replicate any treatments, the addition of other suspects in the verification experiment gives us some built-in replication (also called *hidden* replication in some texts). As seen in Table 13.2, half the experimental runs will be conducted with each

Table 13.2 A two-level factorial experiment with three suspects and eight treatments.

Treatment	Level of Suspect A	Level of Suspect B	Level of Suspect C
1	Low	Low	Low
2	Low	Low	High
3	Low	High	Low
4	Low	High	High
5	High	Low	Low
6	High	Low	High
7	High	High	Low
8	High	High	High

of the suspects at their low and high levels. For example, we use the low level of suspect C on four runs and the high level of suspect C on four runs.

To analyze the results of factorial experiments, we rely on comparing averages and graphical displays.

Brake Rotor Balance

In the brake rotor balance problem described in one of the case studies, rotors required rework if their balance weight was too high. Using a group comparison, the team found two suspect dominant causes, core position in the mold and core thickness. Another suspect came from the chronological link between a change to new tooling in the core-making process and the increase in balance weight rejects. The team planned a verification experiment with the three suspects.

In Table 13.3, we show the two levels of each suspect chosen to capture its full range of variation. The two core-related suspects were expected to produce good results at their nominal levels and deteriorating performance away from the nominal. The team decided to make eight rotors for each of the eight treatments. That is, there were eight runs with eight repeats. No treatment was replicated.

Table 13.3 Suspects and levels for brake rotor verification experiment.

Suspect	Low level	High level
Tooling	Old tooling (4-gang)	New tooling (6-gang)
Core position	Offset (20-thousandths of an inch)	Nominal
Thickness variation	30-thousandths of an inch	Nominal

Table 13.4 Brake rotor verification experiment results.

Treatment	Tooling	Core Position	Thickness variation	Run order	Average balance weight
1	4-gang	Offset	30-thousandths	8	0.56
2	4-gang	Offset	Nominal	1	0.17
3	4-gang	Nominal	30-thousandths	3	0.44
4	4-gang	Nominal	Nominal	7	0.08
5	6-gang	Offset	30-thousandths	2	1.52
6	6-gang	Offset	Nominal	5	0.37
7	6-gang	Nominal	30-thousandths	4	1.34
8	6-gang	Nominal	Nominal	6	0.03

To obtain the cores for the experiment, the team measured thickness variation and sorted cores until they had 16 with high thickness variation and 16 with nominal thickness variation from each set of tooling. The experiment required careful planning since the balance weight of each casting can only be determined after shipping and machining. The castings were tagged for identification after production and tracked through the subsequent process.

The team randomized the casting order of each treatment as given in Table 13.4. The 64 rotors were cast and machined as planned. The experimental plan and the measured balance weights are given in the file *brake rotor balance verification*. The average weight of the eight repeats for each run is given in Table 13.4.

To analyze the data, we plot the weights by treatment in Figure 13.4. We see there are large differences in the balance weights produced by different treatments and relatively little difference within each treatment. We also notice that we have seen roughly the full extent of variation in balance weight given by the problem baseline (the dashed line on Figure 13.4). If this were not the case, we would conclude that we have not found a dominant cause.

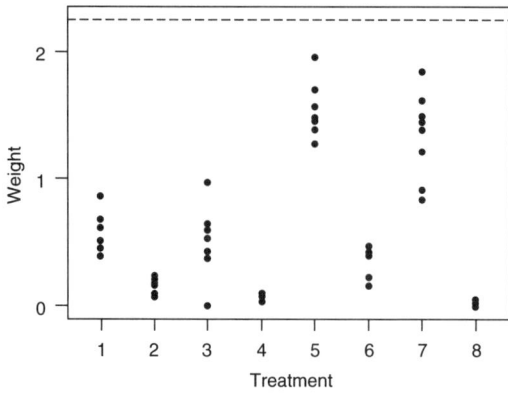

Figure 13.4 Weight by treatment for the brake rotor verification experiment (dashed line gives the full extent of weight variation).

We analyze the experimental results using the within-run average weight for each treatment. We estimate and rank the *effects* of the suspects. A *main effect* due to a particular suspect is the difference in average output (high-low) for that input. For example, using the data in Table 13.4, the main effect for tooling is:

$$\frac{1.52+0.37+1.34+0.03}{4} - \frac{0.56+0.17+0.44+0.08}{4} = 0.50$$

An *interaction effect* measures the change in the main effect for one input as a second input changes. For example, the interaction effect for tooling and thickness is:

$$\frac{1}{2}\left\{\left[\frac{1.52+1.34}{2} - \frac{0.56+0.44}{2}\right] - \left[\frac{0.37+0.44}{2} - \frac{0.17+0.08}{2}\right]\right\} = 0.43$$

The first term on the left side of the equation (except for an extra factor of ½) compares the effect of changing the tooling when the thickness variation is 30 thousandths. The second term measures the effect of changing the tooling when the thickness variation is nominal. Half the difference in the two effects is the interaction.

We can similarly define interactions for three inputs and more. Interactions are important if the effect on the output of changing levels in one suspect depends on the level of another suspect. In terms of finding a dominant cause, we may get extreme values of the output only if both inputs are at their high level. In this instance, we say that the dominant cause is an *interaction* between the two suspects.

In this experiment with eight treatment combinations, we can estimate seven effects. The number of effects estimable is always one less than the number of treatments. We fit a *full model* with all possible effects and construct a Pareto chart (see Appendix F for MINITAB instructions) of the unsigned effects to distinguish between those that are large and small. In Figure 13.5, we see that thickness variation, tooling, and the interaction between thickness variation and tooling are relatively large effects. The vertical dashed line in Figure 13.5 is added by MINITAB based on a test of significance. We ignore this dashed line in our interpretation.

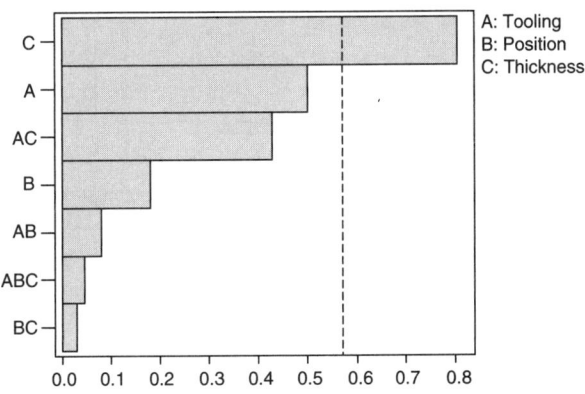

Figure 13.5 Pareto chart of the effects for brake rotor verification experiment.

Table 13.5 Balance weight averages by thickness variation and tooling.

Tooling	Thickness variation		Average
	0.03"	Nominal	
4-gang	0.50	0.12	0.31
6-gang	1.43	0.20	0.82
Average	0.97	0.16	

The effect of core position is small, so we eliminate it as a suspect dominant cause. Since tooling and thickness variation have a relatively large interaction, we need to study the effect of these two suspects simultaneously. In Table 13.5, we calculate the average balance weight for the four combinations of tooling and thickness variation.

We can see the interaction in the table. The effect of changing thickness variation is much greater for the 6-gang tooling. We use a multivari chart, as shown in Figure 13.6, as a convenient way to display the interaction. Note here we plot the individual values as well as averages. From the chart, we see that for the 4-gang tooling, the effect of changing core thickness variation from nominal to the extreme level is moderate. However, for the 6-gang tooling the effect is very large.

In summary, the team found that core position had little effect, and the combination (that is, interaction) of the switch to the 6-gang tooling and thickness variation was a dominant cause. The conclusion that position was not important was a surprise, since that was identified as a suspect in a group comparison (see Chapter 12). This illustrates the potential danger of proceeding without verification. The team could have spent time and effort trying to reduce core position variation with little impact on imbalance.

We finish the story of the brake rotor balance problem in one of the detailed case studies on the CD-ROM.

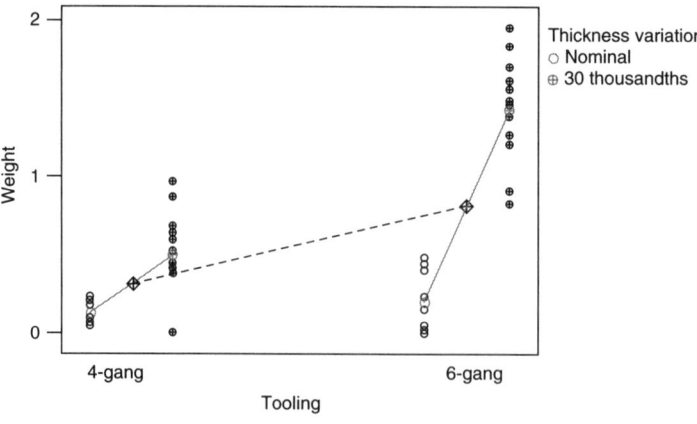

Figure 13.6 Interaction between tooling and core thickness variation.

13.4 FURTHER ISSUES AND COMMENTS

We discussed a number of issues in Section 13.2 relating to experiments to verify a single suspect. We need to consider all of these issues in any verification experiment. Here we consider some further issues relevant in the conduct of an experiment with several suspects.

How Short a List of Suspects Do We Need?

We recommend that you produce as short a list of suspects as possible before undertaking an experiment. If the list is long, the experiment becomes difficult to manage and expensive.

First, if the number of suspects is k, even with only two levels per input, the number of possible treatments is 2^k. For a full factorial design (one replicate of each treatment), the number of runs becomes unfeasible if the number of suspects is greater than about four. We can then use a *fractional factorial experiment* (see Chapter 15 and the supplement to Chapter 15) to reduce the number of runs, but we pay a price in terms of information about interactions. We look at another alternative in the chapter supplement.[4] Second, all of the inputs in the list of suspects are naturally varying and, during each run of the experiment, we need to control the values to define the treatment for the run. With many suspects, the task of simultaneously controlling the inputs becomes very difficult.

We illustrate the dangers of searching for a dominant cause among a long list of suspects using experimental plans with the following example. A team was assigned to reduce scrap due the geometric distortion of rear axle hypoid gear components during heat treatment. The pinion and gear set are critical to provide smooth and quiet transmission of power from the driveshaft to rear axle driveline. In a baseline investigation, the team found the scrap rate was 6%. Using Pareto analysis, they focused on the pinion. They decided to search for a dominant cause without using the recommended method of elimination. Based on existing process knowledge, the team identified five suspect causes: the average size of four pinion gear teeth, variation in tooth size, stem runout before heat treatment, position of the cone in basket (up or down) and the furnace track (left or right). They planned a full factorial experiment using all 32 possible treatment combinations of the five suspects at two levels each. They decided to have two pinions (repeats) per run, so they needed 64 pinions in total.

The team could easily control the last two suspects for each run. However, to find the required eight pinions for each of the eight treatment combinations defined by the first three suspects, the team needed to measure and sort pinions before heat treatment. They continued measuring and sorting for a number of days. Then disaster struck. During the night shift, there was a shortage of parts for the heat treatment process. There was great pressure to sustain volumes, so when members of the night shift found the pinions set aside for the experiment, they processed them. The next day the team decided the proposed experiment was too complicated. They decided to look for a dominant cause with simpler observational investigations.

Can We Verify the Suspects One at a Time?

To verify a dominant cause from a list of suspects, we do not recommend the use of one-at-a-time experiments, where we change only a single suspect at a time. These traditional experiments are a poor choice because they cannot find important interactions between the suspects. For more information on the dangers of one-at-a-time experiments, see Montgomery (2001).

Sometimes, we can test all the suspects simultaneously. In the window leaks example discussed in Chapter 12, the team isolated the primary seal gap and the plastisol application as suspects. To verify, the team selected eight trucks that had passed the aggravated leak test and reinstalled the rear window using different windows and seals, ensuring that the primary seal gap exceeded 1.5 millimeters and that the plastisol score was between 3 and 4. Seven of the eight trucks leaked when they were retested. The conclusions from the verification experiment and group comparison are summarized in Figure 13.7. All the experimental runs came from the upper left quadrant and the results verified that the combination of poor plastisol application (low score) and high primary seal gap resulted in leakers. The team proceeded to improve the plastisol application process by ensuring that all trucks received a plastisol score greater than 4.5.

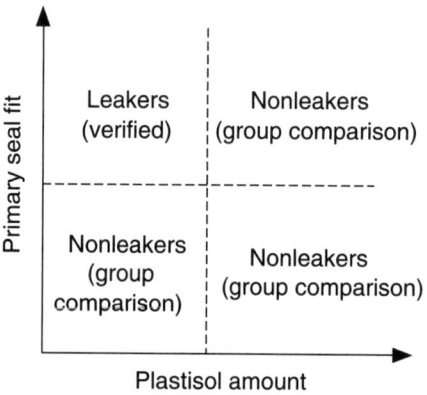

Figure 13.7 Window leaks verification experiment results.

Verification Experiment Summary

We have a list of one or more suspects.

Question

In the current process, are any of the suspects a dominant cause?

Plan

- For each suspect, choose two levels at the extremes of their normal range.
- Define the runs using available information about the time-based family that is home of the dominant cause.

- Select the number of runs. For:
  - A single suspect: use at least three replicates for each level (at least six runs)
  - Two suspects: use a full factorial design and at least two replicates (at least eight runs)
  - Three or more suspects: use a full factorial design with at least one replicate for each treatment
- Determine the number of repeats for each run.
- Randomize the order of the runs as much as possible.
- Make everyone potentially impacted aware of the plan.

Data

Carry out the experiment. Record the output, suspect levels, treatment number, and order for each run. Use a separate row for each repeat.

Analysis

- Plot the output against the treatment number. Add horizontal lines showing the full extent of variation in the output.
- For several suspects, fit a full model and construct a Pareto plot of the main and interaction effects. For large interactions, create a multivari chart with the corresponding inputs.

Conclusion

- If the output does not show the full extent of variation, then none of the suspects is a dominant cause.
- A suspect with a large effect relative to the full extent of variation is a dominant cause.
- If there is a large interaction, the dominant cause involves two (or more) suspects.

We summarize the terminology for experimental plans in Table 13.6.

Table 13.6 Designed experiments terminology.

Term	Meaning
Balance	An experimental design is balanced if there is the same number of replicates for each treatment and the same number of repeats in each run.
(Full) factorial experiment	An experimental plan where all possible combinations of the input levels are used to define the treatments.
Interaction effect	The change in the main effect for one input as a second input changes.

Continued

Table 13.6 Designed experiments terminology. *(Continued)*

Term	Meaning
Levels	The different values of an input characteristic used in an experiment.
Main effect	The change in the average output produced by a change in one input.
Randomization	Radomizing the order of the runs.
Repeat	More than one part (or measurement) made within a run.
Replication	Carrying out more than one run for each treatment.
Run	Assigning the treatment, running the process, and measuring the output.
Treatment	A particular combination of input levels.

Key Points

- We use an experimental plan to verify that one or more suspects obtained using the method of elimination is a dominant cause.
- For each suspect we choose two levels at the extremes of its normal range.
- In the verification experiment, we recommend:

  –For a single suspect: three or more replicates for each level
  –For two suspects: two or more replicates at each of the four treatments
  –For three or four suspects: a full factorial experiment
- We identify a dominant cause by examining main effects and two input interactions.
- In the analysis, we check that the full extent of variation in the output is seen over the runs of the experiment. Otherwise, none of the suspects varied in the experiment is a dominant cause.

Endnotes (see the Chapter 13 Supplement on the CD-ROM)

1. The difference between repeats and replicates is a great source of confusion in the analysis of experimental data. We explore this issue briefly in the supplement to justify the analysis methods we propose.
2. When we identify a suspect cause using the method of elimination, we know that the pattern of variation in the suspect must match that of the output. To verify that the suspect is a dominant cause, we must show that there is no other cause that also matches this pattern of variation. In the

supplement, we explain how randomization and replication can help to reduce the risk of falsely identifying a suspect as the dominant cause. We also explain how we implicitly use blocking, the third fundamental principle of experimental design, to help to avoid this risk.
3. See note 2.
4. In the supplement, we briefly discuss an alternative verification experiment based on a technique called *variables search*.

 Exercises are included on the accompanying CD-ROM

PART IV

Assessing Feasibility and Implementing a Variation Reduction Approach

Opportunity is missed by most people because it is dressed in overalls and looks like work.

—Thomas Edison, 1847–1931

In this final part of the book, we address revisiting the choice of working variation reduction approach, the issues around the assessment, implementation of each approach, and the validation of a solution. The choice of working approach may need to be reconsidered in light of the process knowledge obtained in a search for a dominant cause or in assessing feasibility of a particular approach. We provide detailed how-to directions for assessing feasibility, including consideration of costs, and for implementing each of the seven approaches. We also discuss validating a solution.

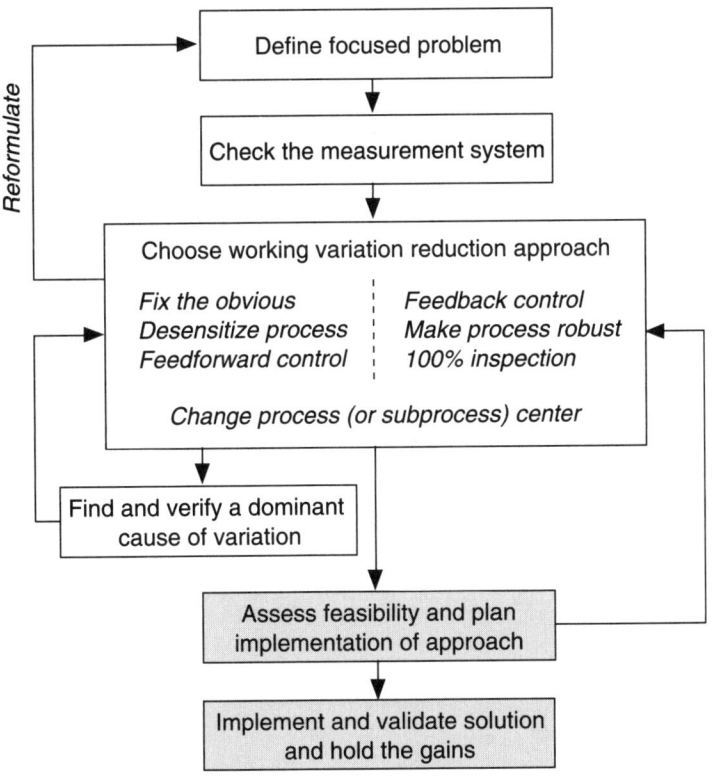

Statistical Engineering variation reduction algorithm.

14

Revisiting the Choice of Variation Reduction Approach

If at first you don't succeed, try, try again.

—Proverb

After conducting a search for a dominant cause, the team should reconsider their choice of working variation reduction approach. In this chapter, we consider the options and how to choose an approach.

If a specific dominant cause has not been found, the team must select one of the three non-cause-based approaches, resume the search for a more specific cause, or abandon the project. If a dominant cause has been found, the team can now consider the feasibility of any one of the variation reduction approaches. They will have accumulated substantial information about the behaviors of the output and the causes that can be used in selecting the approach.

There are four options that directly use the knowledge of the dominant cause:

- Fix the obvious.
- Compensate for the variation in the dominant cause (desensitization or feedforward control).
- Reformulate the problem in terms of the dominant cause and reapply the algorithm (that is, reduce variation in the cause).
- Continue the search for a more specific dominant cause. This is an informal version of reformulation.

The variation reduction approaches that do not require knowledge of a dominant cause should also be considered:

- Use feedback control on the output.
- Make the process robust to variation in the unknown dominant cause.
- Implement or tighten 100% inspection on the output.

Given the knowledge accrued during the search for the cause, the non-caused-based approach may be more or less feasible. The final option is to abandon the project and devote the resources to another problem with a greater likelihood of success.

We explore how to decide among these choices. Since there are few general rules, we present a large number of examples. We consider costs and likelihood of success. There is considerable risk and uncertainty in each choice, but we cannot proceed without making this decision.

14.1 FIXING THE OBVIOUS: IMPLEMENTING AN AVAILABLE SOLUTION

In many problems, once a dominant cause has been identified, the team has sufficient knowledge to determine an obvious solution and move to the Implement and Validate Solution stage of the algorithm. In making this decision, the team must consider possible side effects and the cost of the fix relative to the costs associated with the problem.

The definition of *obvious* depends on the process and the level of process knowledge. The team has an obvious fix if they are confident that it is feasible and there is no need to learn more about the process before it can be implemented. The key issue here is cost. There may be obvious solutions to the problem available at the start of the project (for example, buy new equipment), which are eliminated because of high cost. The team may now identify a lower-cost obvious solution that was not apparent without knowledge of the dominant cause.

There are many examples.

Engine Block Leaks (Center Leaks)

In Chapter 6, we described a project of leaking engine blocks and how the project was focused to three problems related to leaks at three different locations. The team discovered that the dominant cause of the center leakers was core breakage during the casting process. Based on their process knowledge, the team knew they could eliminate or reduce the core breakage by better supporting the core in the mold during the pouring operation. The team added several chaplets (small steel inserts) set in the mold to support the core. They knew that this process change would produce no unfavorable side effects. The change eliminated the problem of center leakers. The extra cost of the chaplets was more than justified by the reduced scrap costs.

Engine Block Leaks (Rear Intake Wall)

The team also found leaks at the rear intake wall of the casting. Using a group comparison (see Chapter 12), they discovered that the dominant cause of this type of leak was wall thickness at a specific location in the left rear wall. Thin walls led to leaks. An obvious solution was to change the mold to make the wall thicker. This process change could be made with a small one-time cost but had negative consequences in terms of the weight of the block. The team made the change and virtually eliminated rear intake wall leakers.

Truck Pull

In Chapter 10 and elsewhere, we described a project to reduce variation in a truck wheel alignment process. The team had access to alignment data that was automatically collected for every truck produced. Each truck was measured on one of four alignment machines. At one point during the project, the team looked at the right caster data, stratified by the alignment machine. The daily averages are plotted in Figure 14.1.

The team was surprised to see the persistent differences among the four gages. The trucks enter the gages haphazardly, so the observed differences must be due to differences in the alignment machines. The alignment machine was a dominant cause of variation in right caster. This cause was not acting during the baseline investigation or the initial investigation of the measurement system. The team took immediate action to recalibrate the four gages to remove the systematic differences. To prevent recurrence, they established a daily monitoring program to correct such differences. There was a small cost associated with the daily check.

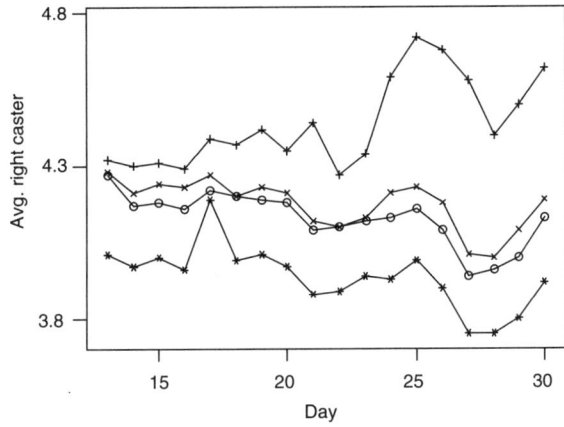

Figure 14.1 Right caster daily averages by alignment machine.

Cylinder Head Shift

In a casting operation, excess side shift variation was a problem. A multivari investigation (see Chapter 11) found the dominant cause of side shift acted in the pattern-to-pattern (cavity-to-cavity) family. The team had the process knowledge to make a one-time adjustment to the dies to shift the pattern averages and reduce the pattern-to-pattern side shift. They then monitored the process to detect the recurrence of the problem before it became material. The value of the variation reduction was greater than the costs of the one-time pattern adjustment and the monitoring.

Fascia Cratering

In the fascia cratering example discussed in Chapter 11, the team found that major cratering occurred on every tenth fascia. They concluded that the dominant cause was the mold release spray. With this knowledge, an obvious (to the team) and immediate short-term solution was to wipe the mold after the spray. This solution had the potential to introduce dirt in the mold and met with resistance from the mold operators. The obvious fix was not feasible. The team decided instead to investigate different mold sprays. That is, they adopted desensitization as the working approach.

Window Leaks

As reported in Chapter 12, a team found that the dominant cause of truck rear window upper ditch leaks was an interaction between the primary seal gap and the plastisol application to the seams. We reproduce the results in Figure 14.2.

The pattern on the plot suggested an obvious fix. The instructions were changed so that the operators where the plastisol was applied, brushed the plastisol to ensure coverage of the critical seals. Because of this change, the rear window upper ditch leaks were totally eliminated. Note that the variation in primary seal gap was not changed. The process was made less sensitive to this variation. The team also decided to pursue another solution based on changing the primary seal to reduce the variation in gap. The operating costs of this proposed solution were less than those of the obvious fix. However, management postponed implementation of this alternative solution because it required robotic application and substantial capital expenditure.

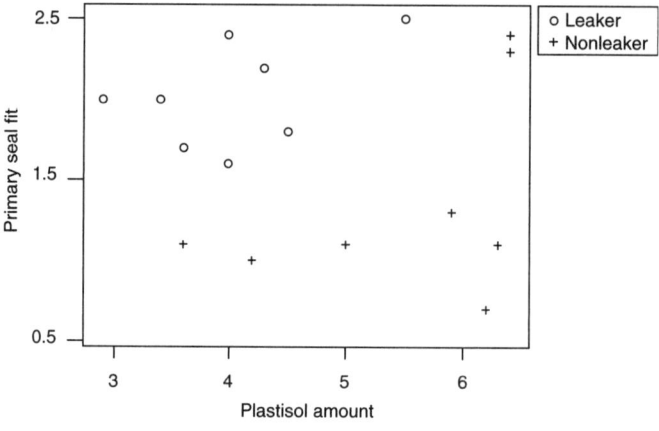

Figure 14.2 Plot of primary seal fit by plastisol amount.

Hubcap Damage

Because of customer complaints, a team was assigned a goal of reducing the incidence of wheel trim and hubcap damage. The team discovered that the dominant cause of broken retaining legs and other damage was a combination of cold weather and contact with curbs. Through comparison with competitors, they found that an obvious fix was to change the hubcap material and design. The team replaced the brittle existing ABS hubcap with a new design made of mineral-reinforced polypropylene. In addition, they increased the number of retaining legs. The increased cost was justified by the elimination of customer complaints.

Assembly Nuts Loose

In an assembly process, a power wrench driven by pneumatic pressure tightens nuts to a specified torque. The dominant cause of loose nuts was found to be occasional drops of pneumatic pressure. An air pressure meter was installed in the air line. An alarm sounded if the pressure dropped below a critical point providing constant inspection of the cause. This is an example of error proofing as defined by Shingo (1986).

Comments

The Fix the Obvious approach is a catchall category that makes use of the other variation reduction approaches. For example:

- 100% Inspection (for example, loose assembly nuts)
- Moving the output center (for example, rear intake wall leakers, cylinder head side shift, truck pull)
- Desensitizing the process (for example, window leaks)

When applying the Fix the Obvious approach, the team should ensure the problem does not recur. In the truck pull, camshaft journal diameter, and cylinder head shift examples, the fix was obvious but all three problems were likely to recur unless a new control or regular maintenance scheme was put in place.

Fix the Obvious is the preferred choice because it applies when there is a clear solution to the problem. There is little uncertainty that the fix will be effective. The main considerations are the cost of the fix relative to the gain and potential negative side effects.

14.2 COMPENSATING FOR VARIATION IN THE DOMINANT CAUSE

If there is no obvious solution that is economical, we can look for changes in the process (that is, changes in fixed inputs) that will eliminate or reduce the effect of the dominant cause. In particular, we are interested in deciding whether we can compensate for variation in the dominant cause through process desensitization (Chapter 16) or feedforward control (Chapter 17).

Refrigerator Frost Buildup

In Chapter 1, we described the problem of frost buildup in refrigerators, where the dominant causes were environmental and usage inputs outside of the control of the manufacturer. The team could not reduce variation in the causes and adopted the Desensitization approach. The team had several design changes in mind that might reduce the effect of the environmental causes of frost buildup. See Chapter 16 for further details.

Iron Ore Variation

A team was charged with reducing variation in iron ore composition as the ore was loaded into ships. This was a reformulated problem from the customer, a steel mill that wanted to reduce variation as the ore was processed. The dominant cause of the variation was the composition of the ore as it came out of the ground. To compensate for this cause, the team decided to look at new ways to create and dismantle stockpiles that were placed in the process to help blend the ore. They could not estimate the costs or benefits without further investigation.

Crossbar Dimension

In Chapter 12, we discussed a problem to reduce variation in a crossbar dimension of a molded part. From the baseline investigation, the full extent of variation was –0.3 to 2.0 thousandths of an inch. The team found that barrel temperature was a dominant cause as shown in Figure 14.3.

At first, the team considered reformulating the problem. Reducing variation in barrel temperature would result in reduced variation in crossbar dimension. Looking closely at Figure 14.3, the team realized there was evidence of a nonlinear relationship between barrel temperature and crossbar dimension. We have added a quadratic fit to the scatter plot to make this conclusion clearer. The variation in crossbar dimension is greater as the barrel temperature varies from 74° to 77° than it is if the barrel temperature ranges from 77° to 80°. The team decided to try to desensitize the process to barrel temperature variation by increasing the average barrel temperature.

The team could not predict the benefits of increasing the average barrel temperature from the data shown in Figure 14.3 since they needed to extrapolate beyond the normal range of barrel temperature. The team could assess the direct cost of the change but they were unsure if the process would gracefully tolerate barrel temperatures much above 80°. That is, there might be negative side effects. Further investigation was required.

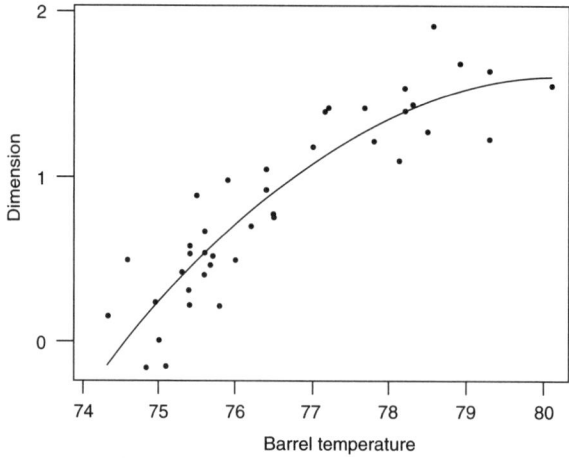

Figure 14.3 Scatter plot of crossbar dimension versus barrel temperature.

Comments

There is uncertainty about finding an effective solution if the team decides to compensate for the effects of a dominant cause. In the refrigerator frost buildup example, the team did not know if they could find affordable changes to the design that would desensitize the refrigerator to changes in the usage and environmental inputs. They could not quantify the benefits until they had investigated the changes. They needed to carry out further process investigations with no certainty of an efficient and effective resolution. Fortunately, they were able to find good design changes. Otherwise, they would have had to absorb the high costs of investigation and reconsider the other variation reduction approaches.

14.3 REFORMULATING THE PROBLEM IN TERMS OF A DOMINANT CAUSE

We can reduce output variation by reducing the variation in a dominant cause. In reformulation, we restart the algorithm with a new problem defined in terms of the dominant cause. We specify the goal for the new problem so that we meet the goal of the original problem. Reformulating the problem is the classic route to variation reduction. Rephrased, we are told to find the "root cause" of the problem and then somehow deal with that cause, hopefully with an obvious fix.

Reformulation is attractive because the problem is moved upstream. This may reduce the cost and complexity of the solution. Reformulation is unattractive because we are replacing one problem with another and we may be no closer to a solution. Note that we may adopt any of the seven variation reduction approaches to solve the reformulated problem. Also, we may reformulate several times before adopting one of the variation reduction approaches.

Sunroof Flushness

Customers were dissatisfied if the sunroof was not flush with the roof of the car. In the sunroof installation process, there was a 90% rework rate due to flushness variation. A team set out to reduce the rework costs.

Flushness was the difference in height between the sunroof seal and the metal roof. It was measured using digital calipers at six points, three at the front and three at the back of the sunroof. A baseline investigation showed that flushness variation was largest at the two front corners. Front corner flushness ranged between –3.5 and 4 millimeters, with an estimated standard deviation of 1.25 millimeters. The team established a goal of reducing the front corner flushness standard deviation to less than 0.5.

Based on engineering knowledge, the team suspected the dominant cause of flushness variation was either crown height or attachment pad height. When the roof panel was adapted to allow installation of a sunroof, six attachment pads were added. The team carried out an investigation using two sets of six vehicles with extreme flushness. For each vehicle, they removed the sunroof module and measured the attachment pad heights and roof crown height at the front and back. The data are given in the file *sunroof flushness input-output*. Here we report the results for left front flushness only. The team used regression analysis

220 Part Four: Assessing Feasibility and Implementing an Approach

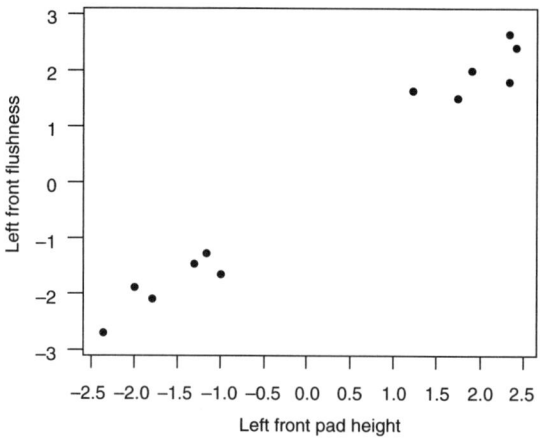

Figure 14.4 Scatter plot of flushness versus left front pad height.

and the scatter plot shown in Figure 14.4 to demonstrate a clear connection between the corner flushness and pad height closest to the corner. Roof crown height was eliminated as a suspect.

The team assumed that pad height was the dominant cause of the variation in flushness without verifying this conclusion. They decided to reformulate the problem in terms of pad height. They carried out a baseline investigation and found the standard deviation in pad height was 1.18 millimeters.

To determine a goal for the reformulated problem, they used the fitted equation from the regression analysis:

```
The regression equation is
left front flushness = -0.126 + 1.05 left front pad height

Predictor        Coef      SE Coef         T         P
Constant      -0.12631     0.09352      -1.35     0.207
left fro       1.04811     0.05015      20.90     0.000

S = 0.3222     R-Sq = 97.8%     R-Sq(adj) = 97.5%

Analysis of Variance

Source           DF          SS          MS         F         P
Regression        1      45.348      45.348    436.87     0.000
Residual Error   10       1.038       0.104
Total            11      46.386
```

The regression analysis corresponds to modeling the output/cause relationship as

$$\text{flushness} = a + b^* \text{ pad height} + \text{noise}$$

where $a + b*$ pad height represents the effect of the dominant cause and *noise* the effects of all other causes. The model constants a (intercept) and b (slope) describe the approximate linear relationship between the output flushness and the dominant cause pad height. Based on the model (see Chapter 2), and assuming that the dominant cause varies independently of the other causes, we have

$$\text{sd}(\textit{flushness}) = \sqrt{b^2 \text{sd}(\textit{pad height})^2 + \text{sd}(\textit{noise})^2}$$

From the regression analysis, we estimate b as 1.05. From the two baseline investigations, we have estimates of the standard deviations of flushness and pad height. Substituting the three estimates into the given equation, we get an estimate for the standard deviation of the noise to be 0.118 millimeters.

Now, we can use the equation to translate the goal of reducing the flushness standard deviation variation to 0.5 into a goal in terms of pad height variation. We use the estimate for the slope b from the regression analysis and the estimated standard deviation of *noise* from the given calculation. Substituting the estimated values for b, the standard deviation of *noise,* and the goal for flushness standard deviation, we get

$$0.5 = \sqrt{1.05^2 \text{stdev}(\textit{pad height})^2 + 0.165^2}$$

Solving gives 0.45 as the goal for pad height standard deviation. To meet the original goal for flushness, in the reformulated problem we need to reduce pad height variation by over 60%.

Battery Seal Failure

Management assigned a team to reduce the frequency of battery leaks. Using a group comparison, the team determined that the dominant cause of leaking was seal strength between the top and bottom of the battery casing. See the results in Figure 14.5. They concluded that stronger seals would eliminate the leak problem. The problem was reformulated to increase the average seal strength. Based on Figure 14.5, the team set the goal to increase seal strength to a minimum of 320 pounds. They did not try to quantify the relationship between the continuous cause, seal strength, and the binary output and hence, they could not quantify the potential gain of increasing the average seal strength.

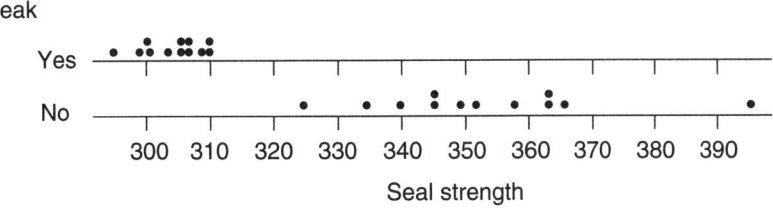

Figure 14.5 Dot plot of seal strength for leakers and nonleakers.

In this example, reformulation is especially useful since the original output is binary and the dominant cause is continuous. It should be easier to find a solution to the reformulated problem than the original since, in any investigation, more information is available from each battery when we measure a continuous as compared to a binary output. Since seal strength is a destructive measurement, this gain is offset by increased costs.

Roof Panel Updings

In a stamping operation, there was a problem with updings (small outward dents in the metal surface) on a roof panel. There was 100% inspection to sort the panels. Those with updings were reworked. The team established the baseline and also found that the measurement system was adequate. Next, using a multivari investigation (see the exercises in Chapter 11), they found that the dominant cause acted in the pallet-to-pallet family. Further investigations showed that updings were caused by particles on the metal blanks.

The team decided to reformulate the problem in terms of the amount of particulate on the steel blanks. Since the dominant cause acted from pallet to pallet, the team was able to use leverage (see Section 9.3) to roughly define the baseline for the amount of particulate. They determined the full extent of variation for particulate amount by counting the total number of particulates on blanks selected from two pallets that had produced stamping with extreme low and high upding counts. This was an inexpensive way to establish the baseline variation compared with the recommended investigation for a new problem as given in Chapter 6.

Comments

We choose to reformulate only after we have considered the other options. Reformulation of a problem is not a solution. That is, we cannot continue to reformulate indefinitely. Eventually one of the variation reduction approaches must be applied. If we can repeatedly find a dominant cause, we can reformulate the problem a number of times. However, the potential benefit arising from solving the reformulated problem is reduced in each iteration by the variation not explained by the identified dominant cause(s). Also, the final cause may be outside local control.

With reformulation, the dominant cause (an input) becomes the output. We then establish a baseline for the new output. We set the goal for the reformulated problem by exploiting what we have learned about the relationship between the cause and the original output. We can set this goal formally, as in the sunroof flushness example, or informally, as in the battery seal failure example.

There is little value in reformulating a problem in terms of a cause that is not dominant. Suppose, as in the sunroof flushness example, we found a cause that is linearly related to the output so that we have the equation

$$\text{stdev}(output) = \sqrt{b^2 \text{stdev}(cause)^2 + \text{stdev}(noise)^2}$$

If the cause is not dominant, then the standard deviation of the noise, the variation due to all other causes, is large fraction of the standard deviation of the output. Even if we completely eliminate the effect of the cause, we do not reduce the output standard deviation substantially.

14.4 CONTINUING THE SEARCH FOR A MORE SPECIFIC DOMINANT CAUSE

Continuing the search for a more specific dominant cause is similar to reformulation, but we proceed less formally. The key is the baseline we use in the search. With reformulation, we determine a new baseline and goal in terms of the identified dominant cause. If we continue the search for a more specific cause, we use the original baseline and continue to measure the original output.

We select this option when the identified dominant cause is vague and there is no obvious solution. In the truck pull example described earlier, the dominant cause of variation was systematic differences among the four gages. This was not a specific cause since we did not identify what was different among the four gages. However, an obvious fix was available.

Here are some examples when the team decided to look for a more specific cause.

Engine Oil Consumption

In the oil consumption example, discussed in Chapter 13, the supplier of the valve lifters was a dominant cause of variation. The two engine assembly plants had different valve lifter suppliers. An immediate fix to reduce customer complaints was to eliminate the valve lifter supplier linked to the excess oil consumption. However, moving to a single supplier was not practical for volume and contractual reasons. Since the identified cause was a large family of inputs including all the specific differences between the two suppliers, the team decided to look for a more specific dominant cause. In particular, the team explored the variation within the valve lifters from the problem supplier to determine a more specific cause of oil consumption.

Camshaft Lobe BC Runout

In the camshaft lobe runout example discussed in Chapter 10 and elsewhere, the problem was excess base circle runout. Through a number of investigations, the team found that a dominant cause of runout variation was heat treatment spindle as illustrated in Figure 14.6. The data for Lobe 12 are given in the file *camshaft lobe runout variation transmission*.

Based on this dominant cause, the team could not see how to apply one of the variation reduction approaches. To keep up with volume requirements, they had to continue using spindle 3. One possible solution was to reduce the runout center for spindle 3. However, the team did not know of an adjuster, and since runout is a measure of variation itself, finding

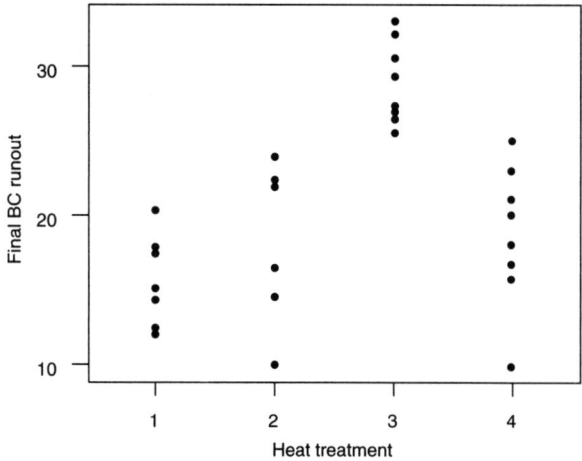

Figure 14.6 Plot of final base circle BC runout by heat treatment spindle.

an adjuster seemed difficult. The team decided to look for a more specific cause. They proceeded by comparing the heat treatment spindles, looking for reasons why they were different. In the end they were unable to find a specific cause. Out of desperation they arranged for overdue maintenance on the heat treatment operation. This somehow eliminated the spindle-to-spindle difference. Next, the team had to worry about how to prevent recurrence of the problem. They decided to monitor the runout after heat treatment on two parts per day from each spindle. Any large spindle-to-spindle differences triggered maintenance of the heat treatment operation.

Crankshaft Main Diameter

In the crankshaft main diameter problem considered in one of the case studies, the dominant cause of final diameter variation was found to be the diameter at an intermediate processing step. There was no obvious fix based on this knowledge. Also, there was no way to desensitize the final diameter variation to diameter variation at the intermediate step. The only processing step between the intermediate and final gages that changed the diameter was lapping to polish the surface. The team opted to look for a more specific upstream cause and continued to use the final diameter baseline.

Comments

We distinguish between reformulation and continuing the search for a more specific cause based on whether or not we determine a new baseline. We are more likely to reformulate in cases where the cause is a different continuous characteristic than the output. If the dominant cause is a discrete characteristic, like machine number or process stream, we typically continue the search for a more specific dominant cause without reformulating.

14.5 DEALING WITH DOMINANT CAUSES THAT INVOLVE TWO INPUTS

When a dominant cause is not a single input, there are a variety of ways to proceed. The choice of the best variation reduction approach depends on the nature of the dominant causes and the problem context. Sometimes we can address such problems by addressing only one of the causes. In other cases, we need to address both causes either simultaneously, or separately, possibly with different variation reduction approaches.

We illustrate using three examples.

Window Leaks

In the truck window leaks problem, the dominant cause of upper ditch leaks was the combination of a large primary seal gap and poor plastisol application to the seams. The proposed and implemented solution was a process change that required operators to brush the plastisol to ensure coverage of the critical seals. With this solution, the team addressed only one of the inputs involved in the dominant cause.

Door Closing Effort

The dominant cause of door closing effort, as discussed in Chapter 10, involved both the component and assembly families. Here there were two large causes acting in separate families. The team proceeded to address each cause (family) separately. Since each cause on its own was not dominant, they could not meet the project goal without addressing both.

Steering Wheel Vibration

In a problem to reduce steering wheel vibration, the dominant causes were found to be imbalances in two transmission components. Both components contributed equally to the problem. The proposed solution was to desensitize the process to both causes by vectoring the two components during the installation so that their imbalances tended to cancel out. Alternately, the team could have reformulated into two problems to address the imbalance in each component separately.

14.6 SUMMARY

The thought process in choosing how to proceed after finding a dominant cause is summarized in Figure 14.7.

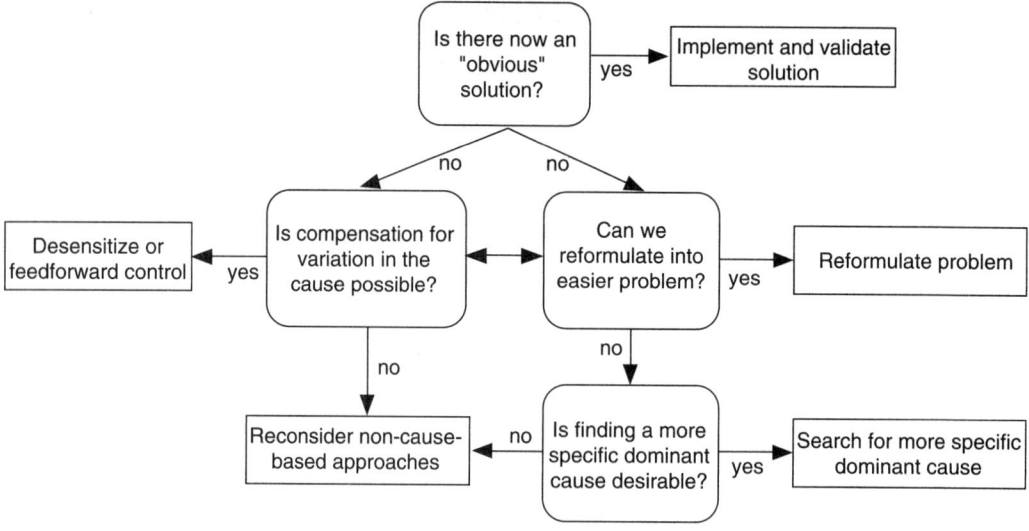

Figure 14.7 Flowchart to help decide how to proceed after finding a dominant cause.

 **Key Points**

- There are three options once a dominant cause has been found:
  – Fix the obvious.
  – Compensate for the cause using desensitization or feedforward control.
  – Reformulate the problem in terms of the dominant cause and reapply the algorithm (that is, reduce variation in the cause).
  – Continue searching for a more specific dominant cause.
- Subject to cost constraints and concerns over side effects, implementing an obvious fix is the most desirable option.
- There is a risk of choosing an approach that will be neither effective nor efficient because of the lack of knowledge. However, we cannot proceed with making a choice.

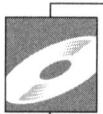 Exercises are included on the accompanying CD-ROM

15
Moving the Process Center

This time, like all times, is a very good one, if we but know what to do with it.

—Ralph Waldo Emerson, 1803–1882

Moving the process center is the only approach that directly addresses an off-target process center. The goal of the approach is to find a way to move the process output center either closer to the target or in the desirable direction if higher or lower is better. An example for a higher is better output is shown in Figure 15.1. We do not need to identify a dominant cause to apply this approach.

The only requirement for moving the process center is an *adjuster;* that is, a fixed process input that can be changed to move the process output center.

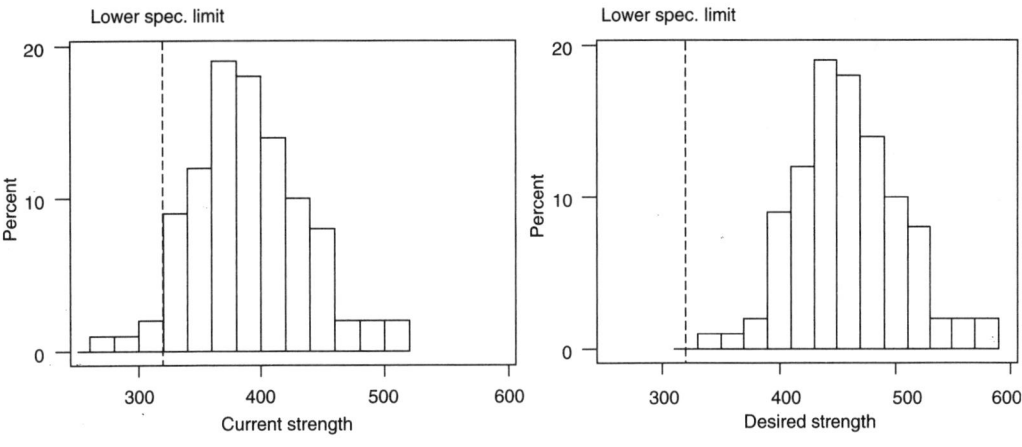

Figure 15.1 Changing the process center.

In many processes, the team will have an available adjuster and thus an obvious fix to the problem. To find an adjuster, we recommend an experimental plan. We use engineering knowledge and experience with the process to choose fixed inputs, called *candidates,* to vary in the experiment. There is a risk that an adjuster will not be found.

The potential costs of moving the process center include the costs of:

- An experiment to find an adjuster, if necessary
- A one-time change to the adjuster
- The ongoing operation of the process at the new setting for the adjuster

To assess the potential benefit of moving the process center, we imagine shifting the baseline process center to a new value.

15.1 EXAMPLES OF MOVING THE PROCESS CENTER

To illustrate, we discuss three examples where the team had to first find an adjuster before they could apply the Move the Process Center approach.

Battery Seal Failure

In the production of automotive batteries, a sealing process joins the top of the batteries to the base. Due to field complaints about leaky seals and high rework costs in the plant, management assigned a team to address the problem. Looking back at several months' production records, the team estimated that the baseline within-plant leak rate was 2.3%. The goal was to reduce this rate to less than 0.5% with the assumption that this would eliminate most of the field failures.

Note that the output is binary. Seals either leak or do not leak. The team suspected that a dominant cause of leaks was low tensile strength of the seal. To assess this idea, the team compared the tensile strengths (in pounds) of 12 leaky seals and 12 good seals. From Figure 15.2, we see that leaky seals are strongly related to low tensile strengths. Based on this conclusion, they

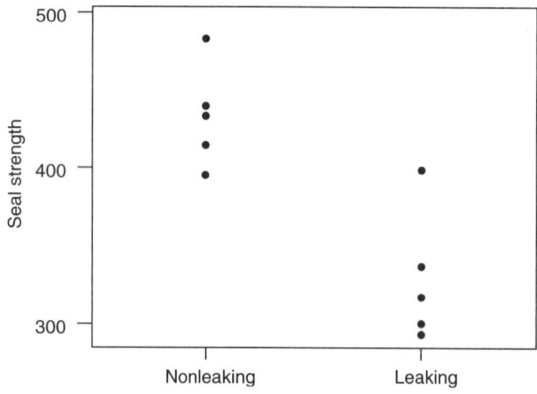

Figure 15.2 Battery seal group comparison results.

reformulated the problem in terms of tensile strength. The tensile strength was measured on a sample of 100 batteries selected from one week's production. The data are stored in the file *battery seal strength baseline* and the results are shown in the histogram in the left panel of Figure 15.1. From the group comparison, the baseline investigation, and engineering knowledge, the team set the minimum acceptable seal strength at 320 pounds, which corresponds to increasing the center of the process output by about 60 pounds. The desired process histogram is shown in the right panel of Figure 15.1. At this point, the team did not attempt to determine if this shift in the process center would meet the initial project goal of reducing the leak rate to less than 0.5%.

The team set out to find an adjuster. Based on engineering and process knowledge, they chose three candidates (fixed inputs) at two levels each. The further apart the levels, the more likely the experiment will detect an effect and the greater the risk of negative side effects. They planned a factorial experiment with eight runs. The candidates and levels are given in Table 15.1.

Table 15.1 Heat seal experiment candidates and levels (existing process levels are indicated by *).

Candidate	Low level	High level
Melt temperature	750°F	800°F*
Melt time	2.5 seconds*	3.1 seconds
Elevator speed	low	high*

In each of the eight experimental runs, five batteries were produced. The runs were conducted in random order. Treatment 7 corresponds to the existing process. The data are given in the file *battery seal strength move center* and in Table 15.2. There were no leaking seals in the 40 batteries produced in the experiment.

Table 15.2 Treatments and seal strength for battery seal experiment.

Treatment	Order	Melt temperature	Melt time	Elevator speed	Seal strength
1	7	Low	Low	Low	413, 505, 489, 452, 465
2	6	Low	High	Low	468, 493, 484, 393, 423
3	8	High	Low	Low	383, 368, 280, 377, 370
4	4	High	High	Low	383, 365, 353, 389, 353
5	1	Low	Low	High	440, 415, 483, 395, 433
6	3	Low	High	High	466, 387, 505, 393, 456
7*	5	High	Low	High	399, 294, 317, 300, 337
8	2	High	High	High	373, 379, 383, 385, 345

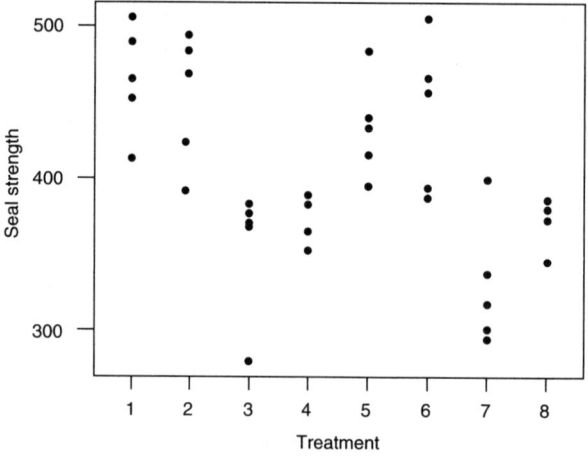

Figure 15.3 Seal strength by treatment.

In the analysis, we first plot the data by treatment as in Figure 15.3. We see, encouragingly, that many treatments have average strength greater than treatment 7, the current operating condition, and many meet the desired minimum of 320 pounds.

To isolate possible adjusters, we use MINITAB to fit a *full model* with all possible main and interaction effects. Then we look at the Pareto plot of the effects in Figure 15.4. Only the effect for melt temperature is large. Also, because none of the interaction effects is large, we conclude that melt temperature is an adjuster.

We can use the main effect plot for melt temperature, given in Figure 15.5, to assess quantitatively the effect of changing melt temperature. Decreasing the melt temperature by 50°F increases the average seal strength by about 90 pounds.

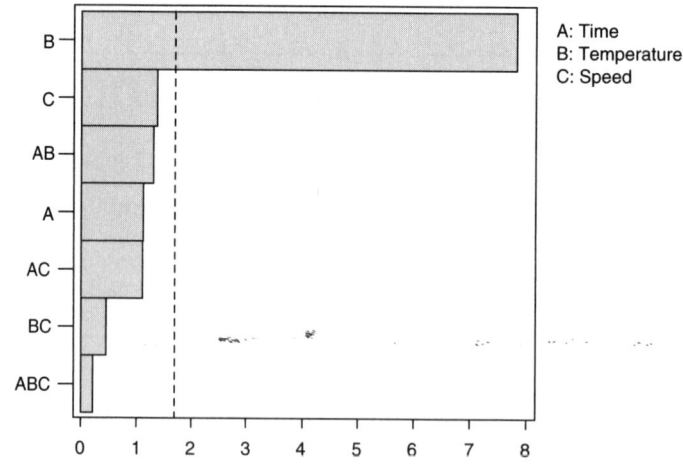

Figure 15.4 Pareto chart of effects for battery seal experiment.

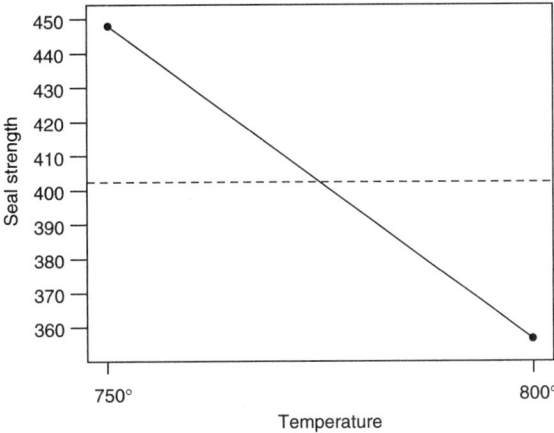

Figure 15.5 Main effect plot for melt temperature.

We do not know from this two-level experiment if the effect of the adjuster is linear. The team decided to use a melt temperature of 770°F, a little over halfway between the low and high levels in the experiment. They did not want to decrease the melt temperature too much for fear of creating other problems related to material flow. To validate the solution, they changed the melt temperature to 770° for one shift and carefully monitored the process for leaks and negative side effects. The results were very promising. Since the cost of the change was negligible, the team implemented the new melt temperature setting and monitored the process for two months. The leak rate was reduced to 0.6%.

Engine Block Leaks (Cylinder Bore)

In Chapter 6, we introduced a problem of leaking engine blocks with three distinct failure modes. The team found that the dominant cause of cylinder bore leaks was the occurrence of *dip bumps* on the barrel cores used to create the bores in the block when it was cast. Each sand core is dipped in core wash, a paintlike substance, to improve the surface finish of the casting. Dip bumps occurred when the core wash ran before drying. The dip bump score for each core was the total number of bumps divided by four, the number of barrels per core. The team reformulated the problem to reduce the average dip bump score and established a new baseline. They did not precisely quantify the relationship between the occurrences of a cylinder bore leak and the dip bump score. The team assumed if they could reduce the average dip bump score to close to zero, then they would eliminate cylinder bore leakers.

The team used a single operator to define and count dip bumps. They did not assess this simple measurement system but assumed that it would add little variation. The team decided not to search for a dominant cause of dip bumps. They felt strongly that dip bumps were inevitable with the current core wash. They decided to experiment with five core washes from different suppliers. They hoped to find a core wash that would reduce the average dip bump score to close to zero.

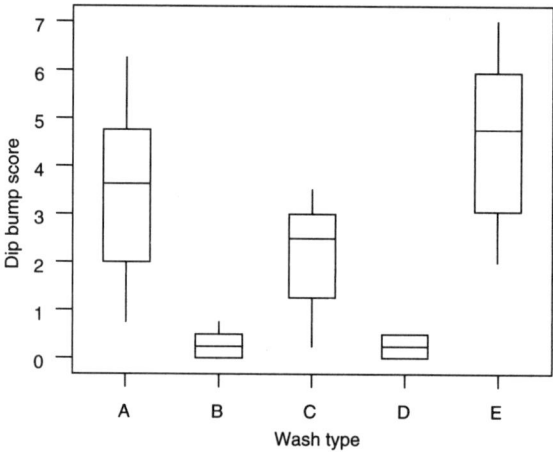

Figure 15.6 Box plots of dip bump score by core wash solution.

In the experiment, for each core wash, 50 cores were processed on each of two days. There were 10 runs, two replicates of each treatment, and 50 repeats for each run. Within each day, the order of treatments was randomized. The dip bump score for each of 500 cores was recorded. The data are given in the file *engine block leaks move center*. The cores from the experiment were recycled. Core wash A is the current wash.

The experimental results are summarized in the Figure 15.6. There were virtually no dip bumps with core washes B and D. The team recommended changing the process to use core wash B since it was cheaper than D.

During the validation of core wash B, the foundry scrapped a half shift worth of engine blocks and shut down the engine assembly line. This disastrous event occurred because, over the longer time frame used in the validation investigation, the new water-based core wash separated and failed to effectively coat the cores. The experiment to assess the different core washes was conducted over such a short time that the separation problem had not occurred. The team had not anticipated that continuous mixing was required with the new wash. Rather than changing the process to incorporate the necessary mixing, the team decided to look again at core wash D (very carefully this time), since it was similar in formulation to the original core wash. In the end, the foundry switched to core wash D and cylinder bore leaks were virtually eliminated.

Differential Carrier Shrink Defect

In the manufacture of a cast iron differential carrier casting, there was a problem with a shrink defect at a boss, a thick area of the casting. The defect was discovered at the drilling operation. The baseline defect rate was about 5%. We first discussed this problem in Chapter 3.

For the purposes of process improvement, a single operator assessed the shrink defect on a scale of 0 to 4, where 0 was the best. Scores of 3 or 4 correspond to a defective casting.

The problem goal was to reduce the average defect score. To establish a baseline and to simultaneously assess the measurement system, the team selected a sample of 200 castings from one day's production. The operator scored each casting twice. The data are in the file *differential carrier shrink defect baseline* and are summarized in Table 15.3.

Table 15.3 Boss shrink defect scores.

		Second measurement				
		0	1	2	3	4
First measurement	0	105	5	0	0	0
	1	3	40	1	0	0
	2	0	3	30	1	0
	3	0	0	0	10	0
	4	0	0	0	0	2

The operator was very consistent in scoring the castings, and the team judged the measurement system to be acceptable. The average score (of the first measurements) in this investigation was 0.75, and 12 out of the 200 castings were defective. The problem goal was to reduce the average score to close to 0, since then there would be almost no defective castings.

The team took the risky decision not to search for a dominant cause. Instead, they adopted the Move the Process Center approach and planned an experiment. The team selected nine candidates as potential adjusters. These candidates included three iron chemistry levels (denoted *A, B,* and *C*), iron temperature *(D)*, pouring time *(E)*, concentration of in-mold alloy *(F)*, two molding inputs from changing the dies that make the sand molds *(G* and *H)*, and squeeze pressure *(J)*. The team selected the low and high levels of each candidate using engineering judgment. We code the two levels of each candidate −1 and +1.

To define a run of the experiment, the team planned to set the candidate levels and operate the casting process for 40 minutes. From past experience with the process, they expected to see castings with the shrink defect within that time period. Within each run, they planned to select 20 castings and score them for shrink. They also planned to randomize the order of the runs.

There are $2^9 = 512$ possible candidate combinations or treatments. It was not feasible to carry out an experiment with this many treatments so the team decided to use a *fractional factorial design* with only 16 runs (see the chapter suppliment[1]). The 16 treatment combinations must be carefully selected from the 512 possibilities. We use MINITAB to make the best selection (see Appendix F). We give the experimental design and results in Table 15.4. The team conducted the experiment without incident and stored the data in the file *differential carrier shrink defect move center*.

Table 15.4 Experimental design and results for differential carrier experiment.

| Treatment | Order | Candidates ||||||||||| Shrink defect score frequency ||||| Average Score |
		A	B	C	D	E	F	G	H	J	0	1	2	3	4	
1	3	+1	−1	−1	+1	+1	+1	−1	−1	+1	7	5	6	2	0	1.15
2	4	−1	+1	+1	+1	−1	+1	−1	−1	−1	5	7	6	2	0	1.25
3	2	−1	+1	−1	−1	+1	+1	−1	+1	−1	1	7	8	3	1	1.80
4	5	−1	−1	+1	−1	+1	+1	+1	−1	−1	6	10	4	0	0	0.90
5	6	−1	−1	−1	+1	−1	−1	−1	−1	+1	4	7	7	2	0	1.35
6	10	−1	+1	+1	−1	+1	−1	−1	+1	+1	8	11	1	0	0	0.65
7	15	+1	+1	+1	+1	+1	−1	−1	−1	−1	0	5	6	9	0	2.20
8	16	+1	+1	+1	−1	+1	+1	+1	+1	+1	4	8	7	1	0	1.25
9	12	+1	+1	−1	+1	−1	−1	−1	+1	−1	1	9	7	3	0	1.60
10	7	−1	−1	−1	+1	+1	+1	+1	−1	+1	7	10	3	0	0	0.80
11	8	−1	+1	+1	+1	−1	−1	+1	+1	−1	12	8	0	0	0	0.40
12	9	−1	−1	−1	−1	−1	−1	+1	+1	+1	4	7	7	2	0	1.35
13	14	+1	+1	−1	−1	+1	−1	+1	−1	−1	4	7	5	3	1	1.50
14	13	+1	−1	−1	−1	−1	+1	−1	+1	+1	1	8	6	5	0	1.75
15	1	+1	−1	+1	−1	−1	+1	+1	−1	+1	1	8	6	4	1	1.80
16	11	+1	−1	+1	+1	−1	−1	+1	−1	−1	6	9	5	0	0	0.95

In Table 15.4, we give the frequency of scores and the average score for the 20 castings in each run. We are looking for a treatment that gives a low average score. Here, due to the discreteness of the scores, a plot of individual casting scores versus treatment is not very informative. Instead, we look at the *performance measure* defined as the average score over each run. In general, a performance measure is a statistic calculated over all the repeats within each run to assess the performance of the process for that treatment. There are promising treatments, such as number 11, where the average score is small.

From Table 15.4, we can see one important property of the selected design. For each candidate, 8 of the 16 runs have the level –1 and the other 8 have the level +1. Due to this balance, we can assess the main effect of any candidate by comparing the averages of these two sets of runs, as we did in Chapter 13.

The price we pay for using a fractional design (rather than a full factorial design) is that we lose information about interactions. We say that certain effects are *confounded*, because we cannot estimate them separately. MINITAB produces a list of effects, called the alias structure, that are confounded when we use MINITAB to generate the design. For the differential carrier experiment, the main effects and two input interactions that are confounded are:

```
Alias Structure (up to order 2)

I
A + F*J
B + G*J
C + H*J
D + E*J
E + D*J
F + A*J
G + B*J
J + A*F + B*G + C*H + D*E
A*B + C*E + D*H + F*G
A*C + B*E + D*G + F*H
A*D + B*H + C*G + E*F
A*E + B*C + D*F + G*H
A*G + B*F + C*D + E*H
A*H + B*D + C*F + E*G
```

In the list, the main effects are labeled by a single letter and the interactions by pairs such as $A*B$. To determine the confounding, we see from the second line, for example, that the main effect of input A is confounded with the two-input interaction between F and J. As a consequence, if the average difference between the eight runs with A at level –1 and A at level +1 is large, the difference may be due to simultaneously changing the levels of inputs F and J or to changing the levels of input A. We cannot separate these two possibilities with the data from the experiment. Because of this confounding, we need to confirm any promising adjuster found in the experiment. Third- and higher-order interactions are also confounded with the given effects. MINITAB will provide the complete confounding (aliasing) structure if desired. We call this experimental plan a *resolution* III design because some main effects are confounded with two-input interactions but not with other main effects. In a resolution IV design, some main effects are confounded with three-input interactions but no two-input interactions; in a resolution V design, some main effects are confounded with four-input interactions but no three-input interactions. Since we assume that the higher the order of an interaction, the more likely it is to be negligible, we want a design with as high a resolution as possible. For a given number of candidates, MINITAB selects, by default, one of the fractional factorial designs with the highest possible resolution.

The Pareto chart of the effects for the full model is given in Figure 15.7. We use the first effect in the list of confounded effects as the label for the string of confounded effects so that, for example, D corresponds to the combined effects of input D, the two-input interaction $E*J$, and other three-input and higher interactions not shown in the confounding list. We see in Figure 15.7 that effects for inputs A, B, D, and G are large relative to the others. If we assume that the two-input and higher-order interactions are negligible, then inputs A, B, D, and G are adjusters.

We give the main effects plots for inputs A, B, D, and G in Figure 15.8. Since a lower average score is better, the desirable levels of the four inputs are -1 for inputs A and B and $+1$ for inputs D and G. The team decided to investigate further the consequences of changing the levels of inputs A and B, the two iron chemistries, to the low level used in the experiment.

To confirm the findings, the team set inputs A and B to their new levels, kept all other inputs at their current levels, and operated the process for four hours. Of the 3108 castings produced, only 12 were scrapped for shrink defect at machining. This scrap rate of 0.38% was a substantial improvement over the original rate of 5%. There were no side effects, so the team adopted the proposed process change.

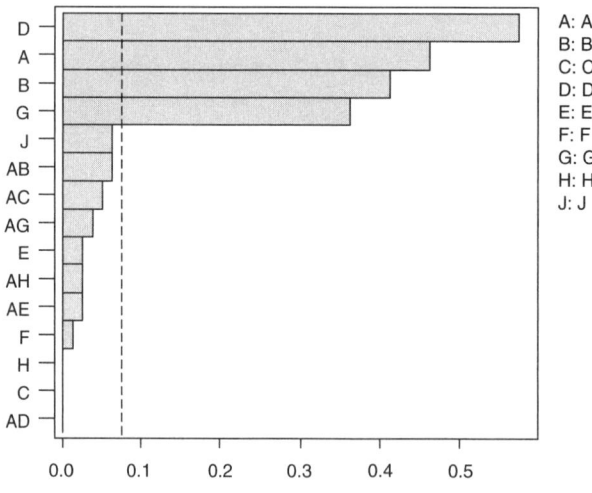

Figure 15.7 Pareto chart of the effects for piston shrink defect experiment.

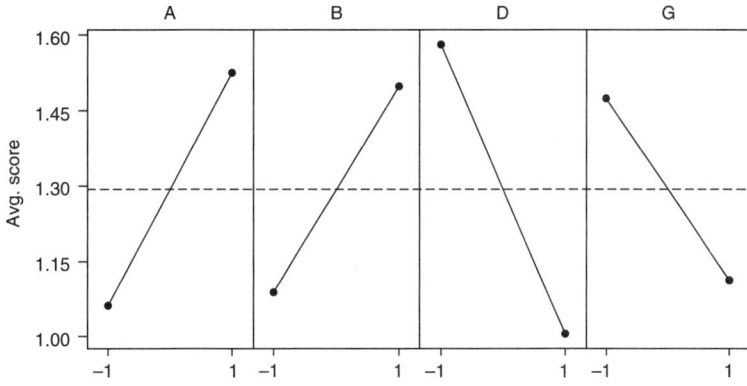

Figure 15.8 Main effects of significant inputs for shrink defect experiment.

Move Process Center Experiment Summary

Use engineering judgment to select candidates as potential adjusters.

Question

Does changing any of the candidates move the process center substantially?

Plan

- Select a study population where we expect to see the full extent of variation.
- Choose two levels for each candidate as extreme as is feasible.
- Define a run for the experiment. If the output is binary, a run must contain many repeats.
- Choose a design with at least eight runs to specify the treatments. For:
  – Three or fewer candidates, use a full factorial design
  – Four or more candidates, select a fractional factorial design with at least resolution III
- Spread the runs across the study population. Randomize the order as much as is feasible.
- Make everyone potentially impacted aware of the plan.

Data

- Carry out the experiment. For a:
  – Continuous output, record the output value, the levels of the candidates, treatment number, and run order, one row for each repeat
  – Binary output, record the proportion of defectives, the levels of the candidates, treatment number, and run order for each run

> **Analysis**
>
> - Plot the output values for each treatment. Look for promising treatments.
> - If there are two or more candidates, use a full model and create a Pareto plot of the effects.
> - Check for possible confounding in the important effects if the design is a fractional factorial.
>
> **Conclusion**
>
> - A candidate with a large effect is an adjuster.
> - For fractional factorial designs, verify an adjuster, if necessary, to break confusion due to confounded effects.

15.2 ASSESSING AND PLANNING A PROCESS CENTER ADJUSTMENT

To successfully implement Move the Process Center we must:

- Identify one or more adjusters, fixed inputs that we can change to move the process center.
- Identify the best level(s) for the adjuster(s).
- Check that the adjustment does not produce substantive negative side effects.
- Estimate the costs of changing the adjuster and ongoing operating costs.
- Estimate the benefit of the adjustment.

If we can accomplish all of these tasks and the benefits outweigh the costs, we adopt the Move the Process Center approach and proceed to the validation stage of the algorithm.

There is some risk that no adjuster will be found. Experiments, such as that used in the differential carrier shrink defect example, can be complex and expensive. In planning such experiments, the candidates and their levels should be selected with great care. There is little value in including an adjuster that is very expensive to change or, if changed, increases operating costs substantially.

Once an adjuster has been identified, further experimentation may be required to determine the appropriate level of the adjuster to move the process center to the desired value. In the battery seal failure experiment, melt temperature was found to be an adjuster. The average seal strength was 356 pounds at 800° and 447 pounds at 750°. A change of 50° in melt temperature produces a change of about 90 pounds in average seal strength. Since the goal was to increase average seal strength by about 60 pounds, the team selected 770° as the new set point for the melt temperature. They assumed a roughly linear relationship between the melt temperature and the seal strength. Alternately, the team could have conducted a second experiment with several levels of melt temperature between 750° and 800° in order to determine the relationship more precisely. Precisely quantifying the adjuster/output relationship is more important if we make repeated adjustments of different sizes, as in feedforward or feedback control.

When changing the process center we need to watch carefully for negative side effects. In the core strength example introduced in Chapter 1, the team found they could eliminate core breakage by increasing core strength. They discovered an adjuster, the amount of resin in the sand mix, which they used to increase the average strength. However, the stronger cores led to more casting defects and the approach was abandoned. We may be able to avoid undesirable side effects by using two or more adjusters simultaneously rather than one adjuster as in the battery seal failure and engine block leaks examples. This is a good reason for searching for several adjusters simultaneously.

For off-target process center problems, we can see the expected benefit of changing the process center using the histogram from the baseline investigation. Based on the problem goal and baseline, we know how far we want to move the process center and the expected benefits of such a move. We also need to assess the costs of changing the adjuster and the operating costs at the new level.

In some problems the two approaches, Move the Process Center and Make the Process Robust, are identical. The differential carrier shrink defect problem is a good example. Whatever the cause of the shrink defect, the change in the fixed chemistry inputs made the process more robust to the cause.

In multistream processes, if we find that the process centers vary from stream to stream, we can reduce the overall variation by moving the centers of each stream to a common target, as shown in Figure 15.9. In the left panel, we show box plots of the output by stream and overall. In the right panel, we show the same plots after we (roughly) align the substream centers. Aligning the substream centers results in a significant reduction in the overall output variation.

In some circumstances, an output center may drift over time. In that case, a one-time adjustment of the process center will not solve the problem over the long term. We may need to use another variation reduction approach, such as feedback control.

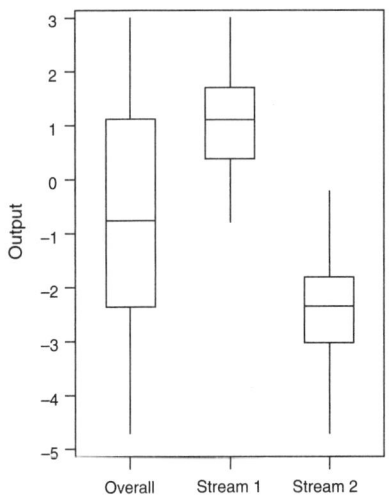 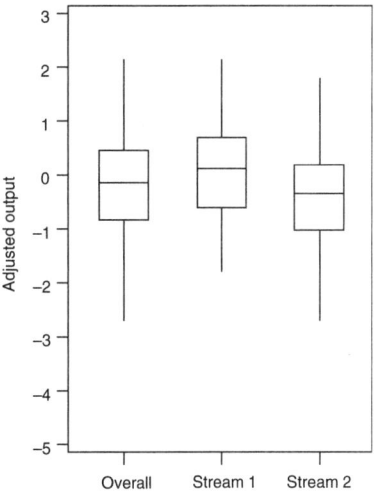

Figure 15.9 The effect of aligning substream centers on output variation.

 **Key Points**

- To move the process center, we need to find one or more adjusters, normally fixed inputs. We change the settings of the adjusters to move the output center.

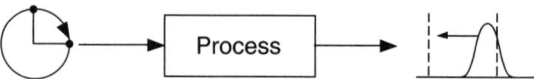

- We use an experimental plan, a full or fractional factorial design, to investigate one or more fixed inputs (candidates) as possible adjusters.
- For processes with multiple streams, we can align the centers of the streams with separate adjustments and hence reduce the overall variation in the process.

Endnote (see the Chapter 15 Supplement on the CD-ROM)

1. Fractional factorial designs are powerful experimental methods for investigating the effects of a large number of inputs simultaneously. In the supplement, we provide an introduction to these designs and references to further work. We will use these designs to determine the knowledge required to implement several of the variation reduction approaches.

 Exercises are included on the accompanying CD-ROM

16
Desensitizing a Process to Variation in a Dominant Cause

Efficiency is doing things right; effectiveness is doing right things.
—Peter F. Drucker

The goal of desensitization is to find and change fixed inputs that flatten the relationship between the output characteristic and the dominant cause (see Figure 16.1). That is, we find an interaction that we exploit to desensitize the process output to changes in the dominant cause.[1] To explore desensitization, we choose a number of fixed inputs to investigate, based on knowledge of the dominant cause and the process. We use an experimental plan to determine if these *candidates* and their new levels will make the process less sensitive to variation in the dominant cause.

With this approach we do not address the dominant cause directly. We continue to live with its variation. By changing the relationship between the cause and the output, we reduce the effect of the cause.

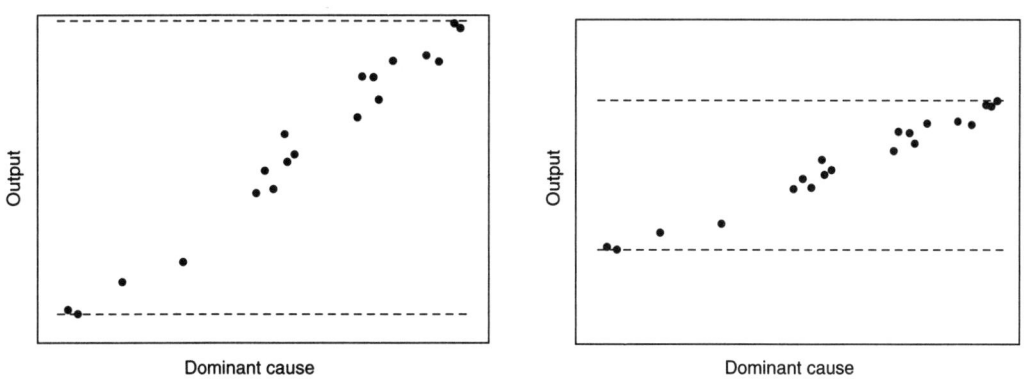

Figure 16.1 Original (left) and new (right) relationship between the dominant cause and output characteristic.

The only requirement for desensitization is to find new settings of the candidates that make the output less sensitive to variation in the dominant cause.

The costs of desensitization include:

- An experiment to find the new process settings that make the process less sensitive to variation in the dominant cause
- A one-time change to the process settings
- The ongoing operation of the process with the new settings

There is no information about whether this approach will be feasible until the experimental investigation is complete. This is a drawback, since the cost of the experimentation may be high and the returns uncertain. The benefit can be assessed using the relationship between the cause and the output. If we could totally eliminate the effect of the dominant cause, the maximum benefit is given by the residual variation in the output due to all other causes.

16.1 EXAMPLES OF DESENSITIZATION

We illustrate the approach with a series of examples in which the complexity of the experimental plan increases.

Engine Block Porosity

We discussed the problem formulation and the search for the dominant cause of excess porosity in a cast-iron engine block in chapters 6, 10, and 13. The team identified low pouring temperature as the dominant cause. During planned and unplanned work stoppages, iron that remained in the pouring ladle cooled since there was no external heat source. In normal production, this was not a problem because the pouring ladle was frequently replenished with fresh hot iron. The team could not address this cause directly. Eliminating or staggering breaks was not possible due to employee resistance, and adding temperature control to the pouring ladles was too costly. They decided to explore desensitization as a possible solution.

The first step was to choose the candidates, fixed inputs that can be changed and might desensitize the process to variation in pouring temperature. Porosity occurs when gases produced during pouring are trapped in the casting. If the pouring temperature is too low, the iron may harden before all the gas can escape. The team selected core wash and core sand composition as candidates. Cores, set in the mold to create spaces in the casting, are dipped in core wash to produce a better surface finish. The cores are made from resins and sand. Both the core wash and the resins generate gas during the pouring process.

The team decided to investigate one new core wash and a single reformulation of the core sand mixture in a factorial experiment. Each candidate had two levels, regular and alternative. There were four treatments as shown in Table 16.1.

To assess the sensitivity of the process to pouring temperature for each treatment, the team used the knowledge that the pouring temperature varied over its full range around the lunch break. For each treatment, three engine blocks were cast just before lunch (high pouring

Table 16.1 Treatments for the engine block porosity desensitization experiment.

Treatment	Sand	Core wash
1	Regular	Regular
2	Regular	Alternative
3	Alternative	Regular
4	Alternative	Alternative

temperature) and another three blocks just after lunch (low pouring temperature). The experiment was replicated over three days so that a total of 72 blocks were measured for porosity. The 12 blocks before and after lunch were divided into three groups of four. Within each group, the team randomized the order of the four treatments. Since temperature increased steadily after lunch, each treatment saw roughly the same range of temperatures. The team managed the randomization easily because the treatments only affected the cores. The data are given in the file *engine block porosity desensitization*.

We start the analysis with a plot of porosity versus treatment number with different plotting symbols for the high and low temperatures, as shown in Figure 16.2. For any treatment, we can see if there is substantially less porosity than in the current process, here given by Treatment 1. All three new treatments are a large improvement over the existing process with smaller average porosity and less variation as the pouring temperature changes.

The team decided to investigate the new core wash further to ensure that there were no side effects such as an increase in other casting defects. There was no cost to this change. The alternative core wash was eventually adopted and the scrap rate due to porosity decreased from about 4% to less than 1%.

In this example, the team designed a full factorial experiment and drew their conclusions from a simple plot. We give three more examples to illustrate some complications that can arise in a desensitization experiment.

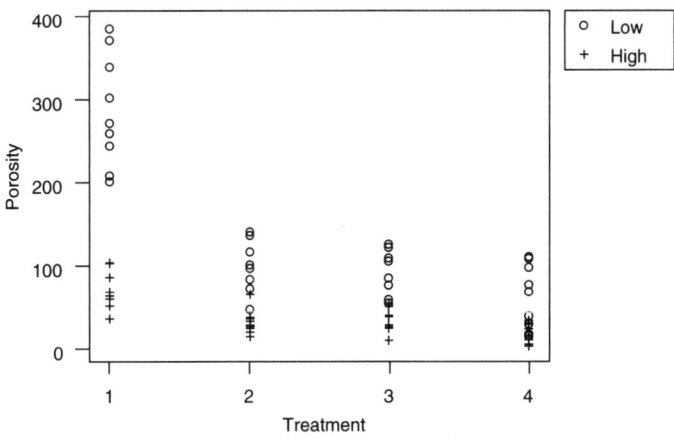

Figure 16.2 Porosity by treatment for high and low pouring temperatures.

Oil Pan Scrap

The baseline oil pan scrap rate was 8%. Applying the method of elimination, the team discovered that a dominant cause of scrap was the amount of lubricant applied to the blanks prior to the draw press. A roll coater with two steel cylinders applied the lubricant. There were no controls on the roll coater and no ongoing measurement of the lubricant amount.

During a verification experiment, the team used a visual five-point score for the amount of oil applied. They specially prepared 80 blanks with lubricant amounts at each of the five levels. The team could control the dominant cause during the experiment but not in normal production. To verify the cause, the team stamped the 80 blanks for each oil level. We see from Figure 16.3 that the scrap rate increased markedly as the lubricant score increased.

There was no obvious cost-effective way to control the amount of lubricant applied to the blanks, so the team decided to try to desensitize the process. There was a lot of uncertainty about whether this approach would be effective, but they decided that it was worth the cost of the investigation in the hopes of avoiding a solution with high capital cost.

Based on knowledge of the cause, the team chose three candidates for the desensitization experiment: lubricant supplier, die temperature, and binder force. After consultation, three lubricant suppliers (the current supplier is labeled A and two other suppliers B and C) provided the lubricant they judged the best suited for the application. Supplier A suggested the lubricant in current use. The team selected the two levels for the other candidates using engineering judgment. They decided to use only the two extreme levels 1 and 5 for the amount of oil on the blanks, the dominant cause.

The team defined a run to be the consecutive stamping of 80 blanks as in the verification experiment. Since the output was binary, each run required sufficient repeats so that some defectives were produced. The team used the scrap rate for each run as a performance measure. The experiment had 24 runs. There were 12 treatments defined by the levels of the three candidates and two runs per treatment, one for each level of the lubricant score. With this plan, each treatment would see the full range of variation in the dominant cause.

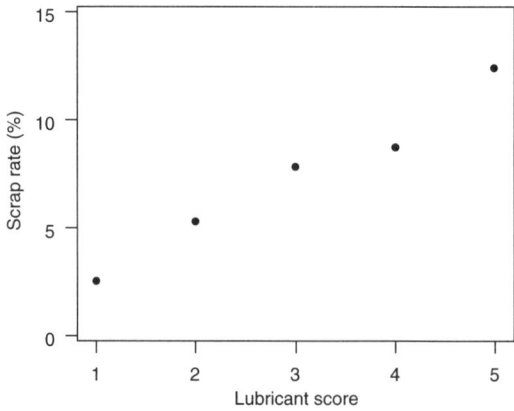

Figure 16.3 Scrap rate by lubricant score.

Table 16.2 Candidate levels and scrap rates for low and high level of lubricant amount.

Treatment	Run order	Lubricant supplier	Die temperature	Binder force	Scrap rate for Lubricant Score 1	Scrap rate for Lubricant Score 5
1	3	A	Low	Low	0.07	0.15
2	5	A	Low	High	0.14	0.16
3	10	A	High	Low	0.06	0.11
4	7	A	High	High	0.15	0.24
5	4	B	Low	Low	0.04	0.01
6	1	B	Low	High	0.10	0.11
7	8	B	High	Low	0.03	0.05
8	12	B	High	High	0.09	0.14
9	6	C	Low	Low	0.03	0.05
10	2	C	Low	High	0.04	0.08
11	11	C	High	Low	0.04	0.04
12	9	C	High	High	0.03	0.07

Since die temperature was difficult to change, all runs with the low temperature were conducted first. Within the low-temperature runs, the six treatments were applied in random order. Once they set the levels of the candidates, the team stamped the two sets of 80 blanks, one set with oil level 5 and one with oil level 1.

We give the experimental plan, including the run order, and the results in Table 16.2 and in the file *oil pan scrap desensitization*.

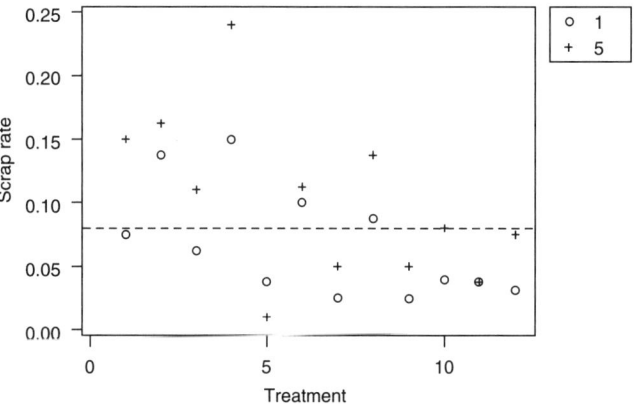

Figure 16.4 Scrap rate for low and high level of lubricant amount versus treatment (dashed horizontal line gives the baseline scrap rate).

246 Part Four: Assessing Feasibility and Implementing an Approach

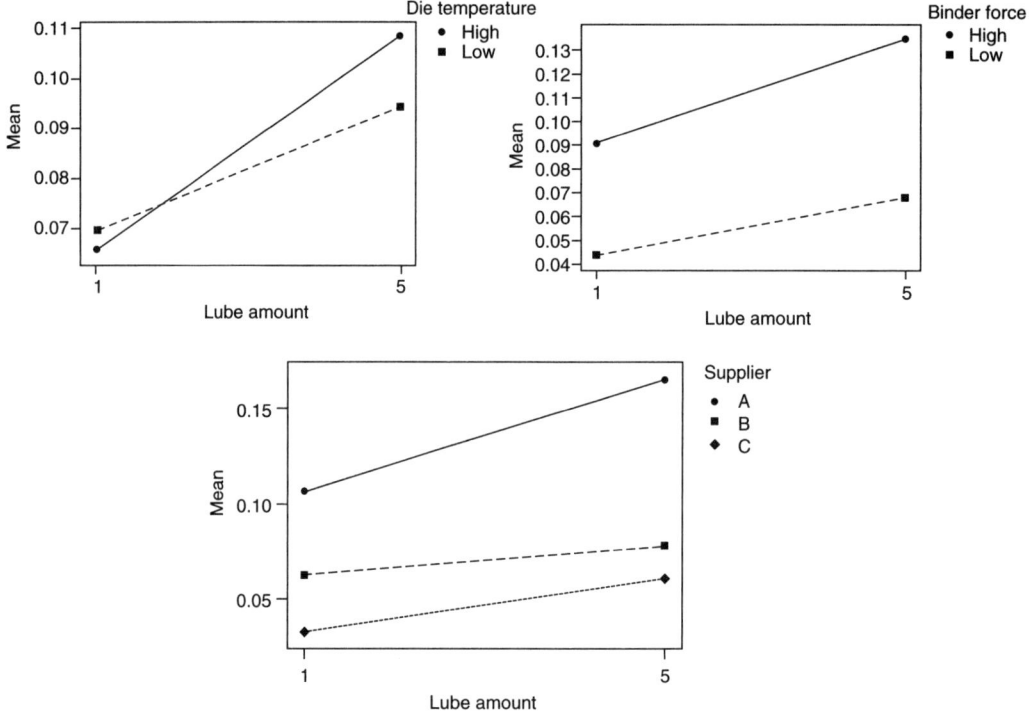

Figure 16.5 Interaction plots for oil pan scrap desensitization experiment.

We start the analysis by plotting the scrap rate by treatment with different plotting symbols for the two lubricant scores. In Figure 16.4, we see that there are promising treatments with low scrap rates for both levels of the dominant cause.

We can look at the effects in more detail using the interaction plots between the candidates and the lubricant score. In Figure 16.5, we see that the scrap rate is:

- Less sensitive to changes in the lubricant amount if the die temperature is low

- Substantially lower with binder force at the low level for both lubricant scores

- Much lower and less sensitive to changes in lubricant amount for lubricant C

Since lubricant C looks so promising, we next consider the interactions between binder force and die temperature by lubricant amount for supplier C only. We subset the data set in MINITAB (see Appendix A). From the interaction plots in Figure 16.6, we conclude that low binder force is the preferred setting for lubricant C to reduce the scrap rate. Die temperature has little effect.

During the investigation, the team measured other important outputs to make sure that changing the process settings would not result in other problems. They found that lubricant C was also better than the original lubricant in terms of steel flow at the deep end of the pan. The team decided to change to the lubricant from supplier C. After the change, the long run scrap rate was reduced to about 3%. There was a small increase in cost that was far outweighed by the savings in scrap costs.

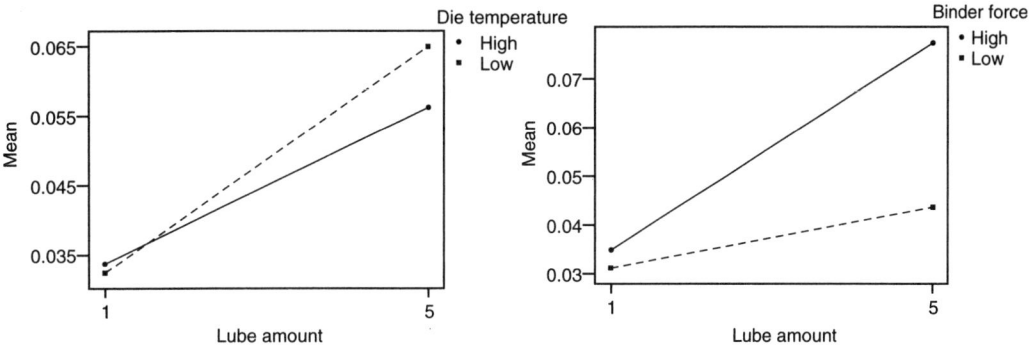

Figure 16.6 Interaction plots of die temperature and binder force by lubricant amount (supplier C only).

Refrigerator Frost Buildup

The manufacturer of a frost-free refrigerator, designed for temperate climates, expanded its market to several tropical countries. Almost immediately, there were many customer complaints about the buildup of frost inside the refrigerator. These complaints had been nonexistent in the traditional market. Management feared that the opportunity for market expansion would be lost. They formed a team to reduce the number of complaints. We discussed this example previously in chapters 1, 3, and 8.

As a first step, the team replaced field measurements by the temperature of the refrigerator cooling plate. They verified that high temperature on the cooling plate led to a buildup of frost under certain conditions. The team set a goal to keep the cooling plate temperature less than 1°C. There was no baseline investigation for either the original problem or for the problem as reformulated in terms of cooling plate temperature. Either investigation would have been very expensive.

Because the problem did not occur in temperate climates, the team knew that the dominant cause must relate to differences in usage or ambient conditions between the tropical and the traditional market. Through further discussion and interviews with field representatives, they decided that the dominant cause of temperature variation on the cooling plate was ambient temperature and relative humidity in combination with the frequency of door openings.

The team could not address the dominant cause directly since it was not under the control of the manufacturer. They decided to pursue desensitization. The goal was to find a refrigerator design that would maintain a consistent low temperature on the cooling plate under varying conditions defined by ambient temperature, relative humidity, and number of door openings per hour.

The team considered four fixed inputs, D1 to D4, related to the refrigerator design. They selected two levels, *original* and *new*, for each candidate, based on engineering knowledge. There were 16 possible prototypes (treatments) using the levels of the four candidates. Due to the high cost of prototypes, the team decided to use a fractional factorial design and build only 8 of the 16 possible combinations. The team used MINITAB to select the design. See the supplement to Chapter 15 for more detail on fractional factorial designs. We show the selected treatments in Table 16.3.

Table 16.3 Eight-run fractional factorial refrigerator desensitization experiment.

Treatment	D1	D2	D3	D4
1	New	New	New	New
2	New	New	Original	Original
3	New	Original	New	Original
4	New	Original	Original	New
5	Original	New	New	Original
6	Original	New	Original	New
7	Original	Original	New	New
8	Original	Original	Original	Original

In this resolution III design, the main effect of input D4 is confounded with the three-input interaction of D1, D2, and D3. Thus pairs of two input interactions such as D1*D2 and D3*D4 are also confounded.

The team set up the testing laboratory with two sets of environmental conditions:

Normal   Conditions to mimic use in temperate climate

Extreme   Conditions to match the worst anticipated usage conditions with high temperature, high humidity, and frequent door openings

For each prototype listed in Table 16.3, the team planned two runs in the laboratory, one at normal and one at extreme environmental conditions. We call this a *crossed design* since we use both levels of the dominant cause for each of the treatments. There are 16 combinations of the 5 inputs (D1 to D4 and environment) out of the 32 possible treatments. We use MINITAB to determine the confounding structure for this fractional design.[2] We assume third- and higher-order interactions are negligible, so we attribute any large differences to main effects and two-input interactions. We reproduce the confounding structure with any effects involving three or more of the fixed inputs (candidates) erased.

```
I
D1
D2
D3
D4
environment
D1*D2 + D3*D4
D1*D3 + D2*D4
D1*D4 + D2*D3
```

```
D1*environment
D2*environment
D3*environment
D4*environment
D1*D2*environment + D3*D4*environment
D1*D3*environment + D2*D4*environment
D1*D4*environment + D2*D3*environment
```

The purpose of the desensitization experiment is to isolate useful interactions between the dominant cause (environment) and one of the candidates (D1 to D4). From the list of confounded effects, we see that the interactions between the four candidates D1 to D4 and the environment are in separate lines and, hence, can be estimated separately. These interactions are the key to desensitizing the cooling plate temperature to changes in the usage conditions. On the other hand, we cannot separate the joint effects D1*D2 and D3*D4 with the environment.

After 30 minutes of cycling, the temperature of the cooling plate was measured. Randomization of the run order was not feasible for time and cost reasons. All eight refrigerators were placed in a special testing room and subjected to one of the environmental conditions simultaneously. This lack of randomization was not a major risk here since all refrigerators are subject to the same environmental conditions. The complete experimental plan and the data are given in Table 16.4. The data are stored in the file *refrigerator frost buildup desensitization* with one row for each temperature measurement and columns indicating the values of D1–D4 and the environmental conditions.

We start the analysis by plotting the cooling plate temperature versus treatment with a separate plotting symbol for the two levels of environment. From Figure 16.7, we see that there are several promising treatments, especially treatment 6.

Table 16.4 Plan and data for refrigerator frost buildup desensitization experiment.

					Environmental conditions	
Treatment	D1	D2	D3	D4	Low (normal)	High (extreme)
1	New	New	New	New	0.7	2.1
2	New	New	Original	Original	2.9	4.8
3	New	Original	New	Original	2.4	9.6
4	New	Original	Original	New	3.8	5.9
5	Original	New	New	Original	1.9	4.0
6	Original	New	Original	New	−0.2	0.1
7	Original	Original	New	New	−0.1	3.5
8	Original	Original	Original	Original	0.2	7.2

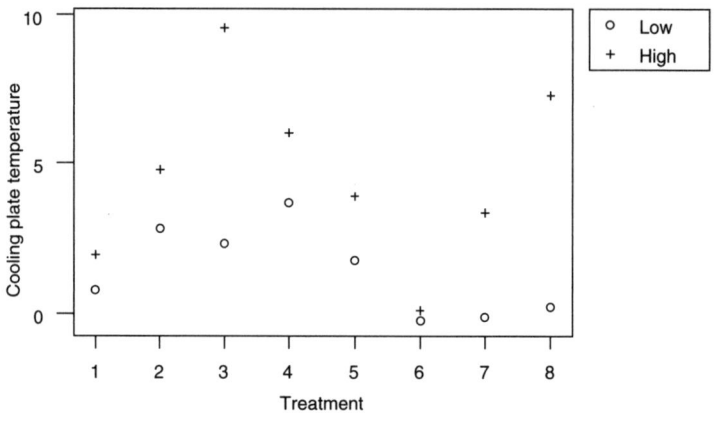

Figure 16.7 Temperature on cooling plate by treatment (different symbols represent output values for different environmental conditions).

In Figure 16.8, we look at the interaction plots for each candidate versus the environmental cause. We hope to find an interaction that gives consistently low cooling plate temperatures for both levels of the environment, as seen in treatment 6.

Using Figure 16.8, the team concluded that they should change D2 and D4 to their new levels. However, changing to the new level of D4 added significant cost. As a result, the

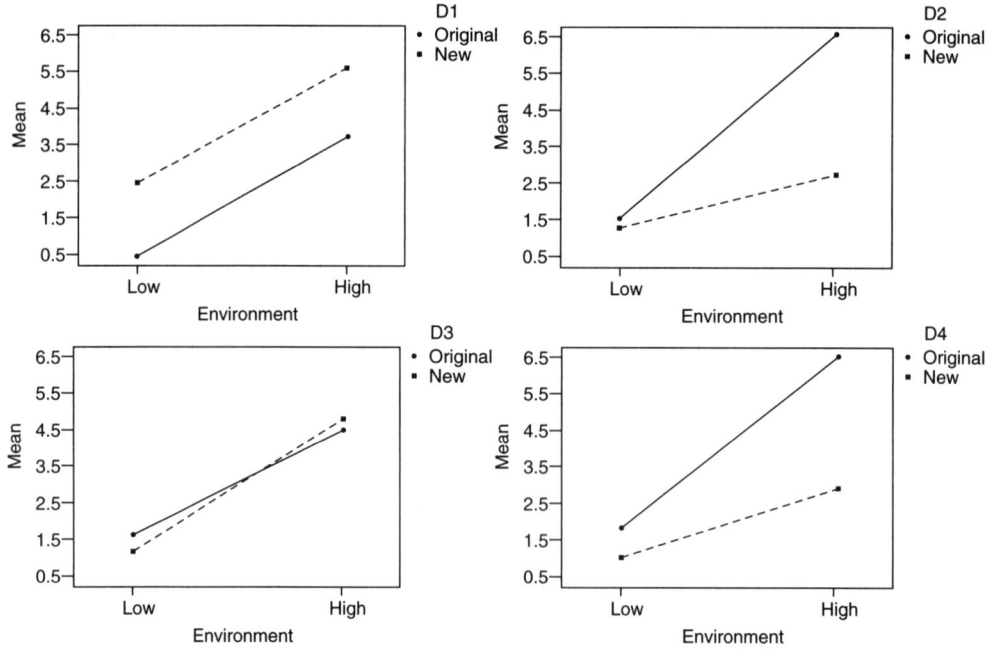

Figure 16.8 Interaction plots of candidates and the environment cause output in cooling plate temperature.

team decided to investigate how much the cooling plate temperature variation could be reduced if only D2 was changed to its new level. This treatment was not included in the original fractional design. When they produced a prototype with only D2 changed and subjected it to both levels of the environmental cause, they found that changing this input alone did not sufficiently reduce the cooling plate temperature and its variation as the environmental input was changed.

Despite the added cost, the team recommended adopting a refrigerator design with both D2 and D4 changed to their new levels. After the change, the frequency of complaints about frost buildup from the tropical market was substantially reduced.

Eddy Current

In a process that produced castings later machined into brake rotors, there was 100% inspection of hardness using a measurement system based on eddy currents. This system was fast and nondestructive. Despite the 100% inspection, there were frequent complaints from the customer, a machining operation, about castings out-of-specification with respect to hardness. The customer used the Brinell method to measure hardness, a standard procedure that is partially destructive and time consuming, not suitable for 100% inspection.

There was a high reject rate at the 100% eddy current inspection station. The operators measured all rejected castings a second time. The plant shipped castings that passed at least one test and scrapped those that failed both tests.

Management assigned a team to reduce scrap costs and customer complaints by reducing casting hardness variation. To establish a baseline, each day for one week, the team collected a haphazard sample of 100 first-time eddy current measurements for a total of 500 hardness measurements. The data are available in the file *eddy current baseline* and are plotted by day in Figure 16.9. The hardness specification is 4.3 to 4.7. The baseline standard

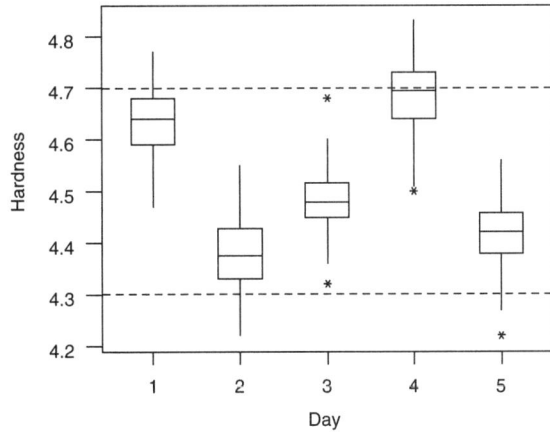

Figure 16.9 Box plots of eddy current hardness measurements by day (dashed horizontal lines give the specification limits).

deviation is 0.1365 and the full extent of hardness variation is 4.52 ± 3*0.14 (average ±3 standard deviation), or about 4.1 to 4.9.

The team assessed the measurement system using a standard gage R&R procedure as described in the supplement to Chapter 7. The measurement system standard deviation was 0.021, so it appeared the measurement system was adequate to proceed to the next stage of the algorithm.

Because of the past complaints from their customer, the team also carried out an investigation to compare the eddy current and Brinell hardness measurement systems. They selected 30 castings with widely varying eddy current measurements, then tagged and shipped these castings to the customer. The customer measured the Brinell hardness for each casting. The data are plotted in Figure 16.10, and given in the file *eddy current Brinell measurement*.

The real problem is now apparent. There is poor correlation between the two measurement systems. Although the earlier gage R&R investigation showed that the eddy current system was repeatable, for any set of castings with the same Brinell hardness, the eddy current system gave widely varying values. The team reformulated the problem to improve the correlation between the two measurement systems. Said in another way, the team decided to reduce the variation in eddy current measurements among castings with the same Brinell hardness.

The team discovered that day-to-day fluctuations in iron chemistry and the level of dirt on the castings were dominant causes of the variation of the eddy current measurement of castings with the same Brinell hardness. They did not look for a more specific cause in the chemistry family. Castings were cleaned by shot blasting. The level of cleaning varied since the shot blast machine did not run using a first-in, first-out protocol. The cleaning times ranged from 5 to 19 minutes.

In regular production, iron chemistry was expensive to control, and it was difficult to remove all the dirt from the casting before measurement. As a result, the team decided to look for a way to run the eddy current measurement system that was less sensitive to varying levels of dirt and iron chemistry.

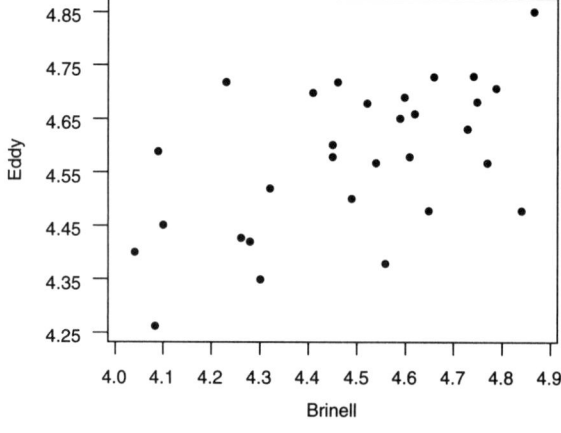

Figure 16.10 Scatter plot of eddy current versus Brinell hardness measurements.

Table 16.5 Candidate levels for eddy current measurement experiment.

Candidate input	Low level	High level
Frequency	200	350
Probe temperature	35	65
Gain	30	40

For the desensitization experiment, the team chose three candidates corresponding to settings on the eddy current system. For each candidate they selected high and low levels on either side of the current settings. The team planned a full factorial with eight treatments using the levels given in Table 16.5.

To ensure that each treatment was assessed under the extreme levels of the two dominant causes, the team selected 16 castings from each of two days' production. Each casting was cleaned for 5 minutes, measured with the eddy current system, cleaned for another 14 minutes, and then measured again. The team measured every casting using all eight treatments at both points in the cleaning cycle. Finally they measured the Brinell hardness of each casting at two different positions on the casting. The team knew that the Brinell hardness measurements had little measurement variation relative to the variation in the eddy current machine. In the analysis, they used the average of the two Brinell values as the true hardness of the casting.

The team hoped that the two days would represent the day-to-day variation in iron chemistry. This was a risky decision. With only two days, the full day-to-day variation in chemistry might not be captured. If the team had found the specific cause within the chemistry family, they could have more easily ensured that each treatment was subject to the full range of the dominant cause. We give the data from the experiment in the file *eddy current desensitization*. There are 16 eddy current measurements and the average Brinell reading for each of 32 castings.

The analysis of this experiment is complicated. The goal in the analysis is to find a combination of levels for the candidates that yields a strong relationship between the Brinell hardness values and the values given by the eddy current system, consistently over the four combinations of the levels of the dominant causes. We do not worry about measurement bias, since if the bias is consistent it can be easily removed.

In the analysis, for each of the eight treatments we plot the hardness as measured by the eddy current system versus the (average) Brinell hardness for all four combinations of the dominant causes. For example, in Figure 16.11, we show the plot for the treatment frequency = 200, temperature = 35, and gain = 30. This plot is typical of all the eight treatments. We see from the Figure 16.11 that there is a weak relationship between the two methods of measurement. For a given Brinell hardness, there is large variation in the eddy current measurements.

The experiment failed to find settings of the candidates that make the eddy current measurement system less sensitive to changes in chemistry and dirt. In all cases, there was no relationship that could be used to predict the Brinell hardness from the eddy current measurements. Had one or more of the treatments shown promise, the team could have conducted a more formal analysis as discussed in the chapter supplement.[3]

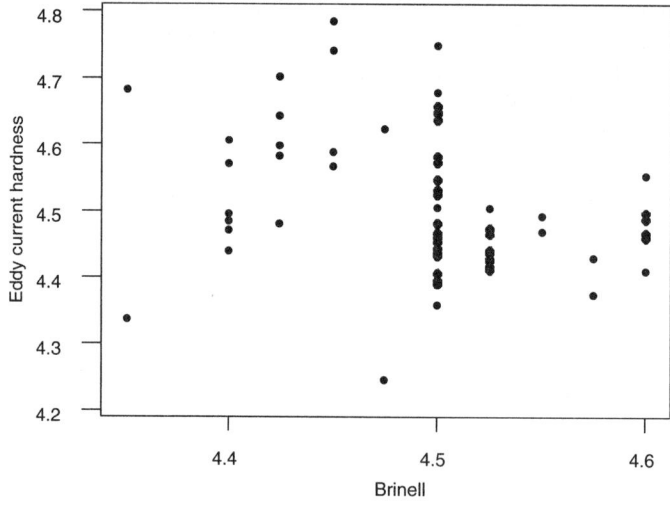

Figure 16.11 Plot of eddy current versus Brinell hardness for freq. = 200, temp. = 35, and gain = 30.

In the end, the team concluded that the experiment was not a failure. They had shown that the eddy current system did not work well and could not be easily improved. There was little value in the 100% inspection. By removing the eddy current system, the foundry was forced to concentrate on reducing variation in the process instead of relying on imperfect 100% inspection.

Desensitization Experiment Summary

Select candidates and levels based on knowledge of the process and the dominant cause.

Question

Which combination, if any, of the candidate levels results in a process less sensitive to variation in the dominant cause?

Plan

- Select a design for the candidates. For:
  - Three or fewer candidates, use a full factorial design
  - Four or more candidates, use a fractional factorial design with resolution III or higher
- Define a treatment number for each combination of the candidates.
- Select two levels for the dominant cause at the extremes of its normal range.
- Use a crossed design where, for each treatment, there are runs for both levels of the dominant cause.
- Define a run of the experiment including treatment assignment and the number of repeats.

- Randomize the order of the runs as much as is feasible.
- Make everyone potentially impacted aware of the plan.

Data

Carry out the experiment. Record the output value, the levels for candidates, the level of the dominant cause, treatment number, and run order, one row for each repeat.

Analysis

- Plot the output for each treatment number. Use a different plotting symbol for the two levels of the dominant cause. Look for promising treatments.
- Construct all cause by candidate interaction plots.

Conclusion

Examine the plots to determine the levels of the candidates (process settings) that make the output less sensitive to variation in the dominant cause.

16.2 ASSESSING AND PLANNING PROCESS DESENSITIZATION

To successfully implement process desensitization, we must:

- Identify new process settings (that is, fixed inputs and their levels) to make the process less sensitive to changes in the dominant cause.
- Move the new process center if not close to the target.
- Check that the change of settings does not produce substantive negative side effects.
- Estimate the costs of changing the settings and the change in ongoing operating costs.
- Estimate the benefit of the process change.

If we can accomplish all these tasks, and the benefits outweigh the costs, we proceed to the validation stage of the algorithm.

In most applications, there is a degree of uncertainty in the choice of candidates. If the experiment fails to find settings that reduce the sensitivity, we may consider other fixed inputs or different levels of the current candidates before we abandon the approach.

Sometimes we may have only partial knowledge of the dominant cause, as in the refrigerator frost buildup example. There, the team did not isolate the dominant cause among ambient temperature, humidity, and usage conditions. In other cases, there may be more than one dominant cause. Then, for the desensitization experiment, we recommend defining a composite cause with two levels that corresponds to the extremes of the identified causes.

The dominant cause varies in normal operation of the process. During a desensitization experiment, however, we need to hold the cause fixed at its low or high level for a run of the experiment. If this is not possible, we may need to resort to a robustness experiment as described in Chapter 19.

We strongly recommend the crossed design to ensure that all interactions between the individual candidates and the dominant cause are separately estimable, even if a fractional design is used for the candidates.

In many instances, we choose not to completely randomize the order of the runs. In the refrigerator example, eight prototypes were built and then simultaneously tested under the two levels of the (composite) dominant cause. Because of high cost, it was not feasible to randomize the order in which each refrigerator was exposed to each level of the cause.

There are some special considerations when planning a desensitization experiment for binary output. There must be enough repeats in each run so that some defectives occur on at least half the runs in the experiment. Otherwise the experiment will provide little useful information. See tables in Bisgaard and Fuller (1995a, 1995b, and 1996) for some guidance concerning the sample sizes necessary. The analysis of an experiment with a binary output can be based on the proportion defect within each run, as in the oil pan scrap example.

We start the analysis of a desensitization experiment with a plot of the individual output values by treatment number, as in Figure 16.2. We use a different plotting symbol for each level of the dominant cause. For complex problems such as the eddy current measurement system, we use scatter plots or other graphical summaries to visually characterize the performance of the process for each treatment. From these summaries, we can assess if desensitization is feasible. In a full factorial experiment, if none of the treatments are promising, we cannot desensitize the process using the selected levels of the candidates. If we have a fractional design, or if one or more treatments appear promising, we look at interaction plots between the individual candidates and the cause as in Figure 16.5.

We may find process settings that are less sensitive to variation in the dominant cause, but that result in an undesirable shift of the process center or some other side effect. In this case, we may look for an adjuster to move the center (see Chapter 15) or we may formulate a new problem to deal with the side effect. In the crossbar dimension example discussed in Chapter 12, barrel temperature was identified as a dominant cause. The team found that increasing the set point for barrel temperature (but not controlling the variation) increased the average size but substantially reduced the variation in the crossbar dimension. They then needed to move the process center and also solve a new problem called *burns*, a visual defect that occurred on some parts molded with the higher barrel temperature. See the exercises for chapters 16 and 18.

We recommend experiments with only two levels for each candidate. If we are successful in finding a candidate that desensitizes the process, we may optimize using a follow-up experiment with the identified candidate at several levels and the dominant cause at two levels. If a candidate is categorical, such as the supplier in the oil pan scrap example, we may use more than two levels in the desensitization experiment.

We have seen many teams proceed directly to the desensitization approach without first identifying a dominant cause. They then conduct an experiment where they change both candidates and suspects. In this way, they hope to identify a dominant cause and desensitize the process at the same time. This is a poor strategy. The experiment will be large and complex since there is little knowledge about which varying input, if any, is a dominant cause. There is also little information to help select the fixed inputs as candidates. It is more effective, both in terms of cost and the likelihood of finding a solution, to search first for a dominant cause (or at least clues about the dominant cause family) using the method of elimination with observational investigations, and then to consider a desensitization experiment.

We sometimes analyze the results of desensitization experiments using statistical models. Model building is an advanced topic that we do not cover here. See experimental design references such as Montgomery (2001) and Wu and Hamada (2000). Additional analysis using models may be necessary if we wish to predict the performance for candidate levels not used in the experiment.

The idea of desensitizing a process to a dominant cause was popularized by Taguchi (1985, 1986). He calls a suspect a *noise factor*. Taguchi (1985) cites several examples, including the famous Ina tile case. In other examples, Taguchi proceeds without knowledge of a dominant cause. This corresponds to the robustness approach described in Chapter 19. We provide a more detailed discussion of the designs and analysis suggested by Taguchi in the supplement to Chapter 19.

Key Points

- A desensitized process is one that is less sensitive (in terms of output variation) to changes in an identified dominant cause.

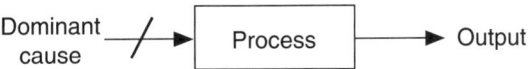

- To find process settings that desensitize the process, we conduct an experiment with one or more candidates. We expose each combination of candidates in the experiment to the full range of the known dominant cause to identify good treatments.
- Within each run of the desensitization experiment, we must be able to control the level of the dominant cause that normally varies.
- Desensitizing a process requires a one-time change to the product design, process design, or the process control plan.

Endnotes (see the Chapter 16 Supplement on the CD-ROM)

1. In the supplement we show how process desensitization exploits a special type of interaction between the dominant cause and one or more of the candidates.
2. The supplement provides more details on crossed fractional factorial experiments used in desensitization experiments and how to determine the confounding structure.
3. In the eddy current example, we concluded from a series of plots that none of the treatments was helpful in making the measurement process less sensitive to dirt and chemistry variation. In the supplement, we consider a formal analysis, useful if one or more of the treatments had shown promise.

 Exercises are included on the accompanying CD-ROM

17
Feedforward Control Based on a Dominant Cause

To improve is to change, to succeed is to change often.
—Winston Churchill, 1874–1965

We use a feedforward controller to reduce the effect of an identified dominant cause of variation. The basic idea is to measure the cause and then predict the output value. If the predicted output is not close to the target value, we make an adjustment to the process to compensate for the predicted deviation. Figure 17.1 is a schematic of the implementation.

For feedforward control to be effective, the requirements are:

- A known relationship between the output and the dominant cause
- A reliable system to measure the dominant cause
- A timely way to adjust the process center

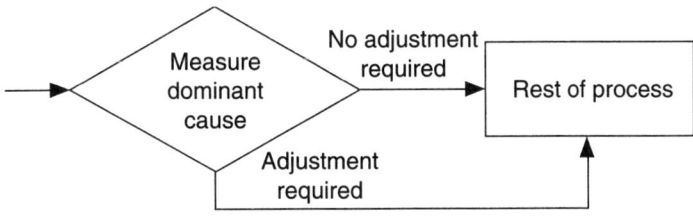

Figure 17.1 Feedforward control schematic.

The potential costs of feedforward control include:

- An investigation to quantify the relationship between the output and the dominant cause
- An experiment to find an adjuster, if not already known
- The ongoing costs of measuring the cause and making the compensating adjustment

We can assess the potential benefit of using feedforward control with the knowledge of the dominant cause and the relationship between the cause and the output. The maximum potential improvement is given by the residual variation in the output due to all other causes.

17.1 EXAMPLES OF FEEDFORWARD CONTROL

We illustrate the use of the feedforward approach with three examples.

Potato Chip Spots

When making potato chips, dark spots on the chips were undesirable to the customers. In the existing process, the dark spot problem was an uncommon but significant concern. A team was assigned to reduce the occurrence of dark spots. Dark spots were measured on a scale from 1 to 10 for each lot of chips. The rating was subjective, but by using the same assessor, the measurement system added little variation. The team used existing production records to quantify the baseline. The lot average score was 1.83 and the full extent of variation was a range of scores from 1 to 8. They also discovered that there was strong time-to-time variation in the occurrence of dark spots.

Stratifying by potato batch and investigating further, they suspected that the sugar concentration in the incoming potatoes was the dominant cause. To verify the suspicion, the team produced chips using five different batches of potatoes chosen to have a wide range of sugar concentrations. From each batch of potatoes, three lots of chips were produced. The resulting data are given in Table 17.1 and plotted in Figure 17.2.

Table 17.1 Potato chip spots data.

Batch	Sugar concentration	Dark spot scores
1	0.3%	1, 1, 2
2	0.4%	2, 1, 3
3	0.5%	4, 5, 5
4	0.6%	5, 4, 4
5	0.7%	5, 6, 7

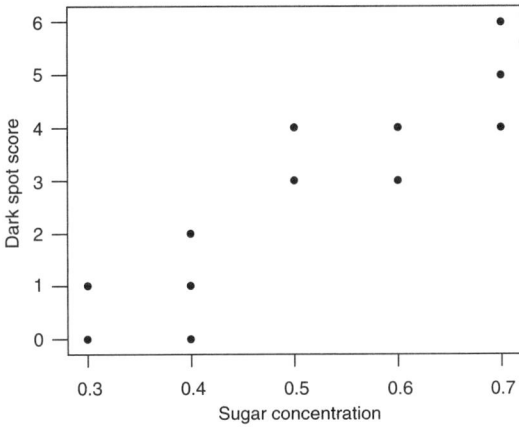

Figure 17.2 Scatter plot of dark spot score versus sugar concentration (%).

The team decided to implement a feedforward control scheme. They noticed that a significant number of dark spots occurred when the sugar concentration was 0.5% or greater. They knew they could reduce the average sugar concentration in a batch of potatoes by storing them for several weeks at temperatures above 13°C. Any batch of potatoes with initial sugar concentration above 0.4% was stored until the sugar concentration decreased. Otherwise the potatoes were processed immediately. The dark spot problem was virtually eliminated (Nikolaou, 1996). There were added storage and logistic costs but no other reported negative side effects due to the storage.

Steering Wheel Vibration

An automobile manufacturer incurred major warranty costs because of customer complaints about steering wheel vibration. A Six Sigma team found that the dominant cause of vibration was imbalance in the transmission. Reformulating the problem in terms of transmission imbalance, the team established a baseline using 200 transmissions selected from two days' production. The data are given in the file *steering wheel vibration baseline*. The distance between the center of gravity and the axis of rotation of the transmission quantifies imbalance, so lower is better. We show the histogram of the baseline data in Figure 17.3. The goal of the project was to reduce the imbalance to less than 2.0 on all transmissions since this would eliminate the vibration problem. In the baseline data, 20.5% of the transmissions exceeded this limit.

The team proceeded to look for a dominant cause without checking the measurement system since they were confident in their ability to measure imbalance. Using a component swap investigation (chapters 11 and 12), the team found that the dominant cause of the imbalance was an interaction between two mating components based on their relative orientation as set in the assembly process. Scaling for relative mass, the imbalance in the two components contributed roughly the same amount to the overall variation. The team also observed in the component swap investigation that the imbalance could be substantially reduced during reassembly by aligning the two components so their individual imbalances were offset.

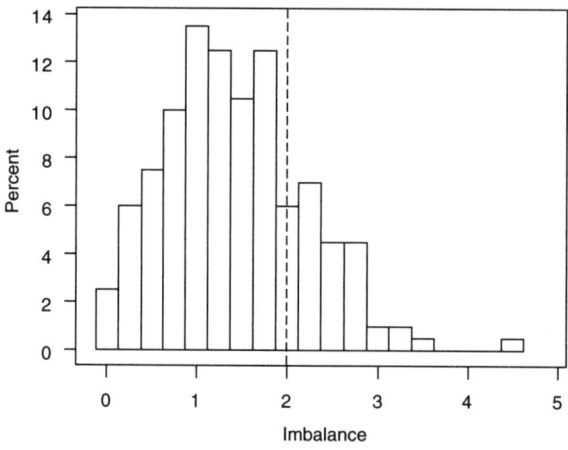

Figure 17.3 Baseline histogram of imbalance.

The team had several choices of approach. They could reformulate the problem again and try to reduce the imbalance in each of the two components. They could implement 100% inspection and rework transmissions with imbalance greater than 2.0, a very expensive option. Instead, they decided to assess the feasibility of feedforward control. In the assembly process, they planned to measure the center of gravity of each component and then assemble the two parts so that there was a maximum offset (that is, the two centers of gravity were set 180° apart). Assembling in this way is called *vectoring*. See Figure 17.4.

To estimate the possible improvement from vectoring, the team measured the center of gravity for 100 pairs of the two components selected over two days to match the baseline investigation. We give the data in the file *steering wheel vibration feedforward*.

The team calculated the imbalance for each sampled pair if vectoring had been used in the assembly. The histogram for these simulated data is shown in Figure 17.5. For these data, 3.5% of the transmissions had imbalance greater than 2.0, a marked improvement over the baseline.

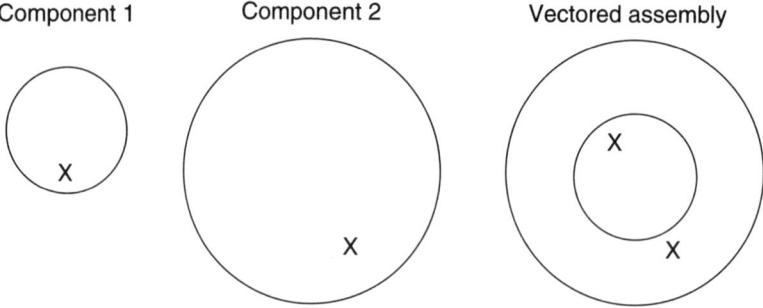

Figure 17.4 Vectoring to reduce imbalance (the X shows the location of the center of gravity for each component).

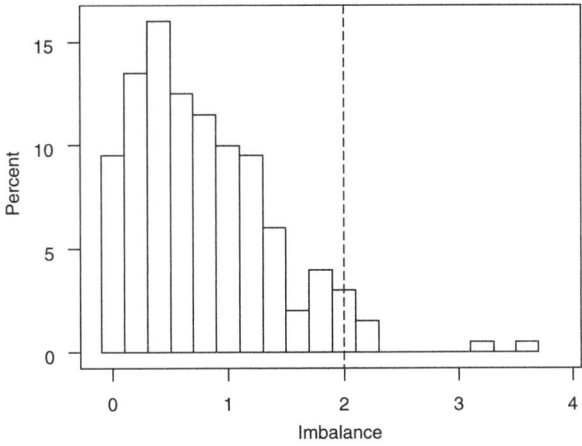

Figure 17.5 Histogram of simulated imbalances for vectored assemblies.

Since vectoring alone cannot meet the goal, the team considered *selective fitting* followed by vectoring, a more complicated and expensive option. The idea was to select a pair of components with similar imbalances and then vector them. There are two steps:

- Sort the second component into bins based on its measured imbalance.
- Measure the first component and select a matching second component from the appropriate bin.

The team needed to determine how best to sort the components (that is, determine the number of bins and their boundaries) and whether sorting was worth the trouble.

To quantify the possible benefits of selective fitting with different bin structures, the team used the data from the feedforward investigation to simulate several scenarios. They explored the effect of sorting by creating two or three bins for the second component. With two bins, the components were divided based on whether their imbalance was less than or greater than 0.75. This is roughly the middle of the range for both components in the sample. With three bins, the boundaries were 0.33 and 0.90. The simulation results are presented in Table 17.2. The benefits of selective fitting are clear. The goal can be reached by using a combination of sorting components and then vectoring. We give a more detailed explanation of how this simulation was done in the supplement.[1]

The team decided to implement feedforward control using selective fitting with two bins and vectoring. They arranged for the off-line measurement of the center of gravity (imbalance and angle) of the second component. The angle was marked on each component. The components were then sorted into two bins depending on whether the imbalance exceeded 0.75. In the assembly process, the center of gravity for the first component was measured. This component was then matched with a mating second component from the appropriate bin and the two selected components were vectored.

Table 17.2 Results of simulating selective fitting.

Assembly method	Average imbalance	Standard deviation of the imbalance	Percentage greater than 2.0
Baseline	1.31	0.83	21%
Vectoring only	0.80	0.61	4.8%
2 bins and vectoring	0.46	0.41	0%
3 bins and vectoring	0.36	0.38	0%

In determining this implementation plan, the team considered a number of logistical issues. They determined that it was feasible to:

- Keep the necessary bin of the second component on the assembly lines.
- Measure the imbalance and direction for the first component as part of the assembly process.

Originally, they planned to have the component suppliers measure, mark, and bin both components. Then, in the assembly process, matching components would have been selected from the appropriate bins. The suppliers resisted and the idea could not be implemented.

The team validated their solution by repeating the baseline investigation. The average overall imbalance of the transmission was 0.59, and fewer than 1% of the assemblies had imbalance exceeding 2.0, not quite reaching the goal but a substantial improvement.

Selective fitting and vectoring added substantial logistic and measurement costs. Occasionally, there were problems because one of the bins for the second component was empty. This is one reason the team had opted for two rather than three bins. The extra improvement from three bins did not seem worthwhile considering the additional cost and complexity.

Truck Pull

In Chapter 1, we described a project to reduce variation in pull, an alignment characteristic of trucks. At an earlier stage of the project than that described in Chapter 1, the team found that the dominant cause of pull variation was the truck frame geometry. They selected feedforward control as a working approach. As part of the search for the dominant cause, 100 frames had been selected from regular production over two weeks. Working with the supplier, the geometry was measured for each frame and reduced to four summaries (left front, right front, left rear, and right rear) thought to affect caster and camber, the measured alignment characteristics that determine pull. The team had the frames built into vehicles using the normal process and measured the caster and camber for each truck. The data are given in the file *truck pull feedforward*. We focus here on left caster.

To establish a relationship between caster and the frame geometry, we fit a regression model to predict caster as a function of the truck geometry. See the supplement to Chapter 11 for more details on regression models. The MINITAB results are:

```
The regression equation is
left caster = -18.6 + 1.24 left front + 0.677 right front + 0.140 left rear +
0.156 right rear

Predictor          Coef       SE Coef        T          P
Constant         -18.6097      0.6100      -30.51     0.000
left front         1.24243     0.04018      30.92     0.000
right front        0.67669     0.03585      18.88     0.000
left rear          0.13953     0.02763       5.05     0.000
right rear         0.15624     0.04162       3.75     0.000

S = 0.1760      R-Sq = 96.4%      R-Sq(adj) = 96.2%

Analysis of Variance

Source            DF        SS         MS         F         P
Regression         4       77.827     19.457     627.82    0.000
Residual Error    95        2.944      0.031
Total             99       80.771
```

Truck frame geometry is a good predictor of the left caster value. We make the prediction by substituting the known frame geometry summaries into the equation:

left caster = –18.6 + 1.24 left front + 0.677 right front + 0.140 left rear + 0.156 right rear

The standard deviation of the left caster for the 100 vehicles in the investigation is 0.903. The residual standard deviation is 0.176. If we could eliminate the effect of the variation in the frame geometry, we could reduce the variation in left caster by a factor of five. There was a similar pattern for the other alignment characteristics.

In reality, a mathematical prediction model was built using knowledge of the geometry of the suspension. The regression model mimicked the mathematical model well.

Since the potential benefit was large, the team recommended implementing feedforward control. For each frame, the supplier measured and bar-coded the frame geometry summaries. Then at the truck assembly plant they:

- Read the geometry summaries from the bar code and used the values to predict caster and camber.

- Built positioning components based on the prediction to adjust the caster and camber.

- Used the custom components to position the suspension components on the frame during the assembly.

After implementing feedforward control, the standard deviation in left caster values, given in the file *truck pull validation*, was 0.25. The full reduction in left caster variation predicted by the regression model was not achieved because there was some error in building and using the positioning components and there were also small errors in the mathematical model.

There was a large benefit for the extra cost in terms of reduced warranty claims. The truck pull variation was reduced by roughly 70% through the use of the feedforward controller.

17.2 ASSESSING AND PLANNING FEEDFORWARD CONTROL

To successfully implement feedforward control, we must:
- Identify an adjuster that can be used repeatedly to quickly move the process center.
- Define a scheme to sample and measure the dominant cause.
- Determine a method to predict the output on the corresponding part.
- Specify an adjustment rule, that is, determine when and how much to adjust the process.
- Assess the sampling, measuring, and adjusting costs.
- Estimate the benefit of the process change.

If we can accomplish all these tasks, and the benefits outweigh the costs, we proceed to the validation stage of the algorithm.

The use of an adjuster for feedforward control is different from its use to move the process center as described in Chapter 15. We expect to make frequent adjustments of different sizes over time to compensate for changes in the dominant cause. Adjusters suitable for a one-time shift of the process may not be useful in feedforward control. In implementing feedforward control, perfect adjustment is typically not possible. For example, when using selective fitting, finding perfectly matched components may be too much effort.

We must be able to measure the dominant cause with an effective and timely measurement system. If there are large errors in measuring the cause, these errors will affect the prediction of the output and hence the adjustment procedure. The measured value of the cause must be available in time to make the adjustment.

In both the potato chip spots and steering wheel vibration examples, we used informal prediction of the output using the measured value of the dominant cause. The simplest formal model is a linear relationship between the cause and the output characteristic. That is

$$\text{predicted output} = \text{intercept} + \text{slope} * \text{dominant cause}$$

Predictions should be accurate for the full range of values normally seen for the dominant cause. Informal methods and simple models may give poor predictions and more complicated models are sometimes warranted.[2]

Feedforward control is effective only if based on a dominant cause. By definition, we cannot predict the output well with a cause of variation that is not dominant. If we base an adjustment on a poor prediction, we may increase rather than decrease the output variation.

If we measure the dominant cause for every unit, we must be able to apply the adjustment to each unit. This can be a difficult task in a complex process, especially if the adjustment takes place far downstream from the measurement of the cause. In the truck pull example, the team solved this difficulty by bar-coding the frames.

The frequency of adjustment depends on how the dominant cause varies. If the dominant cause acts in the part-to-part family, we may need to make an adjustment for every part. If the dominant cause acts in the time-to-time family, we can make adjustments less often, as in the potato chip spots example. Since frequent adjustment adds to the cost and complexity of the process, the feasibility of feedforward control depends on the nature of the variation of the dominant cause. If adjustment costs are large, we may decide to only adjust when the difference between the predicted output and the target is large.

In the truck pull example, we use the residual standard deviation after fitting a regression model relating the output and the dominant cause to estimate the benefits of feedforward control. This standard deviation underestimates the remaining variation because there are likely measurement, prediction and adjustment errors. In cases where the dominant cause acts batch-to-batch, such as in the potato chip spots example, we use the within-batch variation as an optimistic estimate of the benefit. We can sometimes use available process data to simulate the performance of a proposed feedforward controller to assess its benefit, as in the steering wheel vibration example.

Feedforward control is related to feedback control, described in Chapter 18. In both approaches we reduce variation by adjusting the process based on a prediction of the output. The fundamental difference is that, with feedforward control, we predict the output using the dominant cause, while with feedback control, we predict the future output using previously observed output values.

Feedforward control can also be thought of as a way of desensitizing the process to variation in a dominant cause, as discussed in Chapter 16. In the desensitization approach, we searched for process settings that permanently made the process less sensitive to variation in a dominant cause. With a feedforward controller, the desensitization requires repeated adjustments to compensate for the different values of the dominant cause.

There is little published work on feedforward controllers available in the statistical literature. The most comprehensive work is Box and Jenkins (1976). See also Jenkins (1983).

 **Key Points**

- A feedforward controller compensates for the value of a dominant cause by predicting the output from the measured value of the cause and then adjusting the process based on the deviation between the prediction and the target.

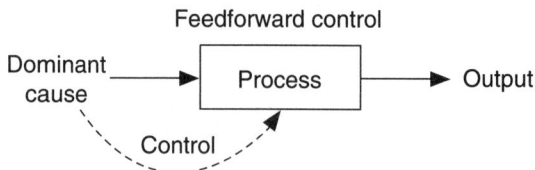

- To implement feedforward control, we require the value of a dominant cause, knowledge of the relationship between the cause and the output in order to make the prediction, and a timely and cost-effective adjuster.
- If the dominant cause acts in the time-to-time family, we need to adjust once, at most, once time period.

Endnotes (see the Chapter 17 Supplement on the CD-ROM)

1. In the supplement, we explain how to assess the potential benefits of vectoring and selective fitting in the steering wheel vibration example using simulation and the available data.
2. In some applications more complicated prediction methods or models are warranted. In the supplement we give references for these methods.

 Exercises are included on the accompanying CD-ROM

18
Feedback Control

Quality is never an accident, it is always the result of intelligent effort.
—John Ruskin, 1819–1900

Feedback control is used to compensate for a predictable pattern in the output characteristic due to known or unknown causes. We reduce variation by predicting the next output value using previously observed output values. Then we adjust the process center to compensate for the deviation between the predicted output and the target. We do not need to identify a dominant cause of variation with this approach. We illustrate how to implement a feedback control scheme in Figure 18.1.

For feedback control to be effective, the requirements are:

- The dominant cause must act in the time-to-time family.
- There must be a timely way to adjust the process center.

The potential costs of feedback control include:

- An investigation to explore the time pattern in the output, if not known
- An experiment to find an adjuster, if one is not known
- The ongoing costs of measuring the output and making the adjustment

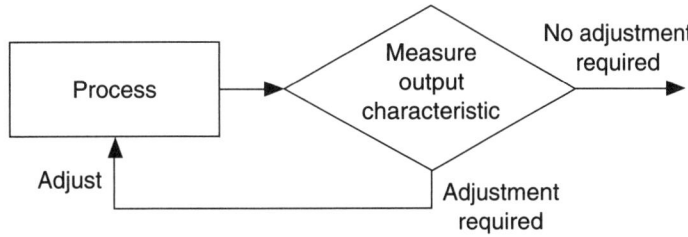

Figure 18.1 Feedback control schematic.

270 Part Four: Assessing Feasibility and Implementing an Approach

We can assess the potential benefits of feedback control once we know the time pattern of variation in the output. With feedback control we can at best hope to eliminate the time-to-time component of the output variation.

18.1 EXAMPLES OF FEEDBACK CONTROL

We illustrate the use of feedback control with five examples.

Parking Brake Tightness

In a vehicle assembly process, there was a problem with excess variation in parking brake tightness that led to considerable rework. To measure tightness the team determined the number of clicks until the brake locked. The specification was 5 to 8 clicks. At final inspection, the plant found parking brakes that were both too tight and too loose.

In the baseline investigation, the tightness ranged between 3 and 9 and there were runs of high and low values. The dominant cause acted in the time-to-time family. Based on this clue and process knowledge, the team speculated that the dominant cause of tightness variation was the length of either the front or axle cables. The two assemblies that include these cables were delivered in batches from a supplier. To search for the dominant cause, the team conducted an investigation where parking brake tightness was measured for three consecutive vehicles for 12 combinations of batches of the two cable assemblies.

The data are given in the file *parking brake tightness multivari*. From Figure 18.2, we see that the dominant cause acts in the front cable assembly and differs from batch to batch. The variation within batches is substantially smaller than the batch-to-batch variation. The baseline tightness values ranged over seven possible values; within a batch of front cable assemblies, the range covered only three values. In the multivari investigation there were six different tightness values, which roughly matches the full extent of variation.

The team did not further explore front cable length as a suspect. They could not measure the length of the cable in the plant since it arrives as part of an assembly. Because the dominant cause was in the batch-to-batch family, they decided to consider feedback control.

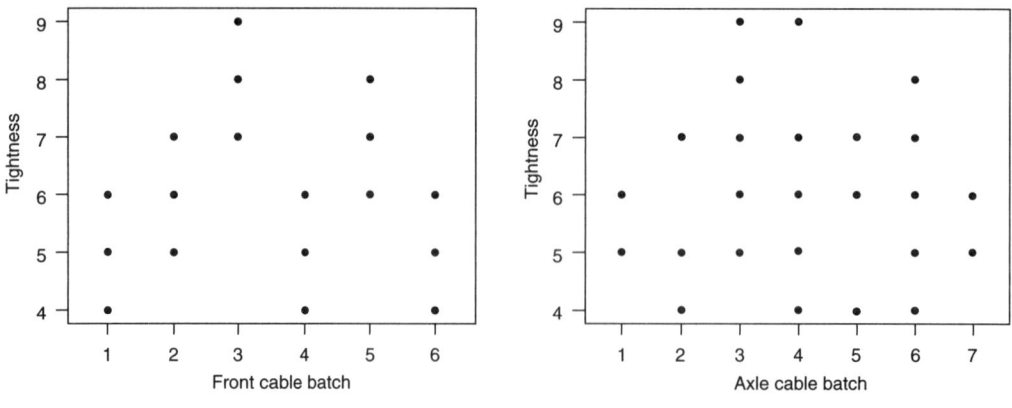

Figure 18.2 Box plot of tightness by front and axle cable batch.

To implement the approach, they needed to find an adjuster. The team knew that changing the depth of an adjustment nut would change parking brake tightness. To calibrate the adjuster, they selected five vehicles of varying tightness to cover the full extent of parking brake tightness variation. The parking brake system on each vehicle had been originally installed with the adjustment nut at a depth of 24 millimeters. The team reinstalled the parking brake system on each vehicle with the nut at the four depths 22, 23, 25, and 26 millimeters and measured the corresponding tightness. The data are given in Table 18.1 and in the file *parking brake tightness adjuster calibration*.

We use MINITAB to fit a regression model to relate the average tightness to the depth with results:

```
The regression equation is
average = -15.0 + 0.880 depth (mm)

Predictor        Coef      SE Coef        T        P
Constant      -15.000        1.666    -9.01    0.003
depth (mm)    0.88000      0.06928    12.70    0.001

S = 0.2191     R-Sq = 98.2%    R-Sq(adj) = 97.6%

Analysis of Variance

Source            DF         SS         MS         F        P
Regression         1     7.7440     7.7440    161.33    0.001
Residual Error     3     0.1440     0.0480
Total              4     7.8880
```

The estimated slope is 0.88, so changing the adjustment nut depth by one millimeter will change the tightness, on average, by about 0.88 clicks.

There was little cost to measuring tightness and to changing the depth of the adjustment nut. Since the dominant cause was in the batch-to-batch family, the team concluded that feedback control was feasible.

To implement the approach, the team decided to measure the tightness of the first five vehicles assembled for each new batch of front cable assemblies. These measurements were made as soon as possible once the parking brake was installed. They made an adjustment

Table 18.1 Parking brake tightness adjuster experiment results.

Vehicle	22 mm depth	23 mm depth	24 mm depth	25 mm depth	26 mm depth
1	2	2	3	4	5
2	2	3	4	5	6
3	4	5	6	7	7
4	7	7	8	9	10
5	8	8	9	10	11
Average	4.6	5.0	6.0	7.0	8.0

if any of these measurements was outside the specification limits (5 to 8). The amount of the adjustment was based on the difference between the target tightness (6.5) and the average tightness for the five vehicles. To increase the average tightness by one click, they increased the depth of the adjustment nut by roughly 1/0.88 = 1.14 millimeters. For ongoing protection, and to ensure that the feedback control was effective, they changed the control plan so that within a batch of front cables, the parking brake tightness was measured on every tenth vehicle. If an out-of-specification vehicle was found, then all vehicles for that batch were inspected and reworked if necessary. After implementation, there was a marked reduction in the amount of rework required because of parking brake tightness variation.

V6 Piston Diameter

We discussed the problem of excess diameter variation in the production of V6 aluminum pistons in chapters 2, 5, 9, and 11. From the problem baseline investigation, the team learned that the process was centered on target with standard deviation 3.32 microns. The problem goal was to reduce the standard deviation to less than 2.0 microns. The team determined that the measurement system was adequate and, using a variation transmission investigation (see Chapter 11), that the home of the dominant cause was the intermediate Operation 270. They reformulated the problem to reduce the variation in the diameter as measured at Operation 270.

At Operation 270, there were two parallel grinders. To better understand the performance of the process, the team measured one piston a minute from each grinder for 200 minutes. They ensured that no adjustments were made to the grinders while the data were collected. The data for stream 1 are given in the file *V6 piston diameter 270*. Diameter is recorded in microns as the deviation from 87 millimeters. Thus, a measured diameter of 87.595 millimeters is recorded as 595. We show the run chart for stream 1 in Figure 18.3.

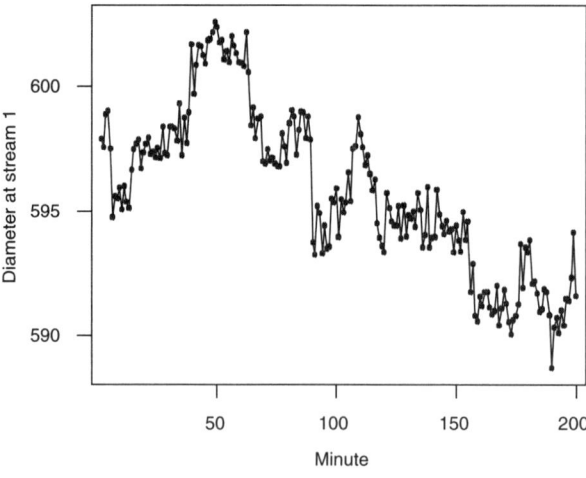

Figure 18.3   Diameter at stream 1 at Operation 270 by minute.

The plot for the second stream is similar. Once the data were stratified by stream, the team saw that the short-term variation in the diameter was small relative to the time-to-time variation.

Based on the run chart, the team decided to look at the feasibility of feedback control separately for each stream. They found this approach attractive because they knew that the operator could make an adjustment at Operation 270 within a short time and with immediate effect.

The team selected an informal feedback scheme similar to precontrol.[1] The rules for each stream were:

- Every 15 minutes, measure the diameter of two consecutive pistons after machining.

- Compare the average diameter to the adjustment limits 592.7 and 600.7 microns.

- If the average falls outside the limits, adjust the process to the target 596.7 microns.

The size of the adjustment, if any, is the difference between the observed average diameter and the target.

The team derived the adjustment limits by working backwards using information about the relationship between the diameter at Operation 270 and the final diameter. To meet the overall goal, they needed the final diameter to be well within the specification limits of 591 ±10 microns.

To assess the potential gain from this approach, the team used the data in the file *V6 piston diameter 270* to simulate the effect of feedback control.[2] They assumed the adjustment was perfect in the simulation, which would not be true in practice. We see the results of the simulation in Figure 18.4.

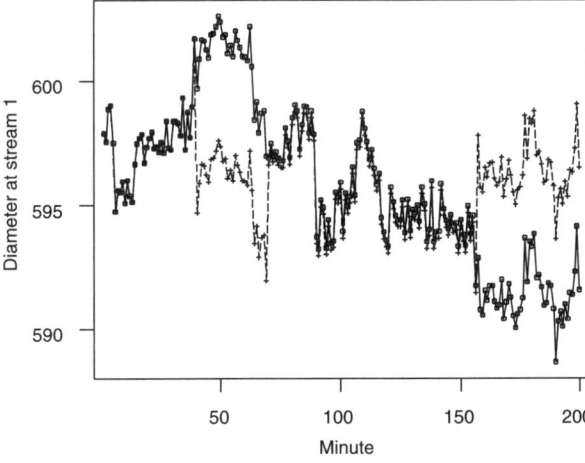

Figure 18.4 Diameter at stream 1 of Operation 270 by minute (solid line gives the original series, dashed line gives the adjusted series).

We estimate the benefit of the proposed feedback scheme by comparing the adjusted and unadjusted series. Using MINITAB, we get:

```
Variable         N       Mean      Median     TrMean     StDev    SE Mean
original        200      5.808      5.548      5.770     3.269     0.231
adjusted        200      6.024      6.078      6.027     1.694     0.120

Variable      Minimum   Maximum        Q1         Q3
original       -1.300    12.586      3.552      7.916
adjusted        1.490    11.703      4.753      7.231
```

Based on the simulation, we expect the feedback controller to reduce the standard deviation of the diameter at Operation 270 from 3.3 to about 1.7. We combine this information with the results from the variation transmission investigation (see the supplement to Chapter 11) to predict that after implementing the feedback controller, the standard deviation in the final diameter would be reduced to

$$\sqrt{0.88^2 (1.7)^2 + 1.224^2} = 1.93 \text{ microns}$$

The team changed the control plan at Operation 270 and trained the operators to use the new control scheme. They also arranged for a periodic process audit to ensure the new scheme was being used.

After implementation, the team repeated the baseline investigation to validate the improvement. The standard deviation of the finished diameter was reduced from 3.32 to 2.20 microns.

V8 Piston Measurement System

In the production of V8 pistons, a 100% inspection gage checked many outputs and rejected pistons not meeting specifications. An auditor retrieved a piston from shipped inventory and remeasured it on the inspection gage. The piston was badly out of specification with respect to one of its diameters, measured at a fixed height on the skirt. The measured diameter was 13.2 microns (measured from the nominal dimension) compared to the specification ±9.0 microns.

Management assigned a team to prevent the shipment of out-of-specification pistons. The team reviewed the calibration procedure for the gage, which was carried out once per day. They conducted a short-term measurement system investigation, measuring 10 pistons three times each in haphazard order within 15 minutes. The data are given in the file *V8 piston diameter short-term measurement*. The results of a one-way analysis of variance from MINITAB are:

```
Analysis of Variance for diameter
Source      DF        SS         MS         F        P
part         9    524.587     58.287    328.07    0.000
```

```
Error        20      3.553    0.178
Total        29    528.140
```

Pooled StDev =   0.4215

The estimate of measurement variation was 0.42 microns. If the diameter was within specification at the original measurement, we cannot explain a measured value 13.2 by the short-term measurement system variation.

The team next decided to assess the stability of the measurement system by measuring the same two pistons every 15 minutes for 12 hours, starting immediately after calibration. The data are given in the file *V8 piston diameter measurement stability* and plotted in Figure 18.5.

Over the 12-hour period, the variation was much larger than expected based on the results from the short-term measurement investigation. The range of diameter values for each piston was around 8 microns. This drift could easily be responsible for the out-of-specification piston found in the audit.

The team immediately increased the frequency of calibration to every two hours. They decided not to look for the cause of the drift but instead implemented a feedback control scheme. Immediately after calibration and then every 15 minutes, the operator measured the same piston and recorded its diameter. If the change from the initial measurement was more than 1.8 microns (in either direction), the process was stopped and the gage was recalibrated. Note that the short-term standard deviation of the difference in two measurements on the same piston is $0.42\sqrt{2} = 0.59$, so a difference of 1.8 microns indicates that the measurement system has drifted.

The changes added to the operating cost of the inspection gage. However, no further out-of-specification pistons were found in audits. There was a positive side effect. The scrap rate due to out-of-specification diameters decreased by about 50%, which more than made up for the loss of cycle time of the gage. Before the change, the gage rejected many

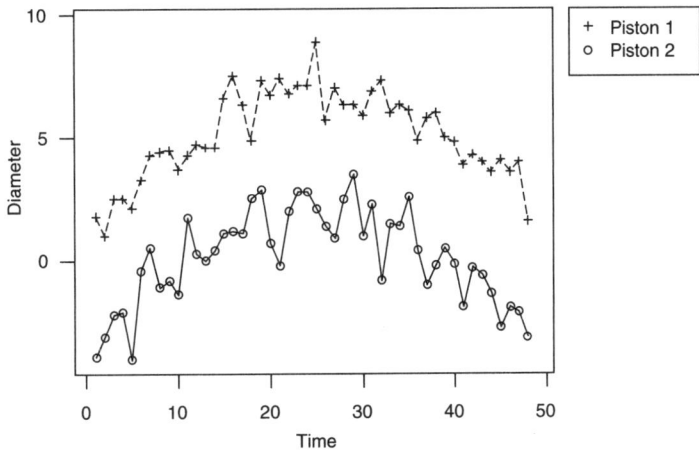

Figure 18.5   Piston diameters versus time.

good pistons. As well, there were fewer adjustments to the upstream machining process that had been driven by the drift in the inspection gage.

Fascia Film Build

We introduced a problem of excess variation in the film build (paint thickness) of fascias in Chapter 3. After finding the dominant cause, the team reformulated the problem in terms of flow rate. To establish a baseline for the new problem, the team recorded the flow rate every minute for three hours. The data are given in the file *fascia film build baseline* and plotted in Figure 18.6. Feedback control is a possibility here since eliminating the drift over time would reduce the variation in flow rate substantially. The range of flow rate values over the short term is less than half the long-term range.

Since the team knew they could adjust the flow rate quickly using a valve, they decided to investigate the possibility of using feedback control to reduce flow rate variation.

This team took a more formal approach. To predict the flow rate at the next minute, they used an *exponentially weighted moving average* (EWMA).[3] To explain this method, let y_t be the measured output at time t (measured in minutes here). Then we predict the next value of the output, denoted $\hat{y}_{t+1}$, using a weighted average of previous values as given in Equation (18.1).

$$\hat{y}_{t+1} = \alpha y_t + \alpha(1-\alpha)y_{t-1} + \alpha(1-\alpha)^2 y_{t-2} + \alpha(1-\alpha)^3 y_{t-3} + \dots , \qquad (18.1)$$

where α (the Greek letter *alpha*) is a constant and $0 < \alpha \leq 1$. In the prediction, the most recent value y_t gets the highest weight α, the next most recent point, y_{t-1}, gets the second highest weight $\alpha(1-\alpha)$, and so on. You can use a bit of probably forgotten high school algebra to show that the weights add to one.

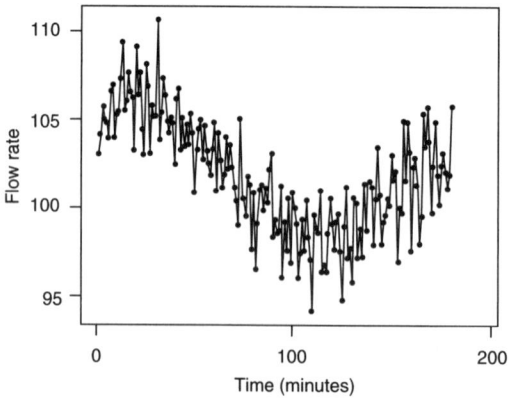

Figure 18.6 Run chart of flow rate.

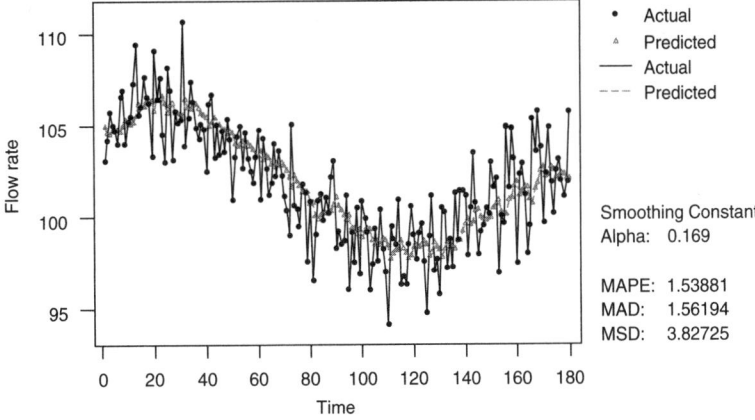

Figure 18.7 EWMA smoothing of paint flow rate.

We use MINITAB (see Appendix C) to estimate the parameter α from the data collected over a period of time when there were no adjustments. For the flow rate data, we get the results shown in Figure 18.7. The estimated value for α is 0.169. We also show the series of predicted values in Figure 18.7 to demonstrate how the EWMA prediction captures the long-term pattern in the series and smoothes the short-term variation.

Once the controller is operational, we start to make adjustments and see the series that includes the effects of earlier adjustments. Let y_t^* be the observed output with the controller in operation. After some algebra,[4] we can show that the appropriate adjustment at time t is $\alpha(y_t^* - \text{Target})$ in the direction of the target. In other words, if y_t^* is greater than the target value, we move the process center down by $\alpha(y_t^* - \text{Target})$. Here we assume adjustments are made after every measurement.

If we make perfect adjustments, we get an estimate of the process standard deviation when the controller is in use from the MINITAB results. For this example, the estimated standard deviation is 1.96 ($\sqrt{MSD}$ in Figure 18.7), a substantial reduction from 3.23, the standard deviation of the unadjusted series, which serves as a baseline here.

To implement this feedback control scheme, the team calibrated the valve and automated the adjustment so that it could be carried out each minute. The standard deviation of the film build was markedly reduced.

Truck Pull

In Chapter 1 and elsewhere, we discussed a problem with excess pull variation in trucks. Pull, an alignment characteristic, is a linear function of left and right camber and caster. The baseline data *truck pull baseline* include the values for left and right caster and camber on more than 28,000 trucks produced over a two-month period. As a result of the problem focusing effort discussed in Chapter 6, the team determined that reducing caster variation was the highest priority. They searched unsuccessfully for a dominant cause of this variation.

The team then decided to investigate feedback control as an approach to reduce the caster variation. We illustrate the assessement of feasibility using right caster. There were similar results for left caster. The target for right caster is 4.5°. A summary of the baseline data is:

```
Descriptive Statistics: r-caster

Variable            N        Mean      Median      TrMean      StDev     SE Mean
r-caster          28258     4.5188     4.5210      4.5192      0.2427    0.0014

Variable         Minimum   Maximum        Q1          Q3
r-caster          3.0440    5.9380      4.3600      4.6780
```

The process is well centered with standard deviation 0.243°. We see in the plot of right caster angle over time (see Figure 18.8) that there is some drift in the process near the middle of the series.

The team knew they could adjust caster using the same custom components that were built for the feedforward control scheme discussed in Chapter 17. Since these components were manufactured and assembled approximately two hours before caster was measured, feedback control based on adjustments after each truck was not feasible. The team decided to investigate the effects of making an adjustment once per shift. The process is run with three eight-hour shifts per day.

As a first step, they looked at a one-way analysis of variance as described in the supplement to Chapter 11. The idea was to separate the baseline variation into two components, variation within shifts and variation from shift to shift. The partial results from MINITAB are:

```
One-way ANOVA: r-caster versus shift

Analysis of Variance for r-caster
Source      DF         SS         MS         F         P
shift      131      456.378     3.484      81.09     0.000
Error    28126     1208.334     0.043
Total    28257     1664.712
```

The estimate of the within-shift standard deviation, labeled MS(Error) in the MINITAB results, is $\sqrt{0.043} = 0.207$. If we could use feedback control to make all the shift averages equal, we would expect a reduction in the baseline standard deviation from 0.243 to 0.207, a modest improvement.

Summarizing the caster shift averages gives:

```
Variable              N        Mean      Median      TrMean      StDev     SE Mean
right caster         132      4.5271     4.5135      4.5269      0.1255    0.0109
shift average

Variable          Minimum   Maximum        Q1          Q3
right caster       4.1909    4.9721      4.4473      4.6154
shift average
```

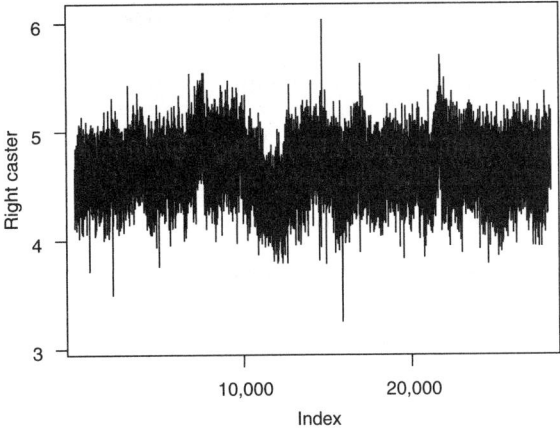

Figure 18.8 Right caster angle over time.

To obtain the shift average summary from the available data, we first store the caster averages by shift (see Appendix B). Across the 132 shifts, the standard deviation of the shift averages is 0.126. We plot the shift averages and the predicted averages from the EWMA in Figure 18.9. The smoothing constant is $\alpha = 0.944$, so the most recent shift is given very high weight in the prediction of the subsequent shift average. If we use feedback based on the EWMA to adjust the process at the start of the shift, then we estimate that the standard deviation of the shift averages will be reduced to 0.077 $\left(\sqrt{0.00598}\right)$, less than half the baseline shift-to-shift variation, 0.126. However, since the within-shift variation is so large, this reduction has little impact on the overall variation.

At this point, the team decided that the potential gain of feedback control would not be worth the cost. Given the two-hour time lag and the imperfect adjuster, they knew they would not be able to achieve the small gain predicted using the baseline data.

The team abandoned the problem because no other approach was feasible.

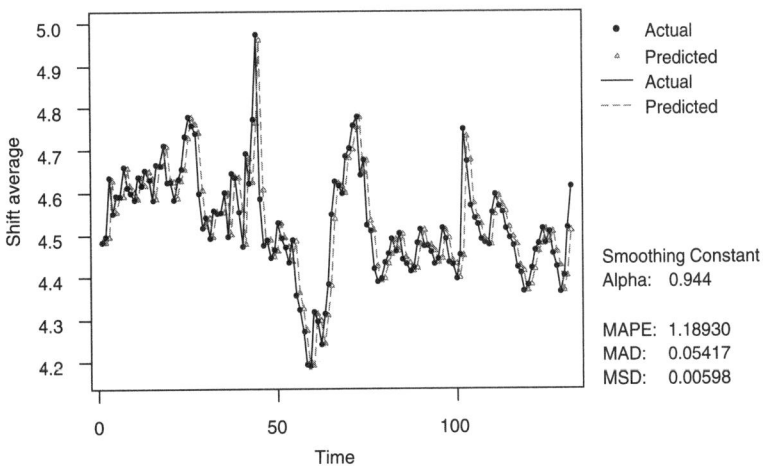

Figure 18.9 Right caster average by shift.

18.2 ASSESSING AND PLANNING FEEDBACK CONTROL

To successfully implement feedback control, we must:

- Identify an adjuster that can be used repeatedly to quickly change the process center.

- Define a scheme to sample and measure the output.

- Determine a method to predict future output values from the current and past output values.

- Specify an adjustment rule, that is, determine when and how much to adjust the process center.

- Assess the sampling, measuring, adjusting costs, and possible side effects.

- Estimate the benefit of the process change.

If we can accomplish all these tasks, and the benefits outweigh the costs, we proceed to the validation stage of the algorithm.

The pattern of variation in the output can be a gradual drift, as in the fascia film build example, or a sudden persistent shift, as shown for machining data in Figure 18.10. Sudden shifts can occur if the output center depends on the setup procedure, the batch of raw materials, tooling changes, and so on. See the chapter exercises for more details on the machining process example.

We observe a pattern of variation with respect to a particular sampling scheme. For example, in the fascia film build example, the team measured the flow rate every minute. The observed pattern would have been different had they measured the flow rate every 15 minutes or every 0.1 seconds. The sampling protocol in the baseline investigation may be sufficient to reveal the time pattern of variation in the output. If not, we suggest a simple multivari investigation with two families, time-to-time and part-to-part. Select the time-to-time family (for example, every 15 minutes, every hour, every shift, and so on) based on the time required to measure the output and to adjust the process.

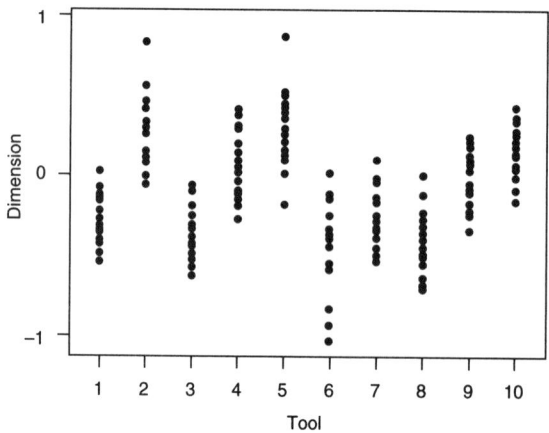

Figure 18.10 A machining process with persistent shifts due to tooling.

We must be able to measure the output quickly compared to the time period that defines the family containing the dominant cause of variation. In an iron foundry, there was a one-hour delay between the time the iron was sampled and the time the chemistry measurements became available. Since the within-hour variation in the process chemistry was relatively large, feedback control was not feasible. In the truck pull example, there was a lag of two hours between the point of adjustment and the measurement of caster. Hence the team considered feedback control schemes that would adjust caster at most every two hours.

We can make similar comments about the action of the adjuster. If the time for the adjustment to take effect exceeds the time period that defines the family containing the dominant cause, we cannot use feedback control effectively.

We recommend determining the pattern of variation before looking for an adjuster since the cost of finding an adjuster may be substantial. If needed, we can search for an adjuster using an experimental plan as described in Chapter 15. Sometimes we need to calibrate a known adjuster, as in the parking brake example, so that we can make adjustments of different sizes to the process center as required.

To specify a feedback control scheme, we need:

- A sampling plan to determine when and how to measure the output

- A rule to predict future output values and to decide what adjustment, if any, is required

- A procedure to make the adjustment

The design of the sampling plan involves a trade-off between the cost of measurement and the ability to predict the future output values. In the V6 piston diameter example, the team decided to measure two pistons every 15 minutes. In the fascia film build example, the team measured flow rate every minute.

To predict future output values, we can use an exponentially weighted moving average or a simple method such as the value of the most recent observation. In the supplement, we discuss the choice of prediction method more fully.[5]

Adjusting after every observation may be undesirable due to the added process complexity and adjustment costs. A simple alternative is to use a dead band or bounded feedback adjustment (Box and Luceno, 1997). With dead bands, we do not adjust the process if the predicted deviation from the target is small. Compared to adjusting after every observation, a dead band scheme results in a smaller decrease in variation and a large reduction in the number of adjustments.

The size of an adjustment was the difference between predicted output and the target in the examples discussed here. However, in cases where a process is subject to regular drift either upward or downward, say due to tool wear, the adjustment may be to the opposite side of the specification limit rather than to the target. This idea is illustrated in the discussion of feedback control in Chapter 3.

Once we have designed the feedback scheme, we can assess the potential benefits in several ways if we have a series of historical data that matches the sampling protocol of the scheme. We can simulate the effect by applying the adjustments to the historical series. If we base the control scheme on an EWMA, we can use MINITAB to estimate the standard deviation of the adjusted process. If we have multivari data with short-term and time-to-time families, we can use a one-way ANOVA to estimate the variation within the short-term family. This is the best we could hope to achieve with a feedback control scheme designed to eliminate the time-to-time variation. In all cases, we assume adjustments are made without error and so, if this is not the case, we overestimate the benefit of a feedback control scheme.

A comprehensive treatment of feedback control from a statistical perspective is given in Box and Luceno (1997) and Del Castillo (2002). There are many variations of feedback control. See Tucker et al. (1993), Box and Jenkins (1994), and Box and Luceno (1997) for further details. Specific examples include acceptance control charts (Duncan, 1986) and precontrol (Shainin and Shainin, 1989; Juran, Gryna, and Bingham, 1979).[6] There is also an extensive engineering literature on feedback control.[7]

If a dominant cause acts part to part, using feedback control will result in increased variation. This is called tampering by Deming (1992, pp. 327–328). The effect of tampering is shown by Deming's famous funnel experiment.

Key Points

- Feedback control is based on predicting future output values using the current and past output values and adjusting if appropriate.
- Feedback control may be effective if the (unknown) dominant cause acts in the time-to-time family.

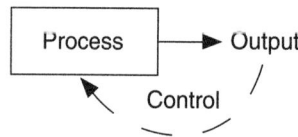

- Feedback control requires an adjuster that moves the process center quickly, has low cost to change, and has small effect on other outputs.
- We may need to calibrate the adjuster with an experiment because we need to make repeated adjustments of different sizes.
- To implement feedback control, we require a sampling plan and a rule to decide when and how large an adjustment to make.

Endnotes (see the Chapter 18 Supplement on the CD-ROM)

1. There are many other informal prediction and adjustment rules. We briefly describe Grubbs' rule, precontrol, and statistical process control charts used for feedback control.
2. We describe how to simulate the benefits of a feedback control scheme using a historical series.

3. In the supplement, we compare EWMA to some other formal prediction methods. We also show that using partial compensation of the current deviation from target to determine the adjustment is appropriate under some assumptions.
4. See note 3.
5. See note 3.
6. See note 1.
7. We describe the connection between the simple feedback control schemes discussed here and engineering proportional-integral-differential (PID) controllers.

 Exercises are included on the accompanying CD-ROM

19

Making a Process Robust

It is a capital mistake to theorize in advance of the facts.
—Sir Arthur Conan Doyle (as Sherlock Holmes), 1859–1930

To make a process robust, we must find changes to fixed inputs that make the process output less sensitive to variation in the *unknown* dominant cause. The robustness approach is similar to desensitization, described in Chapter 16. However, now we do not have knowledge of a specific dominant cause.

The requirements for robustness to be effective are:

- The unknown dominant cause acts in the short-term family of variation.
- New settings of fixed inputs that reduce the effect of the unknown dominant cause resulting in less output variation.

To assess the feasibility of robustness, we choose a number of fixed inputs (candidates) and an experimental plan to determine if changing the levels of these candidates will make the process robust. The first requirement is important because we must define a run to be long enough to see the full extent of variation of the output under the current candidate settings. As a consequence, the dominant cause will vary over its full range within each run. Hence, for every run, we will be able to see if the candidate settings make the process robust to the variation in the unknown dominant cause. If the dominant cause acts in the time-to-time family, it will likely not be feasible to conduct such an experiment since the runs would need to be too long.

The costs of the robustness approach include:

- An experiment to search for the new settings of the candidates
- A one-time change to the process settings
- The ongoing operation with the new process settings

There is no information about whether this approach will be feasible until the experimental investigation is complete. There is a risk of running a high-cost experiment with no return. We cannot assess the benefits of this approach until we find the new settings for the candidates.

19.1 EXAMPLES OF PROCESS ROBUSTNESS

We present three examples of the robustness approach. We selected the third example to demonstrate what can go wrong.

Crossbar Dimension

We discussed the problem of reducing variation in a crossbar dimension in Chapter 12 and the exercises for Chapter 16. The team raised the barrel temperature set point to make the process less sensitive to variation in barrel temperature, the dominant cause. When validating the solution, they showed the variation in the crossbar dimension was substantially reduced but, with the new setting, there was an increase in the frequency of a mold defect called *burn*. They decided to address the burn defect as a new problem. Using a multivari investigation, they showed that the dominant cause of burn acted in the part-to-part family. but the specific dominant cause was not found. They suspected that the defect occurred when the mold cavity filled too fast. In any case, since the suspect dominant cause could not easily be controlled, the team decided to try the robustness approach.

The team planned an experiment with four fixed inputs (candidates): injection speed, injection pressure, back pressure, and screw speed. These candidates were selected because of their influence on fill time and other potential dominant causes in the part-to-part family. They selected two levels for each candidate as given in Table 19.1.

Table 19.1 Candidates and levels for burn robustness experiment (level in current process given by *).

Candidates	Label	Low level	High level
Injection speed	A	slow*	fast
Injection pressure	B	1000*	1200
Back pressure	C	75	100*
Screw rpm	D	0.3	0.6*

The team decided to define a run as five consecutive parts. Since they knew the dominant cause acted in the part-to-part family, they expected it to act within each run. Each run was carried out once the process stabilized after changing the values of the candidates.

The team selected a fractional factorial experiment with the eight runs given in Table 19.2. Since there was no proper baseline investigation, the team assigned the letters to the candidates so that one of the treatments (treatment 5) corresponded to the current process settings (see the Chapter 15 supplement for details). The confounding pattern of the chosen design is given as follows. In the resolution IV design, pairs of two input interactions are confounded.

Alias Structure
A + BCD
B + ACD
C + ABD

D + ABC
AB + CD
AC + BD
AD + BC

The burn on each part was classified into one of four categories of increasing severity. Levels 1 and 2 were acceptable, while levels 3 and 4 resulted in scrap. The order of the runs was randomized. The experimental results are given in Table 19.2 and the file *crossbar dimension robustness*.

Treatments 2 and 3 are promising relative to the current process as given by treatment 5. We plot the burn scores against treatment in Figure 19.1. Because the data are discrete, we add jitter in the vertical direction (see Appendix C). We use MINITAB (see Appendix F) to stack the treatment and burn score columns to create one row for each repeat in the experiment.

Table 19.2 Experimental plan and data for burn robustness experiment.

Treatment	Run order	Injection speed	Injection pressure	Back pressure	Screw speed	Burn scores	Average burn
1	4	Slow	1000	75	0.3	1, 2, 1, 1, 1	1.2
2	8	Fast	1000	75	0.6	1, 1, 1, 1, 1	1.0
3	2	Slow	1200	75	0.6	1, 1, 1, 1, 1	1.0
4	3	Fast	1200	75	0.3	1, 2, 2, 2, 2	1.6
5	5	Slow	1000	100	0.6	1, 3, 2, 2, 1	2.2
6	7	Fast	1000	100	0.3	3, 3, 2, 2, 4	3.4
7	1	Slow	1200	100	0.3	1, 1, 1, 2, 2	2.0
8	6	Fast	1200	100	0.6	2, 2, 4, 3, 2	3.2

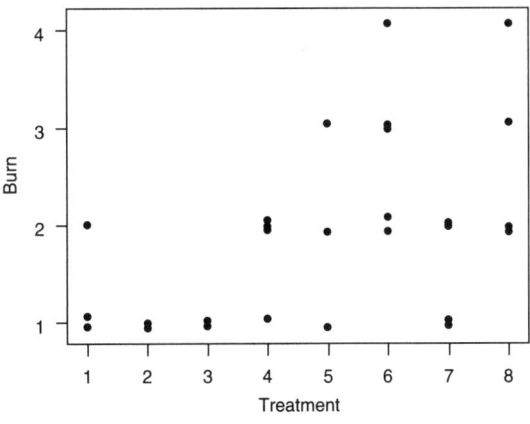

Figure 19.1 Burn by treatment with added vertical jitter.

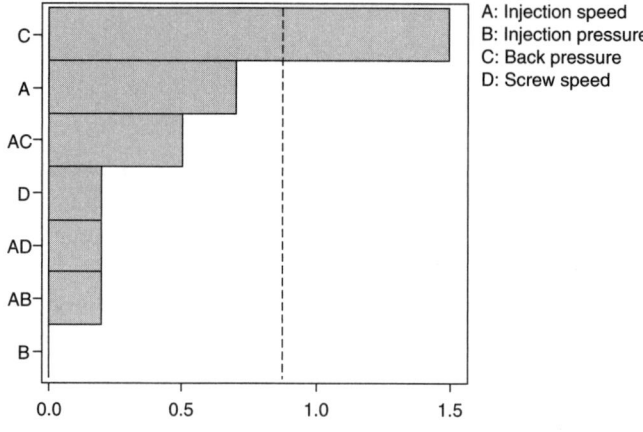

Figure 19.2 Pareto plot of effects on average burn score.

We use the average burn as the performance measure (that is, the *response* in MINITAB) for the analysis. We look for candidate settings that make the performance measure as small as possible.

Fitting a *full model* with all possible effects, we get the Pareto plot of the effects for the average burn score in Figure 19.2. We see that only factor C (back pressure) has a large effect. The team assumed the three-input interaction (*ABD*) confounded with C was negligible.

We show the main effect plot for factor C in Figure 19.3. The low level of back pressure gives lower burn scores on average.

The team decided to reduce the back pressure to 75 and leave the other fixed inputs at their original values. We finish this story in Chapter 21.

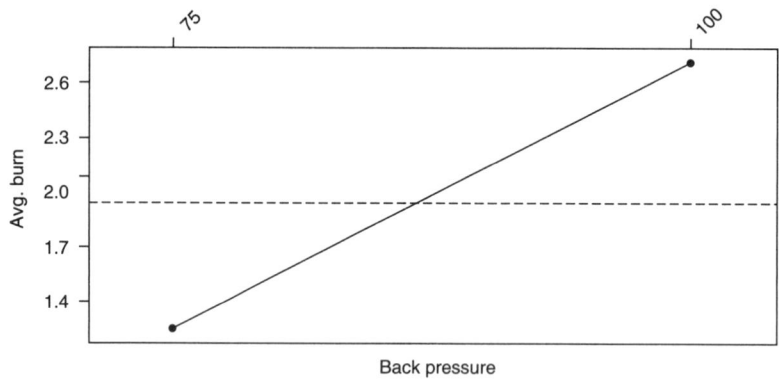

Figure 19.3 Main effect plot for back pressure.

Iron Silicon Concentration

In a project to reduce variation in the silicon concentration of molten cast iron, the team discovered that the existing measurement system was not acceptable. The discrimination ratio was 1.5 and the estimated *stdev(measurement)* was 0.33. The cause of the measurement variation was unknown but acted over the short term.

The measurement process has three major steps:

- Sample the molten iron and pour coins.
- Machine and polish the coins.
- Use a spectrometer to determine the concentration of silicon in the coins.

In the measurement system investigation, the team saw most of the measurement variation in the silicon concentration of coins poured consecutively from the same stream of iron, then prepared and measured together. Rather than search for the dominant cause of the measurement variation, the team decided to try to make the measurement process robust to the unknown cause. They knew that the spectrometer was highly repeatable and hence chose candidates from the other steps of the measurement process. Table 19.3 gives the five candidates and their levels.

Table 19.3 Candidates and levels for iron silicon concentration robustness experiment (level in current process given by *).

Candidates	Label	Low level	High level
Mold temperature	A	300°F	400°F*
Mold design	B	old*	new
Cut depth	C	skin	1/16 inch*
Surface finish	D	lathe*	polish
Sample temperature	E	70°F*	120°F

The definition of a run is a key stage in the planning of a robustness experiment. Since the coin-to-coin variation was dominant, the team decided to sample iron and prepare five coins as quickly as possible for each run. They expected the dominant cause of measurement variation would act within each run.

The team used the resolution III fractional factorial experiment with eight runs and treatments given in Table 19.4. The confounding pattern is:

```
Alias Structure (up to order 3)

A + BD + CE
B + AD + CDE
C + AE + BDE
D + AB + BCE
E + AC + BCD
BC + DE + ABE + ACD
BE + CD + ABC + ADE
```

Table 19.4 Treatments and results for iron silicon concentration robustness experiment.

Treatment	Run order	Mold temperature	Mold design	Cut depth	Surface finish	Sample temperature	Silicon measurements	Log(s)
1	1	300	Old	Skin	Polish	120	2.4, 2.5, 2.3, 2.2, 1.9	−1.55
2	6	400	Old	Skin	Lathe	70	2.8, 2.0, 2.1, 2.6, 2.2	−1.09
3	5	300	New	Skin	Lathe	120	2.2, 2.4, 2.3, 2.2, 2.6	−1.73
4	3	400	New	Skin	Polish	70	2.3, 2.0, 2.3, 2.3, 2.4	−2.03
5	8	300	Old	1/16	Polish	70	2.4, 2.1, 1.8, 2.2, 2.3	−1.51
6	7	400	Old	1/16	Lathe	120	1.9, 2.7, 1.9, 1.9, 2.4	−1.03
7	2	300	New	1/16	Lathe	70	2.5, 2.2, 2.7, 2.5, 2.7	−1.65
8	4	400	New	1/16	Polish	120	2.2, 2.4, 2.1, 2.5, 2.2	−1.79

A resolution III design can estimate main effects, assuming that two input interactions are negligible.

To conduct the experiment, the team manufactured two new molds and selected two old molds. They set the mold temperature to 300° for one of the new molds and to 400° for the second, and similarly for the two old molds. They then quickly sampled iron and poured 40 coins, 10 for each combination of mold design and temperature. Since all coins were produced from essentially the same iron, they assumed that the true concentration of silicon in each coin was the same. Next, the team prepared the 40 coins according to the experimental plan. The data are given in the file *iron silicon concentration robustness* and Table 19.4.

We are looking for a treatment combination that has little variation. As a first step in the analysis, we plot the measured silicon concentration by treatment as shown in Figure 19.4. We see that treatments 4 and 8 look promising since they have relatively little variation.

We summarize the performance of the measurement system for each run using log(s), where s is the standard deviation of the five measurements within each run. We use the log transformation to better meet the assumptions underlying the model for the data. The smaller the within-run variation, the smaller the performance measure. We show the values of the performance measure for each treatment in Table 19.4. As expected, treatments 4 and 8 have the smallest values. From the original investigation of the measurement system, the baseline performance is log (.33) = –1.11.

Figure 19.5 shows the Pareto chart of the effects when fitting a full model. The largest effects are the main effects of candidates *B* (mold design) and *D* (surface finish).

To draw conclusions, we summarize the results using the main effects plots given in Figure 19.6. With the performance measure log(s), smaller is better, so we see that switching to the new mold design and polishing the samples is beneficial.

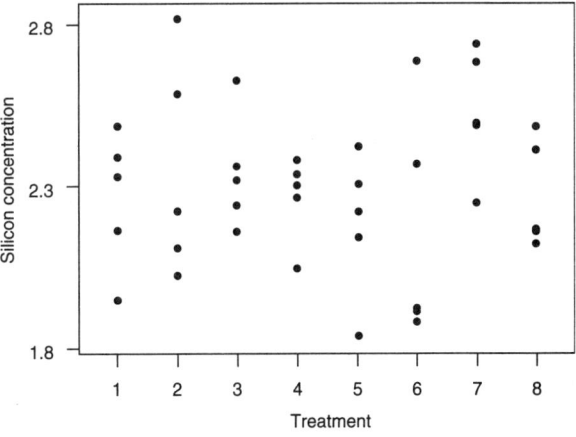

Figure 19.4 Silicon concentration by treatment.

292 Part Four: Assessing Feasibility and Implementing an Approach

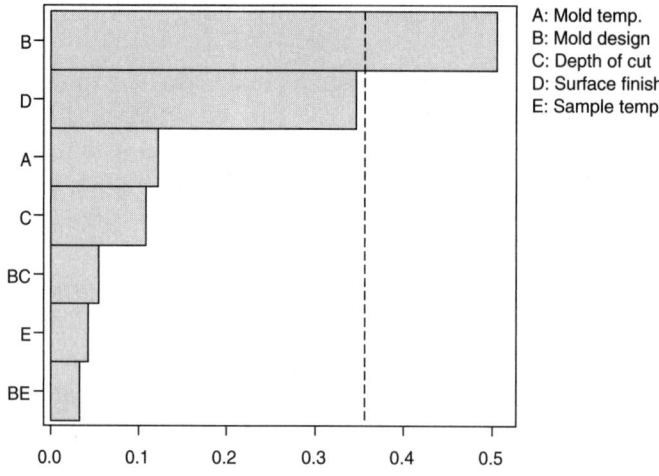

Figure 19.5 Pareto chart of the effects on log(s) for iron silicon experiment.

To validate that the performance of the measurement system would be improved with the new mold design and polished surface finish, the team carried out a simple investigation. They sampled 30 coins from the same stream of iron and measured the silicon concentration using the new levels for the mold design and surface finish and the current levels for the other candidates. Notice that they had not used this treatment in the experiment. The standard deviation of the 30 measurements was 0.15, substantially less then the baseline 0.33. With the new setting we expect the measurement discrimination ratio to increase to 3.3.

The team made a risky decision to use a single batch of iron in the robustness experiment. By using only one silicon concentration, there was a danger that the conclusions from the experiment would not generalize to the range of silicon concentrations seen in the process.

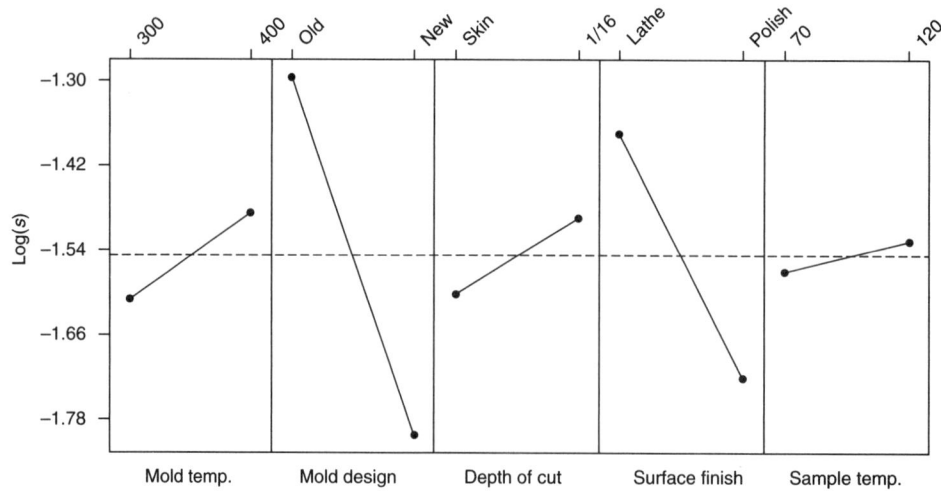

Figure 19.6 Main effects for iron silicon concentration robustness experiment.

Electroplating Pinskip Defect

In a chrome plating process, racks of plastic grills were dipped in a series of chemical baths. The baseline rate of defective grills was 7.3%, measured over several weeks of production. Using Pareto analysis, a team charged with reducing scrap and rework costs found over half the defective grills were due to pinskips, small voids in the plated surface. They decided to try to eliminate the pinskip defectives. The team did not employ the Statistical Engineering algorithm. Instead, they proceeded immediately to the process robustness approach.

They started by brainstorming a list of possible candidates. They selected the six candidates, given in Table 19.5.

Table 19.5 Candidates selected for the pinskip robustness experiment.

Candidates	Name
A	Tank 1 dip time
B	Tank 2 dip time
C	Tank 1 concentration
D	Tank 2 concentration
E	Tank 2 temperature
F	Tank 3 temperature

There was some confusion about the appropriate levels for the candidates, since they all varied somewhat in the normal process. For each candidate, the team decided to use the set point as prescribed in the control plan as one level (−1) and a new set point outside the normal range of variation as the second level (+1). They used their judgment to choose the direction of the change.

For convenience, they defined a run as a rack of 48 grills, all processed simultaneously. The performance measure for each run was the number of defective grills in the rack.

In the past, without much success, the organization had made various attempts to reduce the pinskip defect rate by changing the levels of one fixed input at a time. Now, the team hoped to find a helpful interaction among the candidates. Accordingly, they planned a resolution V experiment with 32 treatments that would allow for the separate estimation of all main effects and two input interactions. This was not possible with six candidates in a 16-run experiment. They planned to conduct the experiment over four days, giving the process time to settle down after changing the candidate levels. The candidates C, D, E, and F (the tank concentrations and temperatures) take the longest time to change, so they ordered the runs to minimize the number of changes of these candidates.

The experimental plan and the number of defective grills for each run are shown in Table 19.6 and are given in the file *electroplating pinskip defect robustness*.

Table 19.6 Electroplating pinskip experimental plan and data.

Treatment	A	B	C	D	E	F	Defectives	Order
1	−1	−1	−1	−1	−1	−1	0	1
2	+1	+1	−1	−1	−1	−1	0	2
3	−1	+1	−1	−1	−1	+1	1	3
4	+1	−1	−1	−1	−1	+1	1	4
5	−1	+1	−1	−1	+1	−1	0	5
6	+1	−1	−1	−1	+1	−1	1	6
7	+1	+1	−1	−1	+1	+1	1	7
8	−1	−1	−1	−1	+1	+1	0	8
9	+1	−1	−1	+1	−1	−1	0	9
10	−1	+1	−1	+1	−1	−1	2	10
11	−1	−1	−1	+1	−1	+1	1	11
12	+1	+1	−1	+1	−1	+1	2	12
13	+1	+1	−1	+1	+1	−1	3	13
14	−1	−1	−1	+1	+1	−1	2	14
15	+1	−1	−1	+1	+1	+1	7	15
16	−1	+1	−1	+1	+1	+1	12	16
17	−1	+1	+1	−1	−1	−1	7	17
18	+1	−1	+1	−1	−1	−1	1	18
19	+1	+1	+1	−1	−1	+1	3	19
20	−1	−1	+1	−1	−1	+1	1	20
21	−1	−1	+1	−1	+1	−1	0	21
22	+1	+1	+1	−1	+1	−1	0	22
23	+1	−1	+1	−1	+1	+1	0	23
24	−1	+1	+1	−1	+1	+1	2	24
25	+1	+1	+1	+1	−1	−1	1	25
26	−1	−1	+1	+1	−1	−1	1	26
27	−1	+1	+1	+1	−1	+1	2	27
28	+1	−1	+1	+1	−1	+1	0	28
29	−1	+1	+1	+1	+1	−1	0	29
30	+1	−1	+1	+1	+1	−1	1	30
31	+1	+1	+1	+1	+1	+1	1	31
32	−1	−1	+1	+1	+1	+1	1	32

The experiment was more time-consuming than expected. The team could not complete the planned eight runs in each day. They found the tank temperatures and concentrations hard to change quickly. In the end, the experiment was conducted over five days.

From Table 19.6, there are many promising treatments with no pinskip rejects in the rack. Using MINITAB and a full model, we get the Pareto chart of effects shown in Figure 19.7. There are three related interactions (*CD*, *CE*, and *DE*) that are large. We present a cube plot in Figure 19.8 for the candidates *C*, *D*, and *E*. The plot gives the average number of grills rejected per rack for the eight combinations of the three candidates.

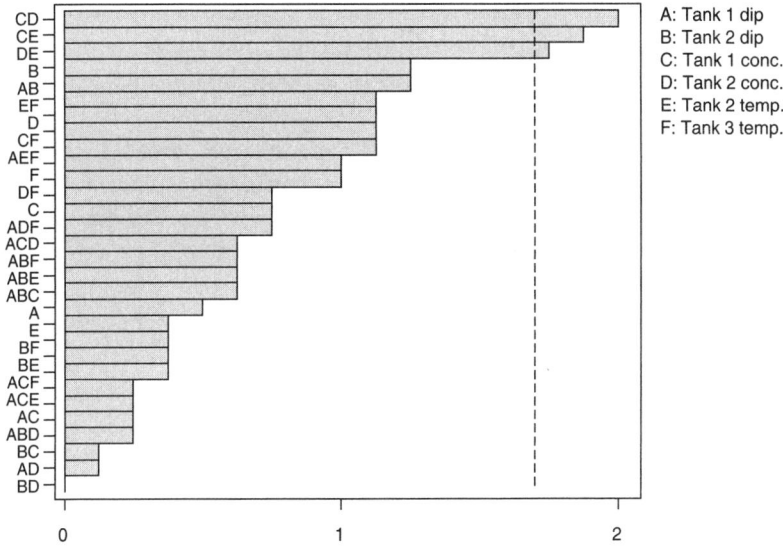

Figure 19.7 Pareto analysis of effects in pinskips experiment.

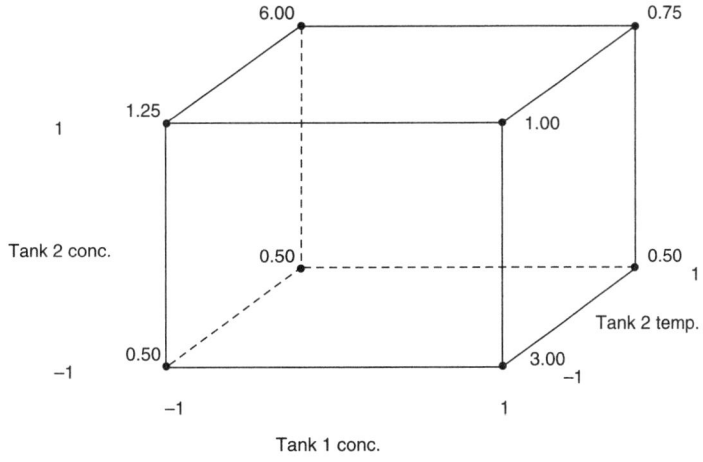

Figure 19.8 Cube plot of average number of pinskip rejects per run for tank 1 concentration, tank 2 concentration, and tank 2 temperature.

The lower the number of rejected grills the better. From the cube plot, the team was surprised to see that the process at the current levels (shown as the lower left-hand point in the plot) had an average of 0.5 rejects per rack, the best observed performance (tied with two other combinations). This was surprising since the current process averaged about 1.6 defective grills per rack.

The team concluded that there was no reason to change the process settings based on the results of the experiment. They decided to abandon the project.

We cannot say for certain what went wrong. One possibility is that none of the selected candidates interacted with the unknown dominant cause. Alternately, if we examine Figure 19.9, a plot of the number of defective grills in the rack versus the order of the runs, we see that there was a large burst of defectives during the middle portion of the experiment. If the dominant cause acts in the time-to-time family, then since each run consisted of a single rack, there was little chance of the cause acting within each run and hence no chance to see if any of the candidates could make the process robust. Defining a run as a single rack of grills is not appropriate if the dominant cause acts in the time-to-time family.

The lesson here is that the team made a number of poor decisions. They should not have jumped directly to the robustness approach. They had little process knowledge to help choose the candidates. They had no assurance that the dominant cause would act within each run or even in the week used for the experiment. They would have been better off first investigating the nature of the process variation and generating more clues about the dominant cause before selecting any particular variation reduction approach.

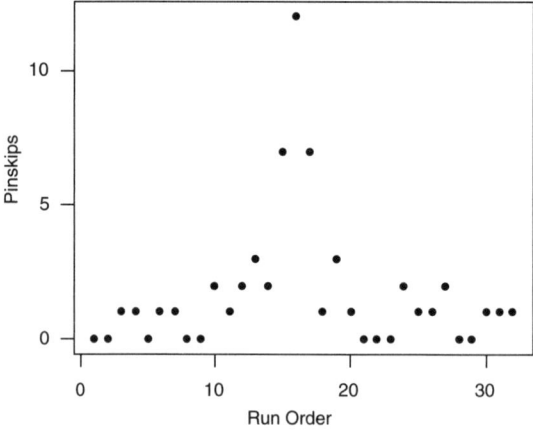

Figure 19.9 Number of defective grills versus run order.

> ## Robustness Experiment Summary
>
> Select candidates and levels based on knowledge of the process.
>
> ### Question
>
> Which, if any, combination of the candidate levels reduces the sensitivity of the process to variation in unknown dominant causes?
>
> ### Plan
>
> - Define a run long enough to see the full extent of variation in the output in the existing process.
> - Determine the number of parts (repeats) to be measured within each run and the performance measure(s).
> - Determine the number of runs.
> - Select a fractional factorial design of resolution III or higher.
> - Randomize the order of the runs as much as is feasible.
> - Make everyone potentially impacted aware of the plan.
>
> ### Data
>
> Carry out the experiment. Record the output values, the levels for candidates, treatment number, and run order, one row for each run. Use a separate column for each repeat.
>
> ### Analysis
>
> - Plot the output by treatment to look for promising treatments. To make this plot, temporarily arrange the data with a separate row for each repeat.
> - Calculate the performance measure(s) across the repeats for each run.
> - Use a full model and a Pareto chart to analyze the performance measure(s) looking for large main and interaction effects.
> - For large effects, construct main and interaction effects plots.
>
> ### Conclusion
>
> Identify the levels of the candidates that lead to the best performance.

19.2 ASSESSING AND PLANNING PROCESS ROBUSTNESS

To successfully implement robustness, we must:

- Identify the fixed inputs and their levels (the new process settings) that make the process robust to variation in the unknown dominant cause(s).
- Move the process center if required.

- Check that the change of settings does not produce substantive negative side effects.
- Estimate the costs of changing the settings and the new ongoing operating costs.
- Estimate the benefit of the change of settings.

If we can accomplish all these tasks, and the benefits outweigh the costs, we proceed to the validation stage of the algorithm.

Since we have little or no knowledge of the dominant cause, we are given little guidance on what candidates to select and how to pick their levels. We recommend using as many candidates as feasible and a fractional factorial design.

The definition of a run is a key step in planning the robustness experiment. To identify settings that reduce the effect of the unknown dominant cause, we need the cause to act within each run. To define a run, we can use knowledge of the time pattern of output variation from a multivari, baseline, or other investigations. In the iron silicon concentration example, the team knew that measuring the same coin several times in a short period of time would show most of the variation in the measurement system. They defined a run to be five consecutive measurements on coins poured from the same iron. In the electroplating pinskip defect example, the team had no knowledge of the time family containing the dominant cause. They specified a run as a single rack of grills. This would have been an appropriate choice if the dominant cause acted within a rack. However, if the dominant cause acted in the rack-to-rack family or slowly over time, then with this definition of a run, the experiment was doomed to fail.

If the time-based family in which the dominant cause acts is unknown, we recommend a multivari investigation before proceeding with a robustness experiment. When the dominant cause acts slowly, we need to have long runs so that the cause acts within each run. Long runs add to the cost and complexity of the experiment, so in this case, determining robust process settings may be infeasible. On the other hand, feedback control may be feasible.

For each run in the experiment, we calculate a performance measure to assess the behavior of the process output within the run. We define the performance measure based on the goal of the problem. There are many possible performance measures.[1] In the iron silicon concentration example, the performance measure was the standard deviation (actually the logarithm of the standard deviation) of the output values measured during the run. The problem goal was to reduce the measurement system variation. In the crossbar dimension example, the team chose the average burn score over the run to measure process performance. By lowering the average score, the team hoped to address the problem goal of reducing the frequency of burn defects. The team could have selected an alternate performance measure such as the proportion of parts scrapped due to burn (as in the electroplating pinskip defect example) in each run. With this choice they would have required longer runs since the output is binary.

We can define and analyze several performance measures (for example, average and standard deviation) within the same experiment. For instance, to make a measurement system more robust, we may simultaneously analyze bias and measurement variation as two performance measures. In the camshaft lobe runout example, introduced in Chapter 1, the goal was to reduce the average and variation in runout. If the team had adopted the robustness approach, in the experiment, they would have calculated both these performance measures within each run. With two or more performance measures we may be forced to make a compromise in the choice of settings.

We need to have enough repeats within each run to get a good estimate of the process performance. It is hard to give a firm rule, but more repeats are better. If measuring the output is expensive we can use a relatively long run but measure a sample of parts within the run.

After the experiment, we can assess the costs and benefits of the approach. We can estimate the performance measure for the new process settings. There are costs associated with changing the candidate levels and the ongoing operating costs at the new levels. We also need to check for negative side effects. Changing candidate levels may shift the center of the process in an undesirable direction. In that case there may be additional costs related to finding an adjuster and operating the process at a different level of the adjuster.

The robustness approach is often selected to reduce the rate of defectives. In the robustness experiment for a binary output, we need a run to be long enough so that each run will likely contain one or more defective parts. If defectives are rare, we may be able to aggravate the process to increase the defect rate for the purposes of the experiment. Then we hope that results obtained under the aggravated conditions are relevant for the standard process. For example, there were field failures of exterior electrical boxes after several years due to corrosion. The team conducted a robustness experiment on the painting process using scored panels in a salt spray chamber (a highly aggravated condition) to see if changing fixed inputs would increase the durability.

There is a strong connection between the robustness approach and the desensitization approach discussed in Chapter 16. In the latter case where we control the known dominant cause in the experiment, we can determine the interactions between the candidates and the cause directly in the experiment. With the robustness approach, we can only observe the interaction indirectly through the performance measure.

There is also a connection between the robustness and moving the process center approaches. In the electroplating pinskip defect example, the output was binary, and the goal of the problem was to reduce the proportion of defective grills. We can view the approach taken as either moving the process center or robustness. In either case, we search for changes to fixed inputs to achieve the goal without knowledge of a dominant cause.

We recommend desensitization over robustness if at all possible. In other words, we recommend first finding the dominant cause of the variation. We have had little success applying the robustness approach. In most circumstances one of the other variation reduction approaches is preferred. Selecting the robustness approach is a last hope.

The idea of analyzing a performance measure such as the within-run standard deviation was first suggested by Bartlett and Kendall (1946). Nair and Pregibon (1988) give a motivation for using $\log(s)$. Taguchi (1986) popularized process robustness and called it *parameter design*.

 **Key Points**

- We make a process robust by changing fixed inputs to reduce the effects of unknown dominant cause(s).

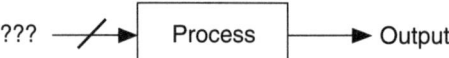

- Selecting candidates for a robustness experiment is difficult due to the lack of knowledge of the dominant cause. We recommend using as many candidates as feasible in a fractional factorial design.
- We must define a run of the experiment to be long enough so that we would see the full extent of variation in the output *if* no process changes were made. We can then be assured that the unknown dominant cause acts within each run.
- We select a performance measure to reflect the goal of the problem. We calculate the value of the performance measure for each run of the experiment and use it as the response in the analysis.
- Making a process robust requires a one-time change to fixed inputs such as the product or process design or the process control plan.

Endnote (see the Chapter 19 Supplement on the CD-ROM)

1. Taguchi popularized the robustness approach. We explore his choice of performance measures, experimental plans, and analysis methods in more detail in the supplement.

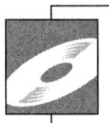 Exercises are included on the accompanying CD-ROM

20
100% Inspection

If you put off everything till you're sure of it, you'll get nothing done.
—Norman Vincent Peale, 1898–1993

The simplest yet most controversial variation reduction approach is 100% inspection. We compare the value of the output characteristic of each part to inspection limits. We then scrap, downgrade, or rework any part with output value outside the inspection limits. Figure 20.1 shows how adding inspection limits reduces the output variation of accepted parts. The inspection limits are tighter than the customer-driven specification limits. To use 100% inspection, we do not need to know the dominant cause or understand the nature of the process variation.

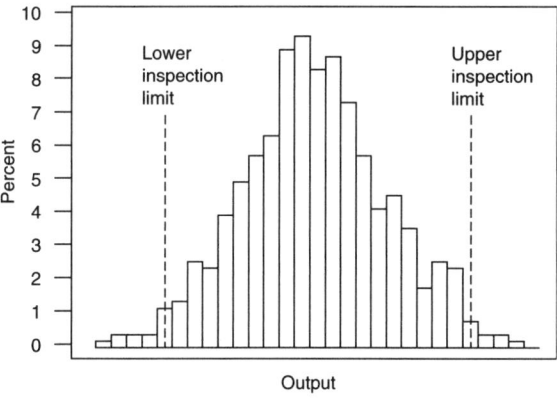

Figure 20.1 Reducing variation by adding inspection limits.

The requirements for 100% inspection to reduce variation are minimal. We need a measurement system for the output characteristic with little measurement bias and variation.

The costs associated with implementing 100% inspection are:

- The cost of measuring every part
- The cost of dealing with rejected parts

We can use the baseline investigation to help assess the second cost. For a continuous output, we use the baseline histogram to estimate the increase in the number of rejects due to adding or tightening inspection limits. 100% inspection is more effective for eliminating outliers than for reducing the standard deviation.

20.1 EXAMPLES OF 100% INSPECTION

We give several examples of the use of 100% inspection to reduce variation. The most common application is to a reformulated problem, where we apply 100% inspection to reduce variation in the dominant cause. In this context, 100% inspection is a form of error proofing.

Manifold Blocked Ports

We introduced the problem of blocked ports in cast-iron exhaust manifolds in Chapter 3. The foundry reacted to a customer complaint that a blocked port had been found on an engine. One step in the current process was the dropping of a ball bearing through the ports of each manifold to ensure that there was no blockage. This manual inspection was sometimes skipped or poorly done.

The team had difficulty establishing a baseline since the defect was so rare. They knew that blocked ports resulted from cracking in the cores during the pouring process, but they had no idea how to determined the cause of core cracking. None of the usual Statistical Engineering tools were helpful, because they could find only two manifolds with a blocked port over a one-month period.

The team decided to automate the inspection to keep the rare defective parts from being shipped. They investigated whether an ultrasound system could correctly classify ports as blocked or not. They were careful in the selection of parts for the investigation. They wanted a sample of manifolds with blocked ports that would demonstrate all conceivable ways a port could be blocked or even partially blocked. This was difficult since blocked ports were so rare. They had only the two manifolds with naturally blocked ports. To get a larger sample, they deliberately created blocked ports in 13 manifolds. They also selected 30 manifolds from regular production that had no blocked ports.

There were three fixed inputs in the ultrasound system. The team planned a full factorial experiment with eight runs. For each run of the experiment, the 45 manifolds were inspected 10 times each and classified as blocked or not. A plan with less chance of study error would use more parts and measure each part once.

For each run, the team counted:

- n1—the number of times a manifold with a blocked port was passed by the system
- n2—the number of times a manifold without a blocked port was rejected by the system

The data are:

Treatment	A	B	C	n1	n2
1	Low	Low	Low	1	7
2	Low	Low	High	2	9
3	Low	High	Low	3	3
4	Low	High	High	1	6
5	High	Low	Low	0	2
6	High	Low	High	0	12
7	High	High	Low	2	13
8	High	High	High	1	7

Since passing a manifold with a blocked port was a serious error, the team considered only treatments 5 and 6 where all manifolds with blocked ports were detected. They selected treatment 5 since it had a lower rate of rejecting manifolds without a blocked port. They changed the operation and control plan so that:

- The ultrasound system inspected all manifolds and automatically rejected any classified as having a blocked port.
- All rejected manifolds were inspected by hand to determine if a port was blocked or not.
- At the start of each shift, one of the manifolds with naturally blocked ports was inspected by the ultrasound system to ensure it was functioning properly.

In this example, defects were very rare, so the costs were limited to the cost of inspection. The inspection cost was less than the loss of goodwill when the customer found manifolds with blocked ports.

Drain Hole Cracks

In a stamping process, the baseline rate of cracks and splits in axle cover drain holes was 3%. There was 100% inspection by the press operators coupled with an off-line audit of a sample from each basket of parts. If a defective part was found in an audit, all baskets filled since the previous audit were contained and reinspected. Despite the inspection effort, there were frequent complaints from the customer about defective stampings.

The team suspected that the dominant cause of the cracks and splits defect was in the incoming steel and thus outside their control. Without further investigation, they decided to 100% inspect parts using digital cameras and image processing to detect the defects. The cost of the inspection equipment was substantial but operating costs were low.

With the new system, there was no need for the audit of inspected parts or for containment. There was an increase in scrap and rework costs. However, there were no further complaints from the customer about cracks and splits.

Broken Tooling

In a machining process, the process engineer discovered that the dominant cause of broken drills in an automated drilling operation was improperly machined parts from upstream in the process. He found that these parts occurred when the production line restarted after a shutdown and decided to use 100% inspection before the drilling operation. He installed a limit switch that detected the poorly machined parts and stopped the line when such a part was found before the drills were damaged. The cost of the inspection was low, and there were substantial savings in tooling and downtime.

This example is classic *error proofing,* 100% inspection applied to the dominant cause.

Rocker Cover Oil Leaks

In an engine assembly plant, management assigned a team to eliminate oil leaks around the rocker cover. The team used repair data to estimate that about 1.4% of the engines had such oil leaks detected at the final test stands.

There were two rocker covers per engine purchased from an outside supplier. The covers were caulked and then bolted to the engine. In the past, the plant had blamed the supplier whenever oil leaks caught the attention of management.

The team decided to look for the dominant cause using a group comparison. They set aside six engines with rocker cover oil leaks. Since each engine had two rocker covers and only one leaked, the team had six leaking and six nonleaking covers. They then measured a large number of characteristics on each of the 12 covers. However, they could not find any characteristic of the rocker cover that separated the leaking and nonleaking sides of the engine. After some thought, they decided to examine the whole process, including the assembly operation, using a component swap investigation. They were surprised to discover that the leaking engines no longer leaked after reassembly. After further investigation, they found the dominant cause was poor torque on the bolts that held the rocker cover to the engine.

They decide to put 100% automated inspection on the torque guns to signal if a rocker cover bolt was not properly tightened. This change virtually eliminated oil leaks around the rocker cover at the final test stands.

20.2 ASSESSING AND PLANNING 100% INSPECTION

To successfully implement 100% inspection, we must:

- Select appropriate inspection limits.
- Measure (with low bias and variation) the output for every part produced.
- Deal with rejects from the inspection.

- Estimate the costs of measurement, loss of volume, and dealing with rejects.
- Estimate the benefits.

If we can accomplish all these tasks, and the benefits outweigh the costs, we proceed to the validation stage of the algorithm.

Many processes already have 100% inspection because the shipment of a defective or out-of-specification part has critical consequences for the customer. For a part such as the exhaust manifold, we consider improving the existing inspection system. In the rocker cover example, the team added a 100% inspection system to the process to check for proper torque.

We can apply 100% inspection if the output can be compared to the inspection limits or standards (sometimes called *boundary samples*) with small measurement error. The key prerequisite is a stable, nondestructive measurement system with low variation and bias.

100% inspection does not eliminate defects or out-of-specification parts. We need to change the control plan to deal with the rejects. In one disastrous example, rejected parts were set aside in a box to be reworked. However, because of poor organization and labeling, the box was shipped to a customer.

Applying 100% inspection to the output is rarely the most cost-effective approach. There is less variation in the parts shipped to the customer, but there are increased costs due to the larger number of rejects. We think 100% inspection is best suited for situations where other variation reduction approaches have not proven feasible or effective. Two potential applications are processes with rare defects or when inspection and rework costs are low. In the manifold blocked port example, the defect was so rare that the team could not discover the dominant cause of the defect using empirical methods. The only feasible approach was to design an effective low-cost 100% inspection system to detect the rare defect.

Applying 100% inspection to a dominant cause is called *source inspection* (Shingo 1986). See Shimbun (1988) for numerous examples. In this case, the costs of dealing with the parts that fall outside the inspection limits on the cause are likely to be small, since we are applying the inspection upstream from the final output. In the rocker cover oil leak example, it was much cheaper to detect and repair the poorly tightened bolts than to deal with the leaks after the engine had been assembled.

We often go directly to source inspection on a dominant cause without reformulating the problem in terms of the cause. This is an example of the Fix the Obvious approach described in Chapter 14.

Most successful applications of 100% inspection use automated measurement. Human inspectors make mistakes at the best of times, and with rare defects, they are unable to remain focused. Using multiple inspectors is not a solution since each inspector may become complacent with the belief that any problem will be found by one of the others.

100% inspection is poorly regarded as a variation reduction approach in the quality improvement literature. For example, one of Deming's 14 points exhorts industry not to rely on mass inspection to "control" quality (Deming, 1992). Despite its unfavorable image, we see 100% inspection applied frequently because immediate process improvement is required and no other improvement approach is feasible without more process knowledge and investigation.

A common modification of 100% inspection is inspection sampling where not every part is measured. One alternative is to define lots that are accepted or rejected based on the quality of a sample taken from the lot. Accepted lots are shipped, and rejected lots are 100%

inspected or otherwise disposed. If we know that lot-to-lot variation is large relative to within-lot variation, inspection sampling can reduce variation. We can improve an existing inspection sampling scheme by redefining a lot, changing the inspection limits, or changing the lot acceptance criteria. Compared to 100% inspection, inspection costs are smaller. However, overall variation will not be reduced to the same degree. If the dominant cause acts part to part, inspection sampling is a poor approach. This is obvious for processes with rare sporadic defectives. Deming (1992, chap. 15) showed that if the process is stable, either no inspection or 100% inspection is optimal. See also VanderWiel and Vardeman (1994).

Key Points

- We use 100% inspection when defects are critical and rare, so that it is difficult and costly to determine a dominant cause, or when the inspection costs are very low.
- 100% inspection applied to a reformulated problem is one form of error proofing.

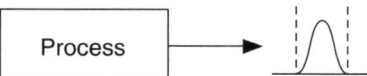

- Most successful applications of 100% inspection rely on automated inspection since human inspectors make mistakes.

21

Validating a Solution and Holding the Gains

An investment in knowledge always pays the best interest.
—Benjamin Franklin, 1706–1790

We have now reached the final stage of the Statistical Engineering algorithm. We have a proposed solution to the problem, a change to one or more fixed inputs of the process. We may propose changes to process settings, the addition or deletion of a process step, the use of a new supplier, changes to the process control plan, and so on. Although this sounds silly, we remind you that the process cannot be improved without making changes to fixed inputs.

There are two remaining tasks:

- *Validate* the proposed solution to see that the goal is met and that there are no substantial negative side effects.

- If the solution is validated, ensure that the process change is made permanent so that the gains are preserved.

21.1 VALIDATING A SOLUTION

We must validate a solution to ensure that under the proposed process changes, the process performance meets the goal of the project or, at the least, that the benefits outweigh the costs. Ideally, we should repeat the baseline investigation to compare the performance before and after the process change. We also need to check carefully for unexpected negative side effects due to the changes in the process.

Depending on the costs and risks, we may initially use a small investigation for a preliminary validation before conducting a full investigation. We do not want to make it too difficult to reverse changes to the process or product until we are confident the expected benefits will materialize. During the validation investigation, we monitor the data as they are being collected. There is no sense in continuing if it becomes clear that the proposed solution is ineffective.

In many problems, we assess the baseline performance of the process using historical data collected over weeks and months. We cannot afford to wait that long to validate the solution. We can use knowledge we have gained in the process investigations to suggest how long we need to observe the original process until we would see the full extent of variation. We plan the validation investigation to last at least that long so that we expect to see the new full extent of variation with the proposed process changes.

We may find a proposed solution inadequate for a number of reasons related to taking shortcuts in the algorithm with the hope of saving time and money. First, the solution may be based on a cause that is not dominant. In that case, the process improvement will be small (see Chapter 2). If the algorithm was followed closely, this cannot occur, since the algorithm requires verification of the dominant cause. Second, we may not meet the goal because of optimistic assumptions. For example, we assume a perfect adjustment method in assessing feedback control that is not realized in practice.

If the solution leads to an improvement that is not sufficient to meet the goal, we need to decide whether or not to continue with the implementation. We can make the decision by reevaluating costs and benefits. To meet the goal, we need to go back to reconsider the possible approaches and look for further changes to fixed process inputs.

If we have reformulated the problem, we must assess the process change against the baseline for the original output. We will have made assumptions about the links between the output and the dominant cause based on uncertain knowledge. When we validate using the original output, we can check that these assumptions were correct.

We give two examples of validation.

Crossbar Dimension

In the production of an injection-molded contactor crossbar, the problem was excessive variation in a crossbar dimension. In the baseline investigation, the team established the full extent of variation of the dimension (measured as the deviation from nominal) as –0.3 to 2.3 thousandths of an inch, with standard deviation 0.46. After some investigation (see Chapter 12), they discovered that barrel temperature was the dominant cause of variation. The team proceeded using the desensitization approach (see the Chapter 16 exercises). They found that increasing the barrel temperature set point reduced the effect of the variation in barrel temperature. Since this change increased the crossbar dimension center, they used a known adjuster to reduce the average dimension to the target value zero.

In a preliminary validation investigation, the team found that increasing the average barrel temperature resulted in burn defects. They did not search for a dominant cause of the burn defect. Instead, they used the robustness approach to find process settings that eliminated the defect while at the same time allowing the increased barrel temperature. See Chapter 19 for further discussion.

The team next proceeded to a full validation with the proposed process settings. In the validation investigation, 300 parts were selected over two shifts. This plan matched the baseline investigation. The team measured the crossbar dimension and inspected each part for the burn defect. The data are given in the file *crossbar dimension validation*. The histogram of the crossbar dimension from the validation investigation, given in Figure 21.1, shows the reduced variation. The standard deviation in the crossbar dimension was reduced to 0.23, and the burn defect occurred on only 2 of the 300 parts. With the new settings, the process performance met the project goal.

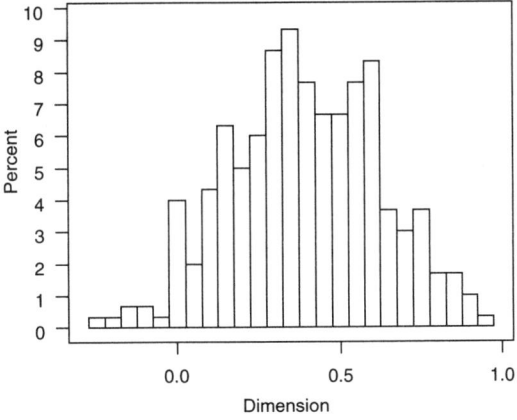

Figure 21.1 Histogram of crossbar dimension in the validation investigation.

Engine Block Porosity

In the engine block porosity example introduced in Chapter 3, the goal was to reduce scrap due to porosity found on a machined surface of an engine block. The baseline scrap rate was about 4%, based on several months of data. There was no formal numerical goal. The team devised a porosity score used throughout the problem solving. They did not conduct a baseline investigation with this new score.

To solve the problem, the team recommended a new core wash supplier. They proceeded with validation in two steps. First, they switched to the new core wash for a single shift and evaluated the results, which were very promising. There were no obvious side effects. They then used the new wash for a week, again monitoring the process carefully for porosity defects and unexpected side effects. The scrap rate due to bank face porosity was less than 1%. The team committed to proceeding with the new core wash. We can see the long-term improvement in Figure 21.2 since the change was made in month 13. The cost savings were several hundred thousand dollars per year.

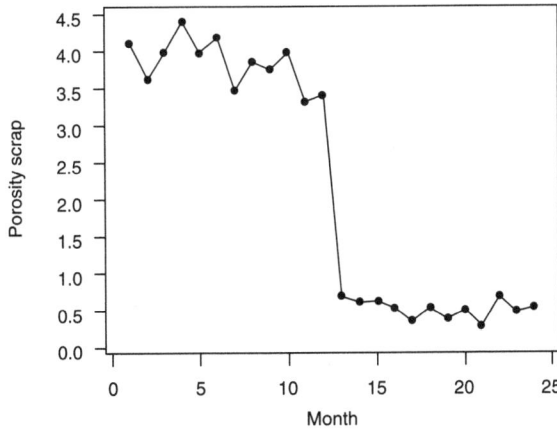

Figure 21.2 Porosity scrap rate by month.

21.2 HOLDING THE GAINS OVER THE LONG TERM

After validating a solution and implementing a process change, we need to ensure that the improvement is preserved over the long term. Most aspects of holding the gains are management issues outside the scope of this book.

In our experience, there are two imminent dangers:

- The recommended process change is not fully implemented or later reversed.

- Other fixed inputs are (later) changed in a way that reduces the effectiveness of the solution.

The difficulty in ensuring the implemented change is not reversed depends on the nature of the change. To change the design of a product or process, we should adhere to formal design change procedures that can be expensive and take considerable time. Such changes are not easily reversed. Changes to the process control plan, on the other hand, are easy to make but also easy to undo or ignore. In these cases, we need to document the changes and ensure that appropriate training is conducted. We also recommend process audits to verify that control plan changes continue to be used.

In many processes, fixed inputs change slowly over time due to wear or aging of equipment. We have discussed several problems where the obvious fix requires maintenance activity. Because at the time the problem has a high profile, the maintenance is carried out and the variation is reduced. Over time, as the original problem is forgotten, maintenance effort decreases and the projected gains slip away. In such cases, the team can increase the chance of holding the gains by implementing a formal monitoring procedure. This is a simple form of feedback control added to the proposed solution. We can use a summary of process performance over time such as a run chart, a regular calculation of process capability, or a control chart (Ryan, 1989; Montgomery, 1996). In all cases, we need a plan to act if evidence from the monitoring suggests the gains are disappearing.

At the end of a project, we need to preserve any new knowledge of process behavior that can be of use to a wider audience. We need to document any process or design changes with the supporting reasons for the change. We do not want improvements undone during future cost-reduction exercises or in the solution to another problem.

We can retain information by:

- Logging design changes in a design guide

- Documenting of all variation reduction projects on a searchable company intranet

Documenting projects and problems is useful but expensive and time-consuming. In our experience most process engineers do not enjoy this activity. See knowledge management books such as Davenport et al. (1997) and O'Dell et al. (1998) for suggestions.

We give three examples that illustrate the issues.

Truck Pull

As described in Chapter 10, in the early phases of the problem to improve the truck alignment process, the team examined right caster data stratified by the four alignment measurement machines that operated in parallel. The team was surprised to see the persistent material differences among averages of the four gages. Because trucks were assigned an alignment gage haphazardly, the gage was a cause of variation in right caster.

The team took immediate action to recalibrate the four gages to remove the systematic differences. To prevent a recurrence, they established a monitoring program to compare the daily averages of each measured characteristic for each gage. If they found significant differences, they recalibrated the four gages. Without such monitoring, there is every reason to believe that the problem would recur.

Fascia Dimension

At a fascia supplier, management assigned a team to address complaints from their customer, a car assembly plant, about difficulties installing the fascias. After consulting with the customer, the team determined that the average of a key dimension was off target and that there were too many large fascias. We discussed this example in the exercises to chapters 6 and 15.

From the baseline investigation, the team estimated the dimension average, measured from nominal, to be 7.3, with a full extent of variation of 2 to 12. After some further investigation, they proposed changing the cure and cycle times in the molding process to move the process center. They made changes to the control plan and carried out a short validation investigation. The average dimension was reduced to 3.1, with a range of 1 to 6. Within a week in actual production, the new control plan was being ignored. The line operators realized that the proposed solution was not as effective as initially thought.

Mistakes were made in this problem. The main difficulty was that the baseline investigation was conducted over too short a time period and did not capture the full extent of output variation. The team also failed to notice an upward drift in the dimension over the course of the baseline investigation. Subsequently, the validation investigation also covered too short a time period. Because of the poor plan and analysis for the baseline investigation, the team was led to an inappropriate approach and solution.

Wheel Bearing Failure Time

Based on a Pareto analysis of warranty claims, management assigned a team to solve the problem of premature failure of wheel bearings within the warranty period of three years. The goal was to reduce the failure rate from 3% to less than 0.3%. Most of the wheel bearing failures occurred in trucks driven primarily on dirt roads in wet and cold conditions. The team used the robustness approach to find a design change to the bearing assembly that increased the average failure time. During the investigation, the team discovered the failure rate had increased after the removal of a dust shield, recommended by an earlier cost-reduction program. No one had examined the potential costly side effects of removing the dust shield, despite a design change protocol that required the assessment of such side effects before the change could be implemented.

21.3 COMPLETING THE IMPLEMENT AND VALIDATE STAGE

We summarize the key tasks necessary to complete the Implement and Validate Solution and Hold the Gains stages of the Statistical Engineering algorithm. We must:

- Implement the solution (that is, change one or more fixed inputs).
- Conduct a validation investigation using the original output to show:
  —The problem goal has been met.
  —There are no substantial negative side effects.
- Establish monitoring and audits to preserve the change, if necessary.
- Document any changes to the product or process design or to the control plan.
- Document the lessons learned.

After completing all these tasks, we can celebrate our success!

Key Points

- We need to validate process changes to ensure the long-term reduction of process variation.
- In the solution validation, watch for unexpected negative side effects on other output characteristics.
- To have a lasting impact, the implemented change needs to be made permanent.

Exercises are included on the accompanying CD-ROM

References

Abraham, B., and J. Ledholter. 1983. *Statistical Methods for Forecasting.* New York: John Wiley & Sons.

Agrawal, R., 1997. Variation Transmission in Multistage Industrial Processes. Statistics Ph.D. thesis, University of Waterloo, Ontario, Canada.

Agrawal, R., J. F. Lawless, and R. J. MacKay. 1999. "Analysis of Variation Transmission in Manufacturing Processes, Part II." *Journal of Quality Technology* 31:143–154.

Amster, S., and K. L. Tsui. 1993. "Counterexamples for the Component Search Procedure." *Quality Engineering* 5:545–552.

Asher, R. K., Sr. 1987. "Statistical Process Control." *Products Finishing* 51:20–22.

Aström, K. J. 1970. *Introduction to Stochastic Control.* Mathematics in Science and Engineering Series. Vol. 70. New York: Academic Press.

Aström, K. J., and B. Wittenmark. 1989. *Adaptive Control.* Reading, MA: Addison-Wesley.

Automotive Industry Action Group (AIAG). 1995a. *Measurement Systems Analysis.* Second edition. Southfield, MI: AIAG.

Automotive Industry Action Group (AIAG). 1995b. *Statistical Process Control.* Second edition. Southfield, MI: AIAG.

Automotive Industry Action Group (AIAG). 1998. *Quality Systems Requirements QS-9000.* Third edition. Southfield, MI: AIAG.

Bartlett, M. S., and D. G. Kendall. 1946. "The Statistical Analysis of Variance: Heterogeneity and the Logarithmic Transformation." *Journal of the Royal Statistical Society. Series B*, 8:128–138.

Bhote, K. R., and A. K. Bhote. 2000. *World Class Quality.* Second edition. Saranac Lake, NY: American Management Association (AMACOM).

Bisgaard, S., and H. Fuller. 1995a. "Sample Size Estimates for Designs with Binary Responses." *Journal of Quality Technology* 27:344–354.

———. 1995b. "Analysis of Factorial Experiments with Ordered Categories as the Response." *Quality Engineering* 8: 199–207.

———. 1996. "Correction to 'Sample Size estimates for Designs with Binary Responses' (1995V27 p344-354)." *Journal of Quality Technology* 28: 496.

Box, G.E.P. 1988. "Signal to Noise Ratio, Performance Criteria, and Transformations (with discussion)." *Technometrics* 30:1–40.

———. 1999. "Statistics as a Catalyst to Learning by Scientific Method Part II: A Discussion." *Journal of Quality Technology* 31: 16–29.

Box, G.E.P., W. G Hunter, and J. S. Hunter. 1978. *Statistics for Experimenters: An Introduction to Design, Data Analysis and Model Building.* New York: John Wiley & Sons.

Box, G.E.P., and G. M. Jenkins. 1976. *Time Series Analysis: Forecasting and Control.* Revised edition. Englewood Cliffs, NJ: Prentice-Hall.

Box, G.E.P., G. M. Jenkins, and G. C. Reinsel. 1994. *Time Series Analysis: Forecasting and Control.* Third edition. Englewood Cliffs, NJ: Prentice-Hall.

Box, G.E.P., and T. Kramer. 1992. "Statistical Process Monitoring and Feedback Adjustment: A Discussion." *Technometrics* 34:251–285.

Box, G.E.P., and A. Luceno. 1997. *Statistical Control by Monitoring and Feedback Adjustment.* New York: John Wiley & Sons.

Boyles, R. A. 2001. "Gage Capability for Pass-Fail Inspection." *Technometrics* 43: 223–229.

Brassard, M. 1988. *The Memory Jogger: A Pocket Guide of Tools for Continuous Improvement.* Second edition. Metheun, MA: Goal/QPC.

Breiman, L., J. Friedman, C. J. Stone, and R. A. Olshen. 1984. *Classification and Regression Trees.* Boca Raton, FL: Chapman and Hall/CRC.

Breyfogle, F. W., III. 1999. *Implementing Six Sigma: Smarter Solutions Using Statistical Methods.* New York: John Wiley & Sons.

Burdick, R. K., and G. A. Larsen. 1997. "Confidence Intervals on Measures of Variability in R&R Studies." *Journal of Quality Technology* 29:261–273.

Cameron, A. C., and P. K. Trivedi. 1998. *Regression Analysis of Count Data.* Cambridge, England: Cambridge University Press.

Canadian Oxford Dictionary: The Foremost Authority on Current Canadian English. 2002. Edited by Katherine Barber. Oxford, England: Oxford University Press.

Cleveland, W. S. 1979. "Robust Locally Weighted Regression and Smoothing Scatterplots." *Journal of the American Statistical Association* 74:829–836.

Coleman, D. E., and D C. Montgomery. 1993. "A Systematic Approach to Planning for a Designed Industrial Experiment." *Technometrics* 35:1–26.

Cotter, C. S. 1979. "A Screening Design for Factorial Experiments with Interactions." *Biometrika* 66:317–320.

Davenport, T. H., and L. Prusak. 1997. *Working Knowledge: How Organizations Manage What They Know.* Harvard, MA: Harvard Business School Publishing.

Deming, W. E. 1992. *Out of the Crisis.* Seventh printing. Cambridge, MA: MIT Center for Advanced Engineering Study.

De Mast, J., 2003. "Quality Improvement from the Viewpoint of Statistical Method." *Quality and Reliability Engineering International* 19:255–264.

———. 2004. "A Methodological Comparison of Three Strategies for Quality Improvement." *International Journal of Quality and Reliability Management* 21(2): 198–213.

De Mast, J., K.C.B. Roes, and R.J.M.M. Does. 2001. "The Multi-Vari Chart: A Systematic Approach." *Quality Engineering* 13:437–448.

De Mast, J., W.A.J. Schippers, R.J.M.M. Does, and E. Van den Heuvel. 2000. "Steps and Strategies in Process Improvement." *Quality and Reliability Engineering International* 16:301–311.

Del Castillo, E. 2002. *Statistical Process Adjustment for Quality Control.* New York: John Wiley & Sons.

Duncan, A. J. 1986. *Quality Control and Industrial Statistics.* Fifth edition. Homewood, IL: Richard D. Irwin.

Evans, J. R., and W. M. Lindsay. 1993. *The Management and Control of Quality.* Second edition. St. Paul, MN: West.

Farnum, N. R. 1994. *Modern Statistical Quality Control and Improvement.* Belmont, CA: Duxbury.

Feder, P. I. 1974. "Some Differences Between Fixed, Mixed, and Random Effects Analysis of Variance Models. *Technometrics* 6:98–106.

Goldratt, E. M. 1992. *The Goal.* Second revised edition. Great Barrington, MA: North River.

Grubbs, F. E. 1954. "An Optimum Procedure for Setting Machines or Adjusting Processes." *Industrial Quality Control*, July. Reprinted in the *Journal of Quality Technology* 15(4):186–189.

Hahn, G. J. 1984. "Experimental Design in the Complex World." *Technometrics* 26:19–31.

Hamada, M., and J. A. Nelder. 1997. "Generalized Linear Models for Quality Improvement Experiments." *Journal of Quality Technology* 29:292–304.

Harrington, H. J. 1987. *The Improvement Process: How America's Leading Companies Improve Quality*. New York: McGraw-Hill.

Harry, M. J. 1997. *The Vision of Six Sigma: A Roadmap for Breakthrough*. Fifth edition. Phoenix, AZ: TriStar.

Harry, M., and R. Schroeder. 2000. *Six Sigma: The Breakthrough Strategy Revolutionizing the World's Top Corporations*. New York: Doubleday.

Hastie, J. H., and R. J. Tibshirani. 1990. *Generalized Additive Models*. Boca Raton, FL: Chapman and Hall/CRC.

Hosmer, D. W., and S. Lemeshow. 2000. *Applied Logistic Regression*. Second edition. New York: John Wiley & Sons.

Hoyle, D. 2001. *ISO 9000 Quality Systems Handbook*. Fourth edition. Woburn, MA: Butterworth and Heinemann.

Ingram, D. J. 1993. "A Statistical Approach to Component Swapping." *American Society for Quality Control (ASQC) Quality Congress Transactions*, Milwaukee, WI, 85–91.

Ishikawa, K. 1982. *Guide to Quality Control*. Second revised edition. Tokyo: Asian Productivity Organization.

Jenkins, G. M. 1983. "Feedforward-Feedback Control Schemes." In *Encyclopedia of Statistical Sciences*, edited by S. Kotz and N. L. Johnson. New York: John Wiley & Sons.

Juran, J. M. 1988. *Juran on Planning for Quality*. New York: Free Press.

Juran, J. M., and F. M. Gryna, Jr. 1980. *Quality Planning and Analysis*. Second edition. New York: McGraw-Hill.

Juran, J. M., F. M. Gryna, and R. S. Bingham, eds. 1979. *Quality Control Handbook*. Third edition. New York: McGraw Hill.

Kalbfleisch, J. G. 1985. *Probability and Statistical Inference*. Vol. 2, *Probability*. Second edition. New York: Springer-Verlag.

Kotz, S., and N. Johnson. 2002. "Process Capability Indices: A Review 1992–2000." *Journal of Quality Technology* 34:2–53.

Kume, H. 1985. *Statistical Methods for Quality Improvement*. Tokyo: Association for Overseas Technical Scholarships (AOTS).

Lawless, J. F., R. J. MacKay, and J. A. Robinson. 1999. "Analysis of Variation Transmission in Manufacturing Processes, Part I." *Journal of Quality Technology* 31:131–142.

Ledolter, J., and A. Swersey. 1997a. "Dorian Shainin's Variables Search Procedure: A Critical Assessment." *Journal of Quality Technology* 29:237–247.

Ledolter, J., and A. Swersey. 1997b. "An Evaluation of Pre-control." *Journal of Quality Technology* 29:163–171.

Lewis, J. P. 2002. *Fundamentals of Project Management: Developing Core Competencies to Help Outperform the Competition*. Second edition. Saranac Lake, NY: American Management Association (AMACOM).

Liberatore, R. L. 2001. "Teaching the Role of SPC in Industrial Statistics." *Quality Progress*, July, 89–94.

Logothetis, N. 1990. "A Perspective on Shainin's Approach to Experimental Design for Quality Improvement." *Quality and Reliability Engineering International* 6:195–202.

Mackertich, N. A. 1990. "Pre-control vs. Control Charting: A Critical Comparison." *Quality Engineering* 2:253–260.

McCallagh, P., and J. L. Nelder. 1989. *Generalized Linear Models.* Second edition. Boca Raton, FL: Chapman Hall/CRC.

Mease, D., V. N. Nair, and A. Sudjianto. 2004. "Selective Assembly in Manufacturing: Statistical Issues and Optimal Binning Strategies." *Technometrics* 46:165–175.

Meyer, R. D., D. M. Steinberg, and G.E.P. Box. 1996. "Follow-up Designs to Resolve Confounding in Multifactor Experiments (with discussion)." *Technometrics* 38:303–332.

MINITAB. 2000a. *MINITAB User's Guide 1: Data, Graphics, and Macros.* Release 13 for Windows. State College, PA: MINITAB.

———. 2000b. *MINITAB User's Guide 2: Data Analysis and Quality Tools.* Release 13 for Windows. State College, PA: MINITAB.

Montgomery, D. C. 1996. *Introduction to Statistical Quality Control.* Third edition. New York: John Wiley & Sons.

———. 2001. *Design and Analysis of Experiments.* Fifth edition. New York: John Wiley & Sons.

Montgomery, D. C., E. A. Peck, and G. G. Vining. 2001. *Introduction to Linear Regression Analysis.* Third edition. New York: John Wiley & Sons.

Nair, V. N. 1992. "Taguchi's Parameter Design: A Panel Discussion." *Technometrics* 34:127–162.

Nair, V. N., and D. Pregibon. 1988. "Analyzing Dispersion Effects from Replicated Factorial Experiments." *Technometrics* 30: 247–257.

Nelson, L. S. 1985. "Sample Size Tables for Analysis of Variance." *Journal of Quality Technology* 17:167–169.

Neter, J., M. H. Kutner, C. J. Nachtsheim, and W. Wasserman W. 1996. *Applied Linear Statistical Models.* Fourth edition. Chicago: Irwin.

Nikolaou, M. 1996. "Computer-aided Process Engineering in the Snack Food Industry." *Proceedings Chemical Process Control Conference* (CPC-V), Tahoe City.

Odeh, R. E., and M. Fox. 1975. *Sample Size Choice: Charts for Experimenters with Linear Models.* Second edition. Homewood, IL: Irwin.

O'Dell, C. S., N. Essaides, and N. Ostro. 1998. *If Only We Knew What We Know: The Transfer of Internal Knowledge and Best Practice.* New York: Free Press.

Oldford, R. W., and R. J. MacKay. 2001. Stat 231 course notes, Fall 2001, University of Waterloo, Ontario.

Parmet, Y., and D. M. Steinberg. 2001. "Quality Improvement from Disassembly-Reassembly Experiments." *Communications in Statistics: Theory and Methods* 30:969–985.

Phadke, M. S. 1989. *Quality Engineering Using Robust Design.* Englewood Cliffs, NJ: Prentice-Hall.

Prett, D. M., and C. E. Gracia. 1988. *Fundamental Process Control.* Boston: Butterworths.

Quinlan, J. 1985. "Product Improvement by Application of Taguchi Methods." *Third Supplier Symposium on Taguchi Methods*, American Supplier Institute, Dearborn, MI, 367–384.

Robinson, G. K. 2000. *Practical Strategies for Experimenting.* New York: John Wiley & Sons.

Ross, P. J. 1988. *Taguchi Techniques for Quality Engineering: Loss Function, Orthogonal Experiments, Parameter and Tolerance Design.* New York: McGraw-Hill.

Ryan, B. F., B. L. Joiner, T. Ryan, Jr. 2000. *Minitab Handbook.* Fourth edition. Belmont, CA: Duxbury.

Ryan, T. P. 1989. *Statistical Methods for Quality Improvement.* New York: John Wiley & Sons.

Satterthwaite, F. E. 1954. "A Simple, Effective Process Control Method." Rath & Strong Inc. Report 54-1, Boston, MA.

Scholtes P. R. 1998. *The Team Handbook: How to Use Teams to Improve Quality.* Madison, WI: Oriel.

Seder, L. A. 1950a. "Diagnosis with Diagrams: Part I." *Industrial Quality Control,* January, 11–19.

———. 1950b. "Diagnosis with Diagrams: Part II." *Industrial Quality Control,* February, 7–11.

———. 1990. "Diagnosis with Diagrams." *Quality Engineering* 2:505–530 (reprinted from original in *Industrial Quality Control,* 1950).

Senge, P. M. 1990. *The Fifth Discipline: The Art and Practice of the Learning Organization.* New York: Currency/Doubleday.

Shainin, D., and P. Shainin. 1988. "Better Than Taguchi Orthogonal Tables." *Quality and Reliability Engineering International* 4:143–149.

———. 1989. "Pre-control Versus $\bar{X}$ & R Charting: Continuous or Immediate Quality Improvement?" *Quality Engineering* 1:419–429.

Shainin, R. D. 1992. "Technical Problem Solving Strategies: A Case Study." 46th *Annual Quality Congress Proceedings*, American Society for Quality Control (ASQC), Milwaukee, WI, 876–882.

———. 1993. "Strategies for Technical Problem Solving." *Quality Engineering* 5(3):433–448.

Shimbun, N. K., ed. 1988. *Poka-yoke: Improving Product Quality by Preventing Defects.* Cambridge, MA: Productivity Press.

Shingo, S. 1986. *Zero Quality Control: Source Inspection and the Poka-yoke System.* Stanford, CT: Productivity Press.

Snee, R. D. 2001. "My Process Is Too Variable: Now What Do I Do?: How to Produce and Use a Successful Multi-vari Study." *Quality Progress*, December, 65–68.

Spiers, B. 1989. "Analysis of Destructive Measuring Systems," *Forty-third Annual Quality Congress Proceedings*, American Society for Quality Control (ASQC), Milwaukee, WI, 22–27.

Steiner, S. H., and R. J. MacKay. 1997–1998. "Strategies for Variability Reduction." *Quality Engineering* 10: 125–136.

Taguchi, G. 1986. *Introduction to Quality Engineering, Designing Quality into Products and Processes.* Tokyo: Asian Productivity Association.

———. 1987. *System of Experimental Design: Engineering Methods to Optimize Quality and Minimize Costs.* White Plains, NY: UNIPUB/Kraus International.

Taguchi, G., and Y. Wu. 1985. *Introduction to Off-line Quality Control.* Nagaya, Japan: Central Japan Quality Control Association.

Taylor, W. A. 1991. *Optimization and Variation Reduction in Quality.* New York: McGraw-Hill.

Tippett, L.H.C. 1934. *Applications of Statistical Methods to the Control of Quality in Industrial Production.* Brighton, UK: Manchester Statistical Society.

Todd, R. H., D. K. Allen, and L. Alting. 1994. *Manufacturing Processes Reference Guide.* New York: Industrial Press.

Traver, R. W. 1985. "Pre-control: A Good Alternative to $\bar{X}$-R Charts." *Quality Progress*, September, 11–14.

Traver, R. W. 1995. *Manufacturing Solutions for Consistent Quality and Reliability.* New York: American Management Association.

Tukey, J. W. 1959., "A Quick, Compact, Two-sample Test to Duckworth's Specifications." *Journal of Quality Technology* 1:31–48.

VanderWiel, S. A., and S. B. Vardeman. 1994. "A Discussion of All-or-None Inspection Policies." *Technometrics* 36: 102–109.

Wheeler, D. J., and R. W. Lyday. 1989. *Evaluating the Measurement Process.* Second edition. Knoxville, TN: SPC.

———. 1990. *Understanding Industrial Experimentation.* Second edition. Knoxville, TN: SPC.

———. 1992. "Problems with Gage R&R Studies." *American Society for Quality Control (ASQC) Quality Congress Transactions*, Milwaukee, WI, 179–185.

Wu, C.F.J., and M. Hamada. 2000. *Experiments: Planning, Analysis, and Parameter Design Optimization.* New York: John Wiley & Sons.

Zaciewski, R. D., and L. Nemeth. 1995. "The Multi-vari Chart: An Underutilized Quality Tool." *Quality Progress* 28:81–83.

Index

Note: All page references following CD refer to text on the compact disk.

A

Abraham, B., CD-288
accelerated test, 71, 107, 109
adjuster, 40, 266, 227
    substream alignment and, 20, 239
aggravation, use of, 71
Agrawal, R., CD-235
alias stucture, 235
alias effects. *See* confounding
Amster, S., CD-253
analysis of variance (ANOVA), 95, 150, 157, 162, CD-229, CD-333
    gage R&R and, CD-215
    multivari investigations and, CD-239–46
    one-way, CD-333–35
    replicates and repeats, CD-263
    two or more inputs, CD-335–36
Analysis step of QPDAC, 51, 57
    checklist, 65
ANOVA. *See* analysis of variance
Asher, R. K. Sr., CD-236
assembly family, 141–43, CD-237
assembly nuts loose, 217
assembly versus component investigation, 141
    summary, 144
Aström, K. J., CD-295
attribute, 53, 64
    baseline, 73
    combined, CD-194–95
    confidence interval for, CD-201–3
    stratification of, CD-193–94

Automotive Industry Action Group (AIAG)
    assessing variation, CD-209, CD-217
    gage R&R investigation, CD-214
    Measurement Systems Analysis manual, 91, CD-211, CD-215
    QS-9000 Manual, 46, CD-211–14
    SPC Manual, CD-203
available data, 125
average
    baseline attribute, 74
    confidence interval and, CD-202
    connection to modd, 23
    histogram and, 19
    MINITAB summary, CD-311–13

B

balance, 194, 208
Bartlett, M. S., 299
base circle (BC) runout. *See* camshaft lobe runout
baseline investigation, 73–80
    confidence intervals, CD-201–3
    effect of outliers, CD-200
    full extent of variation, 76
    importance of, 42–43
    patterns over time, 112–13
    process stability and, CD-203–4
batch-to-batch family, 156, 270
battery seal failure. *See* examples
BC runout. *See* camshaft lobe runout

bell-shaped histograms, 19, 75, CD-187–89
  Gaussian models and, 23
best subsets regression, CD-342–44
Bhote, A. K., 46, CD-234, CD-253–54,
  CD-259, CD-265
Bhote, K. R., 46, CD-234, CD-253–54,
  CD-259, CD-265
binary characteristic, 15
binary output
  difficulties with, 4
  dominant causes for, 17
  full extent of variation, 87
  group comparisons and, 179–82
  logistic regression for, CD-261
  measurement system assessment,
    CD-205–9
  multivari investigations, 162
  sample size and, 5, 77
  summaries for, 22
  use of aggravation, 71
  verification experiments, 199
Bingham, S., 282
Bisgaard, S., 256
block bore diameter, *See* examples
blocking, CD-264
bottle label height. *See* examples
boundary samples, 305
Box, G.E.P., 51, 194, 267, 281–82, CD-198,
  CD-260, CD-267, CD-288, CD-290,
  CD-295, CD-298, CD-333
box plots, CD-319
Boyles, R. A., CD-209
brainstorming, CD-225–26
brake rotor balance. *See* examples
Brassard, M., CD-225
Breakthrough Cookbook, CD-191
Breyfogle, F. W., CD-191
broken tooling, 304

C

Cameron, A. C., CD-261
camshaft journal diameter. *See* examples
camshaft lobe angle error, 21
camshaft lobe runout. *See* examples
candidate
  adjuster, 228, 238
  for desensitization, 241, 254
  for robustness, 285, 297

capability ratios, 19, 73, CD-183,
  CD-194
casting thickness. *See* examples
categorical characteristic, 15, 22
cause-and-effect diagrams, CD-227
cause of variation, 16
  classification, CD-182
  dominant, 16
center points, use of, CD-275–76
characteristics
  calculating derived, CD-304
  classification, 15
  definition, 14
  numerical summaries, CD-311–15
Cleveland, W. S., CD-288
Coleman, D. E., 200
common cause, CD-182
component family, 141, 143, 167–76
  use of leverage, CD-237
component-swap investigation, 167–76
  alternative plans, CD-257
  summary, 174–75
  using three groups, CD-254
concentration diagrams, 140
conclusion step in QPDAC, 58
  checklist, 65
confidence intervals, CD-201–3
  sample size, 77
confounding, 235
  aliased effects and, CD-272
  elimination of, CD-264
continuous characteristic, 15
  numerical summary, CD-311
continuous output
  baseline sample size, 77
  dominant cause and, 17
  input/output investigation, 183–192
contrast matrix, CD-268
control chart. *See* run chart
  acceptance control chart, 282
  as a baseline, 81
  dominant cause and, CD-182
  feedback control and, CD-290
  process monitoring, 310
control factors, CD-182, 278
controlled processes, adjustments to,
  CD-294
convenience sampling, 54
cost/benefit analysis, 47, 49, 217

costs
    100% inspection, 38, 302
    desensitization, 33, 255
    feedback control, 36, 280
    feedforward control, 34, 266
    fix the obvious, 214
    move the process center, 40, 228
    robustness, 298
    sampling protocols and, CD-197
Cotter, C. S., CD-253
count data, CD-261
crankshaft main diameter. *See* examples
credit card defect. *See* examples
crossbar dimension. *See* examples
crossed design, 248, 256, CD-278–81
    summary, 254
cross tabulation, CD-314–15
cube plot, 295
customer, definition, 14–15
cylinder head rail damage, 140–41
cylinder head scrap. *See* examples

D

data storage in MINITAB, 78, CD-304
Data step in QPDAC, 51, 57
    checklist, 65
Davenport, T. H., 310
Define Focused Problem Stage, 42, 69–87
Del Castillo, E., 282, CD-289
De Mast, J., CD-192
Deming, W. E., 11, 51, 282, 305–6, CD-198
desensitization approach, 32, 241–58
    feedforward control vs., 267
    fractional factorial experiments, CD-278–81
    implementation, 255–57
    investigation summary, 254–55
    mathematical representation, CD-277–78
    robustness vs., 299
designed experiment
    definition, 55–56
    for desensitization, 254
    for moving the process center, 237
    for robustness, 297
    for verification, 206
    fractional factorial, CD-267, CD-278
    full factorial, 200

one-factor-at-a-time, 205
    terminology, 194, 208
design resolution interpretation, CD-274
destructive measurement system assessment, CD-209–11
diagnostic tree, 118, 123
differential carrier shrink defects. *See* examples
disassembly-reassembly investigation, 141, 144. *See* component swap investigation
discrete characteristics, 15
discrimination ratio, 96–97, 103
    comparison to other criteria, CD-214
    destructive measurement systems and, CD-210
DMAIC (Define, Measure, Analyze, Improve, Control), CD-191–92
documentation of projects, 310
dominant cause, 16, CD-180
    binary output, 18
    importance of, 26
    involving two or more inputs, 225, CD-180
    obvious fix, 30–32, 214–17
    problem reformulation, 219–23
    specificity of, 223–24
    variation reduction approaches and, 43–44
    verification of, 193–209, CD-263–65
door closing effort. *See* examples
downstream family. *See* variation transmission investigation
draftsman plot, CD-257–59, 323–24,
drain hole cracks, 303–4
Duncan, A. J., 282

E

eddy current measurement. *See* examples
effects in experiments, 203, 208. *See also* main effect, interaction effect
    estimate of, 203, CD-269
    ranking, 203
80/20 rule. *See* Pareto principle
electric box durability, 107
electric motor failure, CD-65
electroplating pinskip defects. *See* examples
elimination, method of. *See* method of elimination
engine block leaks. *See* examples
engine block porosity. *See* examples

engine oil consumption. *See* examples
error proofing, 39, 304
errors, 54–55, 64, CD-178. *See* measurement error; sample error; study error
Evans, J. R., CD-225
examples
    assembly nuts loose, 217
    battery seal failure, 221–22, 228–31, 238–39
    block bore diameter, CD-45–46, CD-54, CD-67, CD-243–44
    bottle label height, 119, 124–25, CD-51–52
    brake rotor balance, 46, 201–4, CD-3–11, CD-48
    broken tooling, 304
    camshaft journal diameter, 90, 90–96, 98–101, 155–59, CD-249–50
    camshaft lobe angle error, 21
    camshaft lobe runout, 5–6, 18–21, 109–10, 133–36, 138–140, 223–24, 298, CD-47, CD-59–60
    casting thickness, CD-239–42, CD-246–49
    crankshaft main diameter, 7–8, 36, 224, CD-23–38
    credit card defect, CD-206–9
    crossbar dimension, 183–84, 193, 196–97, 218, 256, 286, 308–9, CD-51, CD-64
    cylinder head rail damage, 140–41
    cylinder head scrap, 151–55, 215
    differential carrier shrink defect, 232–37
    door closing effort, 142–43, 225
    drain hole cracks, 303–4
    eddy current measurement, 251–54, CD-281–83
    electric box durability, 107
    electric motor failure, CD-65–66
    electroplating pinskip defect, 293–96
    engine block leaks, 4–5, 48, 69–70, 180, 214, 231–32, CD-53–54
    engine block porosity, 32–33, 136–38, 145, 242–43, 309, CD-52, CD-57–58, CD-230–33
    engine oil consumption, 120–21, 125–26, 194–197, 223
    fascia cratering, 159–61, 193, 215–16
    fascia dimension, 311, CD-45, CD-61
    fascia film build, 112, 276–77, 280–81
    fascia ghosting, 61–63

furnace switch, 38
headrest failure, 167–70
hubcap damage, 216–17
hypoid gear distortion, 205
iron ore variation, 218
iron silicon concentration, 288–92, CD-53
machining dimension, CD-67–68
manifold blocked port, 13–14, 107, 302–3
manifold sand scrap, 188–90, CD-57
molded plastic casing, 39
nylon bond strength, CD-58
oil consumption, 195–97
oil pan scrap, 244–47
paint film build, 8–9, CD-54–55, CD-68–70, CD-75
parking brake tightness, 112–13, 270–72
potato chip spots, 260–61, 267
power window buzz, 142, 170–74
precision shaft diameter, CD-68
pump noise, 85–86
refrigerator frost build-up, 9–10, 32, 217, 247–51, CD-64–65
rocker cover oil leaks, 304
rod thickness, 71–80, 132–33, CD-13–21, CD-229–30
roof paint defect, 165–66
roof panel updings, 222, CD-45, CD-52, CD-55–56
sand core strength, 7, 108–9, CD-56, CD-62–63
seat cover shirring, CD-70–72
sheet metal, CD-61–62
sonic weld, CD-63–64
spark plug connection, CD-56
speedometer cable shrinkage, 111
steering knuckle strength, CD-58–59
steering wheel vibration, 34, 225, 261–64, CD-285–87
sunroof flushness, 219–21, CD-56–57
tin plate strength, CD-209–10
transmission shaft diameter, 110–11, CD-54
truck pull, 2–4, 81–85, 185–88, 215, 223, 264–67, 277–79, 311, CD-42–43, CD-49–50, CD-66–67, CD-75–76
V6 piston diameter, 21–23, 52–61, 118–19, 121–23, 128, 163–65, 272–74, CD-60–61, CD-67, CD-233, CD-250–52

V8 piston diameter, 274–76, CD-47–48
valve train test stands, 31, 125–26
weatherstrip torsional rigidity,
 CD-72–74
wheel bearing failure time, 46–47, 109,
 311–12
window leaks, 180–82, 192, 206, 216, 225,
 CD-325
exhaust manifold. *See* manifold
experimental plans, 55–56, 64. *See* designed
 experiments
experimental run, 194, 208
 robustness experiment and, 298
exponentially-weighted moving average
 (EWMA), 276–77, 281, CD-292–95,
 330–31
exponential smoothing, CD-330–31
external causes, CD-182

F

factorial experiment, 200
 fractional, CD-267
factors, definition, 200, CD-267
family of causes, 118
 comparison of two families, 131,
  CD-229
 comparison of three or more families, 149,
  CD-239
 haphazard effect and, CD-246–50
 investigation type and, 125
family of variation. *See* family of causes
Farnum, N. R., CD-214
fascia cratering. *See* examples
fascia dimension. *See* examples
fascia film build. *See* examples
fascia ghosting, 61–63
Feder, P. L., CD-245
feedback control, 34–36, 112–13, 269–83,
 CD-289–95
 implementation, 280–82
 informal, CD-289–90
 simulation, CD-290–92
feedforward control, 33–34, 259–68
 implementation, 266–67
 selective fitting and, 263, CD-285–87
fitted line, CD-195–96, 341–42
fixed input, 15–16, CD-182
Fix the Obvious approach, 30–32, 214–17

flowchart. *See* process map
focused problems. *See* Define Focused
 Problem stage
fold-over experiment, CD-275
Fox, M., CD-197
fractional factorial experiment, CD-267
 crossed designs, 248, 256, CD-278–81
 design resolution and, 235, CD-274
 moving the process center and, 233
F-tests, CD-234
Fuller, H., 256
full extent of variation, 75–76, 87
 comparing component families, 167
 input/output relationship and,
  190–91
 leverage and, 128
 measurement system investigation
  and, 93
 method of elimination and, 126–28
 multivari investigation and, 161
 observing less than, 145
 variation transmission and, 166
 robustness and, 285
 suspect dominant cause and, 194, 198
full factorial experiment, 200. *See* designed
 experiment
 component-swapping in, CD-254–55
full model, 203
furnace switch, 38

G

gage reproducibility and repeatability (R&R),
 CD-214–17
gains, holding the, 310–12
Gaussian model, 23, CD-187–89
goal
 baseline and, 73
 project vs. problem, 42, 49, 69, 86
Gracia, C. E., CD-295
graphical summaries, CD-317–31
group comparison, 179
 matrix scatter plot, CD-257–59
 paired comparisons versus,
  CD-259–60
 summary, 181–82
Grubbs, F. E., CD-289
Grubbs' rule for feedback, CD-289
Gryna, F. M., Jr., 46, 282, CD-192

H

Hahn, G. J., 200
Hamada, M., 257, CD-261, 267, 281
haphazard effect, 150, 153, 157, CD-246–50, CD-328
haphazard family. *See* haphazard effect
haphazard sampling, 54, 111
Harrington, H. J., 14
Harry, M., CD-191
headrest failure. *See* examples
hidden replication, 200
histogram, 19, 20, CD-317
    idealized, 23–24
Hosmer, D. W., CD-261
Hoyle, D., 46
hubcap damage, 216–17
Hunter, G., CD-267
Hunter, J. S., CD-267
hypothesis testing, 148, 191, 199, CD-234–35
hypoid gear distortion, 205

I

indicator variables, 190, CD-261
infrastructure, supportive, 47
Ingram, D., CD-235, 253
input characteristic, 15
input/output relationship investigation, 183–92, CD-181
    summary, 191
inspection limits, 301, 304
inspection, 100%, 38–39, 301–6
interaction effect, 203, CD-181
    confounding in fractional factorials, CD-272
interaction plots, 246, CD-249
internal cause, CD-182
investigation summary
    assembly versus components, 144
    baseline, 80–81
    component swap, 174–75
    desensitization, 254–55
    group comparison, 181–82
    input/output relationship, 190–91
    measurement, 97–98
    moving process center, 237–38
    multivari, 161–62
    robustness, 297

    variation transmission, 166–67
    verification, 206–7
iron ore variation, 218
iron silicon concentration. *See* examples
Ishikawa, K., CD-225
isoplot investigation, CD-218–19

J

Jenkins, G. M., 267, 282
Johnson, N., CD-183
Juran, J. M., 16, 46, 48, 282, CD-179, CD-192

K

Kalbfleisch, J. G., CD-180
Kendall, G., 299
Kotz, S., CD-183
Kramer, T., CD-290
Kume, H., 131

L

Lawless, J. F., CD-235
least squares, CD-195
Ledolter, J., CD-265, CD-288, CD-290
Lemeshow, S., CD-261
levels for inputs in experiments, 194
    dominant cause in desensitization, 254
    move process center, 237
    verification of suspects, 198
leverage, 128
    comparing assembly and component families, 141, CD-237
    component swap process, 167, 176
    outliers and, CD-200
Lewis, J. P., 48
Liberatore, R. L., CD-198
Lindsay, M., CD-225
linearity, definition, CD-212
local optimization, 48
locally weighted scatterplot smoother (LOWESS), CD-288
logistic regression, CD-261
logistics in planning investigations, 56, 63
Logothetis, N., CD-192
LOWESS (locally weighted scatterplot smoother), CD-288

low-frequency problems, 182
Luceno, A., 281–82, CD-295

M

machining dimension, CD-67
MacKay, J., 29, 51
Mackertich, N. A., CD-290
macros in MINITAB, CD-306–9
main effects, 203, CD-269
    confounding in fractional factorial designs, 236
    plot, 230, CD-354
management role, 47–49
manifold blocked port. *See* examples
manifold sand scrap. *See* examples
matrix scatter plot, CD-257–59, 323–26
mean, 23–24
Mease, D., CD-285
measurement bias, 91
    estimation of, 98–101
    linearity and, CD-212
    relative bias, 101
measurement error, 55, 64, 90–91
    confidence intervals and, CD-198, CD-202
measurement system, 89–104, CD-205–23
    attributes of, 90–91, CD-211
    binary, CD-205–9
    destructive, CD-209–11
    gage R&R investigation, CD-214–17
    improvement of, 101–2
    investigation summary, 97–98
    isoplot assessment, CD-218–19
measurement variation, 91
    dominant cause, CD-221
    effect of, CD-220–21
    interpretation of, CD-220
method of elimination, 117, 121
    comparison to other methods, CD-225
    dominant cause and, 121
    failure of, CD-227
    implementation, 124–30
    iterative nature, 124
Meyer, R. D., CD-275
MINITAB, CD-299-356
    ANOVA, CD-333–37
    best subsets regression, CD-342–44
    data storage, CD-301–3
    derived characteristics, CD-304
    exponential smoothing, CD-329–30
    fitted line plots, CD-341–42
    graphical summaries, CD-317–31
    macros, CD-306–9
    numerical summaries, CD-311–15
    patterned data, CD-303
    regression analysis, CD-339–45
model building, CD-260, 287–88
molded plastic casing, 39–40
Montgomery, D. C., 81, 194, 200, 206, 257, 310, CD-260, CD-267, CD-275
Move the Process Center approach, 39, 227–40, CD-6
multiple regression model, CD-260, CD-288
multivari investigation, 135–38, 149–63
    analysis of variance for, CD-239–50
    chart, 150, CD-326–29, 327–28
    summary, 161

N

Nair, V. N., 299, CD-297
Nelder, J. A., CD-261
Nelder, L., CD-261
Nelson, L. S., 13, CD-197
Nemeth, L., 150
Neter, J., CD-197, CD-287, CD-333, CD-342
Nikolaou, M., 261–64
noise factors, 257, CD-182
    fractional factorial experiments, CD-278
    output variation and, 17
    robustness to variation, 36–37
nonparametric tests, CD-234
nylon bond strength, CD-58

O

observational plan, 55–56, 64
obvious fix, definition of, 214
Odeh, R. E., CD-197
O'Dell, C. S., 310
oil pan scrap. *See* examples
Oldford, R. W., 51
operations swap, CD-235–36
organizational learning, 47, 49
outlier, 57, CD-198
output characteristics, 15

output drift, CD-236
owners of processes, 15

P

paint film build. *See* examples
paired comparison, CD-259–60
parameter, 23
parameter design, 299
Pareto, Vilfredo, CD-179
Pareto analysis, 38, 48, 72
    chart, CD-179
    effects in factorial experiments, 203, CD-275, CD-352
Pareto principle, CD-179–89
    causes and, 16, 26
parking brake tightness. *See* examples
Parmet, Y., CD-253
part-to-part family, 135–38, 149, 153
    feedback control, 282
    feedforward control and, 267
    systematic vs. haphazard variation in, 150
patterned data, CD-303
patterns over time, 112–13
performance measure
    binary measurement systems and, CD-205
    binary output, 74
    robustness experiments, 36, 298, CD-297
Phadke, M. S., CD-297
PID (proportional, integral, derivative) controllers, CD-295
Plan step in QPDAC, 51
    checklist, 65
Pooled StDev, CD-229–32, 335
potato chip spots. *See* examples
power window buzz. *See* examples
precision of measurement system, 91
precision shaft diameter, CD-68
precontrol, CD-289
Pregibon, D., 299
Prett, D. M., CD-295
problem baseline. *See* baseline investigation
problem
    focusing, 69–87
    hierarchy of, 73
    reformulation, 219–23
    vs. project, 69

process capability. *See* capability ratios
process center adjuster. *See* adjuster
process
    controlled, CD-294–95
    holding gains, 310–12
    knowledge of, 51–66
    language of, 13–15
    owners of, 15, 63
    repeatability of, 14
    robustness, 36–37
    stability, CD-203–4
    subprocesses and, 13
    variation transmitted through, 27
process center
    alignment, 31
    average and, 19
    moving, 39–40, 227–40, CD-267–76
    specification limits and, 227
    variation and, 20
process certification, 46
process improvement team, 42, 48
process language and definitions, 13–15
process maps, 13
project, 69
pump noise. *See* examples

Q

QPDAC (Question, Plan, Data, Analysis and Conclusion) framework, 51–58, 124–25
    checklist, 65
Question step of QPDAC, 51–53
    checklist, 65
Quinlan, J., 111

R

randomization, 195, CD-264
randomized sequencing, CD-236
random sampling, 54, CD-196–98
rare defects, 182, 302–3
    group comparison, 182
    100% inspection, 302–3
reformulation of problems, 219–23
refrigerator frost buildup. *See* examples
regression analysis, 183, CD-259
    best subsets, CD-342–44
    with count data, CD-261

interpretation of, 191
logistic, CD-261
MINITAB, CD-339–46
variation transmission investigations, 165, CD-250–53
relative bias of measurement systems, 101
repeatability, CD-211–12
repeats, 194, 208
  binary output, 199
  vs. replicates, CD-263
replicates, 194, 208, CD-263
  hidden, 200
replication. *See* replicates
reproducibility, definition, CD-211–12
resolution, 235, CD-274
Robinson, G. K., 200
robustness approach. 36–37, 110–11, 285–300
  investigation summary, 297
  Taguchi method, CD-297–98
rocker cover oil leaks, 304
rod thickness. *See* examples
roof paint defects, 165–66
roof panel updings. *See* examples
root mean squared deviation (RMSD), CD-184–85
Ross, P. J., CD-192
row/column format, CD-301–3
run chart, 21, CD-318
run. *See* experimental run
Ryan, B. F., 310, CD-299
Ryan, T. P., 81, 194

S

sample error
  avoidance of, 124
  confidence intervals and, CD-202
  definition, 54, 64
  description, CD-198
  sample sizes and, 55, 76
  sampling protocols and, CD-197
sampling protocol, 54–55
  measurement variation and, 93
  order preservation and, 144–45
  sample error and, 60
sand core strength. *See* examples
Satterthwaite, F. E., CD-290
scarcity of effects principle, CD-275

scatter plot
  adding jitter, CD-321–22
  fitted lines and, CD-195
  function, CD-320–26
  labeling points, CD-322
  matrix, CD-257–59
  numerical attributes and, CD-195–96
Schoeder, R., CD-191
Scholtes, P. R., CD-191, 225
Scholtes algorithm, CD-191–92
seat cover shirring, CD-70
Seder, L. A., 150
selective fitting, 34, 263–64, CD-285–87
Senge, P. M., 49
Shainin, D., 282, CD-253, 265, 290
Shainin, P., 282, CD-265, 290
Shainin, R. D., CD-191
sheet metal, CD-61–62
Shimbun, N. K., 305
Shingo, S., 39, 217, 305
side effects, 30, 239
signal-to-noise ratio, CD-298
Six Sigma, CD-191
smoothers, predictions, CD-288
Snee, R. D., 150
sonic weld, CD-63
source inspection, 39, 305
spark plug connection, CD-56
special causes, CD-182
speedometer cable shrinkage, 111
sporadic problems, 46
stability, definition, CD-212
stack heights, 24–27
standard deviation, 19
  combining, 25–27
  confidence intervals and, CD-203
  data summaries, CD-311–13
  dominant cause and, CD-180
  effect of outliers, CD-199
  estimation precision, 77
  Gaussian models and, CD-188
Statistical Engineering Variation Reduction Algorithm, 41–49
statistical process control (SPC), 310, CD-182
statistical significance, 191, CD-234–35
steering knuckle strength, CD-58
steering wheel vibrations. *See* examples
Steinberg, M., CD-253

Steiner, S. H., 29
stoplight control, CD-289–90
stratification, 131–34
study error
    avoidance of, 71, 124
    definition, 54, 64
    description, CD-198
study population, 54, 64, 198
subprocess, 13
subsetting data, CD-304–5
sunroof flushness. *See* examples
supplier, definition, 15
suspect, 121, 124, 208
    choosing levels for, 198
    interactions between, 203
    verification experiments for, 193–209
Swersey, A., CD-265, 290
systematic sampling, 54, CD-196–98

T

Taguchi, G., 257, 299, CD-182, 192, 253, 278, 297–98
target population, 52
target process, 52
Taylor, W. A., CD-192
team. *See* process improvement team
time-based families. *See* multivari investigation
    comparison of, 135–38
    order preservation, 144–45
    traceability, 144–45
time series models, CD-288
tin plate strength, CD-209–210
Tippett, L.H.C., CD-253
Todd, R. H., 46
traceability, 144–45
transmission shaft diameters. *See* examples
Traver, R. W., CD-290
treatments, 200, 208
Trivedi, K., CD-261
trivial many, CD-179
truck pull. *See* examples
Tsui, L., CD-253
t-tests, CD-234
Tukey, J. W., CD-234

U

unit, definition, 14, 64
upstream family. *See* variation transmission investigation

V

V6 piston diameter. *See* examples
V8 piston diameter. *See* examples
validation of a solution, 307–9
valve train test stands. *See* examples
VanderWiel, S. A., 306
Vardeman, B., 306
variables search, CD-265
variation
    approaches to reducing, 29–40
    causes of, 15–23, CD-182
    definition, 1
    displaying, 18–23
    explaining changes in, CD-232–33
    quantifying, 18–23
    reduction algorithm, 41–50
    unequal levels of, 146
variation transmission investigation, 163–67
    regression analysis for, CD-250–53
    summary, 166–67
varying inputs, 15–16
vectoring, 262
verification of a dominant cause, 193–209
    investigation summary, 206–7
vital few, CD-179

W

weatherstrip torsional rigidity, CD-72–74
wheel bearing failure time. *See* examples
Wheeler, D. J., 194
window leaks. *See* examples
Wittenmark, B., CD-295
working approach flowchart, 114, 226
Wu, C. F. J., 257, CD-267, 281

Z

Zaciewski, R. D., 150